Understanding Stellar Evolution

AAS | IOP Astronomy

AAS Editor in Chief

Ethan Vishniac, John Hopkins University, Maryland, US

About the program:

AAS-IOP Astronomy ebooks is the official book program of the American Astronomical Society (AAS), and aims to share in depth the most fascinating areas of astronomy, astrophysics, solar physics and planetary science. The program includes publications in the following topics:

- GALAXIES AND COSMOLOGY
- INTERSTELLAR MATTER AND THE LOCAL UNIVERSE
- STARS AND STELLAR PHYSICS
- EDUCATION, OUTREACH AND HERITAGE
- HIGH-ENERGY PHENOMENA AND FUNDAMENTAL PHYSICS
- THE SUN AND THE HELIOSPHERE
- THE SOLAR SYSTEM, EXOPLANETS, AND ASTROBIOLOGY
- INSTRUMENTATION, SOFTWARE, LABORATORY ASTROPHYSICS AND DATA

Books in the program range in level from short introductory texts on fast-moving areas, graduate and upper-level undergraduate textbooks, research monographs and practical handbooks.

For a complete list of published and forthcoming titles, please visit iopscience.org/books/aas.

About the American Astronomical Society

The American Astronomical Society (aas.org), established 1899, is the major organization of professional astronomers in North America. The membership (~7,000) also includes physicists, mathematicians, geologists, engineers and others whose research interests lie within the broad spectrum of subjects now comprising the contemporary astronomical sciences. The mission of the Society is to enhance and share humanity's scientific understanding of the universe.

Understanding Stellar Evolution

Henny J.G.L.M. Lamers
University of Amsterdam
Emily M. Levesque
University of Washington

IOP Publishing

Media content for this book is available from Book information at https://doi.org/10.1088/978-0-7503-1278-3.

ISBN 978-0-7503-1278-3 (ebook)
ISBN 978-0-7503-1279-0 (print)
ISBN 978-0-7503-1280-6 (mobi)

DOI 10.1088/978-0-7503-1278-3

Version: 20181101

AAS–IOP Astronomy
ISSN 2514-3433 (online)
ISSN 2515-141X (print)

British Library Cataloguing-in-Publication Data: A catalogue record for this book is available from the British Library.

Published by IOP Publishing, wholly owned by The Institute of Physics, London

IOP Publishing, Temple Circus, Temple Way, Bristol, BS1 6HG, UK

US Office: IOP Publishing, Inc., 190 North Independence Mall West, Suite 601, Philadelphia, PA 19106, USA

Contents

Preface

This e-book is based on the lecture notes of the graduate class "*Understanding Stellar Evolution*" developed and taught by Henny Lamers during the spring quarter at the University of Washington in Seattle from 2004–2014 and now taught by Emily Levesque. The course consisted of twenty 80-minute lectures. The lecture notes have developed over the years due to many comments, questions, and suggestions by graduate students and by requests from colleagues at the University of Washington.

Goal of this e-book: Our goal is to give students a "feeling" and "physical understanding" of stellar structure and evolution. We therefore try to avoid long, purely mathematical derivations and explanations. Although these may be straightforward and appealing to mathematically inclined students, they often obscure the physical processes behind them. Instead, we try to explain basic concepts in simple, sometimes intuitive, physical terms. When needed, we provide references to where the more rigorous explanations and derivations can be found. From the reactions of the students, we sense that this approach is appreciated.

References: Each chapter contains a short list of references. In a few cases, we refer to the original papers in which a new concept was introduced; however, in most cases, we refer to a more current paper, often a recent review paper, where the topic and its consequences are explained more clearly. We believe that this is more useful for the students, particularly as these reviews often cite the original paper. Wherever possible, the list of references contains URLs for easy access to the relevant paper.

History: When a formula or concept is named after a person (e.g. the Schwarzschild criterion, the Chandrasekhar limit, or the Hayashi track), some information about that person is given. We hope that in this way students appreciate the historical development of stellar evolution theory.

Figures: The figures were made by Emily Levesque. Many of the figures are taken from the published literature, but are often simplified to show the essential information more clearly. We make extensive use of color figures. When evolution or stellar structure diagrams are shown, we try to use a consistent color scheme: red for hydrogen and H-fusion, blue for helium and He-fusion, and green for carbon and C-fusion. We are grateful to the many authors and journals for permission to use figures from their publications, especially to Onno Pols for the many figures from his lecture notes.

Consulted books: Henny Lamers is grateful to his Utrecht colleague Onno Pols for his lecture notes on "*Stellar structure and evolution*" and "*Binary evolution*," with Frank Verbunt. These notes have been a great help, especially for explaining the complicated late evolution phases of low-mass stars. Other consulted books include the excellent and extensive book *Physics, Formation and Evolution of Rotating Stars* (Maeder, 2009), *Stars and Stellar Evolution* (de Boer and Seggewiss, 2008), *Evolution of Stars and Stellar Populations* (Salaris and Cassisi, 2006), *An Introduction to Stellar Structure and Evolution* (Prialnik, 2004), *Stellar Interiors: Physical Principles, Structure, and Evolution* (Hansen and Kawaler, 1994), and *Stellar Structure and Evolution* (Kippenhahn and Weigert, 1990).

Proofreading: We thank colleagues and students for comments on individual chapters. Colleagues: Carlo Abate, Conny Aerts, Eric Agol, Norbert Langer, Andre Maeder, Max Pettini, Onno Pols, Maurizio Salaris, Gertjan Savonije, Jorick Vink, and Ralph Wijers.

Students of the Astronomy Department of the University of Washington: Giada Arney, Hannah Bish, Iryna Butsky, Russell Deitrick, Trevor Dorn-Wallenstein, David Fleming, Margarte Lazzarini, Andrew Lincowski, Jacob Lustig-Yaeger, Rodrigo Luger, Eddie Schwieterman, Grace Telford, Michael Tremmel, Spencer Wallace, Kolby Weisenburger, Matthew Wilde, and Diana Windemut. A special thanks goes to Chris Suberlak for painstakingly finding all the errors and typos in the penultimate version of the lecture notes. We hope he found them all!

Versions: The first version of these lecture notes was handwritten by Henny Lamers. I am grateful to Marion Wijburg at Utrecht University and Rachel Beck and Brandi Okano at the University of Washington for deciphering my scribbles and the notes' many equations.

Thanks to students: Many thanks goes to the many graduate students who took this course over the years. We thank them for their discussions, comments, and suggestions about this class, the notes, and the exercises. We learned a lot from your questions during the lectures!

Henny Lamers, Bilthoven, the Netherlands, and Seattle, US, 2016–spring 2017
h.j.g.l.m.lamers@uu.nl
Emily Levesque, Seattle, US, fall 2016–spring 2017
emsque@uw.edu

Comments for teachers.
These lectures are based on a four-credit class (ASTR 531). The course is given in a spring semester of 10 weeks with two 80-minute lectures per week. Enrolled students are expected to spend 12 hours per week on this class. This book is slightly more extended than the original lecture notes. At the request of students, we added sections on binary evolution.

Throughout this book you will find questions that we asked the students during the class. They are indicated by "Q?". The students were encouraged to answer these questions in class, which sometimes resulted in a class discussion where different arguments were compared.

Suggested homework questions for each chapter are given at the end of the chapter. The answers are available upon request, to teachers only. The answers can be obtained from the authors: Henny Lamers at h.j.g.l.m.lamers@uu.nl and Emily M. Levesque at emsque@uw.edu.

For readers of the printed version of this book we have prepared an extensive index list, which can be obtained from the authors upon request.

To help teachers using this book, a set of lecture slides is provided. The slides were developed by Emily Levesque and cover the full set of content presented in the text. For ease of use we have made these available in both Keynote (v6.6.1) and Powerpoint (v15.14) format.

Author biographies

Henny J.G.L.M. Lamers

Henny J.G.L.M. Lamers is Professor Emeritus in Astrophysics at Utrecht University and the University of Amsterdam. He studied at the Universities of Nijmegen and Utrecht and was a postdoc at Princeton University. His research concentrated on the study of stellar winds, mass-loss, and its effect on the evolution of massive stars. In the last 10 years he studied the dynamics and survival of star clusters.

In 1976 he received the NASA Group Achievement award and in 1980 the Schmeits Award for Astronomy in the Netherlands. In 1989 he held the Honorary Brittingham Chair of the University of Wisconsin. He has been teaching in different countries and was nominated for teacher of the year by students in 2003.

He has given more than 500 public lectures and is the author or editor of nine professional astronomy books and four popular astronomy books. He is a member of the Netherlands Royal Academy of Arts and Sciences, and an honorary member of the American Astronomical Society. Asteroid 21563 is named Henny Lamers.

Emily M. Levesque

Emily M. Levesque is an Assistant Professor of Astronomy at the University of Washington in Seattle. Her primary research area is massive stellar astrophysics, with a focus on evolved massive stars such as red supergiants and their potential role as cosmological tools. Her research accolades include a 2017 Alfred P. Sloan fellowship in physics, the 2014 Annie Jump Cannon research prize from the American Astronomical Society, and the 2012 Robert J. Trumpler Award from the Astronomical Society of the Pacific. She was both an Einstein and Hubble postdoctoral fellow at the University of Colorado, and received her PhD in astronomy from the University of Hawaii and her S.B. in physics from the Massachusetts Institute of Technology.

Henny J.G.L.M. Lamers and Emily M. Levesque

Chapter 1

Stars: Setting the Stage

Everybody knows what stars are: twinkling light spots in the night sky. In reality, they are huge spheres of gas at high temperature that emit a quantity of light that we can hardly imagine. The properties of stars are described by quite simple laws of physics. To understand the stars and their evolution, it is important to get a "feeling" for their properties, so that they are not just described by astronomically large numbers. To this purpose, we will first consider the Sun and compare it with the Earth and other objects on a human scale. After that, we describe the structure and evolution of stars in a nutshell, as an introduction to the main goal of this book: the understanding of what stars are, what their structure is, why they shine, why they are stable, how they are born, how they evolve, and how they die.

1.1 The Sun: Our Star

To get a feeling for the masses, sizes, temperatures, and ages of stars, we first consider our star: the Sun. We will use approximate values here; more accurate values are listed in Appendix A. Here and throughout the rest of the book, we will express quantities in c.g.s. units (centimeter, gram, second), as is usual in astronomy. Some conversion factors are listed in Appendix A.

- **The mass of the Sun is 2×10^{33} g**. This is 3×10^5 times as large as the Earth mass of 6×10^{27} g.
- **The radius of the Sun is 7×10^{10} cm**. This is 110 times the Earth radius of 6400 km. This implies that the 1.3 million Earths could fit inside the Sun.
- **The mean density of the Sun is 1.4 g cm^{-3}**, which is 0.25 times the Earth density of 5.5 g cm^{-3}. The mean density of the solar gas is even higher than the density of water, which is 1 g cm^{-3}.
- **The luminosity of the Sun is 4×10^{33} ergs s^{-1}**. This is a huge amount of energy. According to the web, there were about 1.3 billion (1.3×10^9) cars in 2016. Assume that a car uses on average 1 liter of gasoline per 10 km and that the

doi:10.1088/978-0-7503-1278-3ch1

energy contained in 1 liter of gasoline is 3×10^{14} erg (30 MJ). If all cars in the world drove 100 km per day for 2.8 million years, they would use the same amount of energy that the Sun emits in one second!

- **The surface temperature of the Sun is about 6000 K**. This is about three times higher than the temperature of burning natural gas, which is about 2000 K. The hottest flames on Earth, i.e., those of burning dicyanoacetylene in oxygen, have a temperature of 5300 K.
- **The age of the Sun and the solar system is 4.5 billion years**. A very slow snail with a speed of 1 cm per hour would be able to go around the earth 9000 times during that time.

1.2 The Chemical Composition of the Sun and Stars

The chemical composition of the outer layers of the Sun and stars has been determined spectroscopically. This reflects the composition of the interstellar clouds from which the stars are formed. The composition is roughly the same for all stars in the disk of the Milky Way. The abundances of the 10 most abundant elements in the solar atmosphere, compared to hydrogen by number, are listed below.

Table 1.1. The Abundance of the 10 Most Abundant Elements in the Atmosphere of the Sun and Stars in the Milky Way (Cox 2000).

Nr	Element	Z	m (AMU)	Abund	Nr	Element	Z	m (AMU)	Abund
1	H	1	1.0079	0.00	6	N	7	14.007	−3.95
2	He	2	4.0026	−1.01	7	Mg	12	24.305	−4.42
3	O	8	15.999	−3.07	8	Si	14	28.086	−4.45
4	C	6	12.011	−3.44	9	S	16	32.066	−4.79
5	Ne	10	20.180	−3.91	10	Fe	26	55.847	−4.46

Note. The atomic mass is expressed in atomic mass units: 1 AMU = 1.66054×10^{-24} g. The abundance by number of element E relative to H is expressed in $\log (N_E/N_H)$.

This composition can also be expressed in terms of mass fractions as

$$\mathrm{H} : \mathrm{He} : \mathrm{C+N+O+Ne} : \mathrm{rest} = 0.70 : 0.28 : 0.016 : 0.003. \quad (1.1)$$

Put another way, 1 g of gas contains 0.016 g of C + N + O + Ne and 0.003 g of more massive elements. The composition may be different for stars in other galaxies; it depends on the amount of star formation that occurred in a galaxy in the past. For instance, for stars in the Small Magellanic Cloud, the total mass fraction of all elements, apart from H and He, is about five times smaller than that in Galactic stars.

1.3 The Structure of Stars

Due to stellar evolution, a star becomes highly structured, both chemically and physically. For discussing the evolution of stars, it is useful to distinguish several regions in stars.

- **The core** is the central region of a star where fusion occurs or has occurred.
- **The shell zone** indicates that fusion occurs or has occurred in a shell or shells around the core.

- **The envelope** is the region between the shell zone and the atmosphere.
- **The atmosphere** is the region of optical depth $10 < \tau < 1$ from which radiation can escape.
- **The photosphere** is the top of the atmosphere, where the optical depth $\tau \lesssim 1$ at the wavelength where most of the light is emitted.

(The optical depth τ describes the probability that photons can escape: this probability is $e^{-\tau}$.)

Many stars are surrounded by extended layers above the photosphere. The following regions can be distinguished.

- **The chromospheres and corona** are the regions above the photosphere where the temperature rises far above the photospheric temperature. This is typical for stars with convective envelopes, which generate shocks and magnetic fields in the upper atmosphere; therefore, only cool stars have chromospheres and coronae. The Sun has a chromosphere of $\sim 10^5$ K and a corona of $\sim 10^6$ K.
- **The wind** is the region above the photosphere from which gas escapes into the interstellar medium. The velocity that a stellar wind reaches is typically a few times the escape velocity at the stellar surface. In some evolutionary phases, the wind carries away so much mass that the resulting mass loss from the star affects its evolution. The Sun has a stellar wind with a velocity of ~ 600 km s^{-1} and a mass-loss rate of $\sim 10^{-14}$ $M_\odot$ yr^{-1}, where the symbol $M_\odot$ indicates a solar mass. This is less than the mass-loss rate due to the generation of energy by nuclear fusion: $(dM/dt)_{\text{fusion}} = -L/c^2 \approx -7 \times 10^{-14}$ $M_\odot$ yr^{-1}. The H-fusion lifetime of the Sun is $\sim 10^{10}$ yr, so the Sun loses only $\sim 10^{-3}$ of its mass during this period. In later evolution phases of the Sun, the mass-loss rate will increase drastically.

1.4 Stellar Evolution in a Nutshell

Stars are spheres of gas that are held together by their own gravitational attraction. During much of its lifetime, a star is remarkably stable. This implies that at each depth within a star the inward pull by gravity is exactly counterbalanced by the outward force that results from gas pressure plus radiation pressure. Pressure only produces a force if there is a gradient, so there must be a strong temperature and density gradient in the star to counterbalance gravity, with the stellar center having a much higher density and temperature than the atmosphere. The gradient in temperature results in an outward flow of energy from the center to the surface. This energy leaves the star at the photosphere and is radiated into space. Stars must therefore radiate and continuously replenish this energy loss. This implies that stars must have an internal energy source that can sustain the temperature gradient during their long lifetimes. This energy source is nuclear fusion in the center of the star, where the temperature and density are high enough for fusion to occur.

If a nuclear fusion reaction stops because the required element is exhausted, gravity forces the star to contract. This releases potential energy (if the star did not contract, its interior would cool due to energy loss; this would reduce the gas pressure, which would no longer be able to withstand the pull of the gravity, and the interior would shrink anyway). Part of the released potential energy is used to supply the energy flow from the center to the outside. The remainder is used for raising the internal temperature. When the central temperature is high enough, the next fusion reaction may occur, converting the product of the previous phase into more massive elements. This cycle of nuclear fusion and contraction can go on until all elements that are available for fusion are exhausted.

Surprisingly, throughout their evolution, most stars maintain a perfect hydrostatic balance between the inward pull by gravity and outward force by the pressure gradient. This balance is maintained even as some layers of a star may be in turmoil on a smaller scale (for instance, by convection).

When nuclear energy sources are exhausted, a star can no longer provide the necessary energy to keep the gas pressure high enough to withstand gravity. Gravity finally wins! In low-mass stars, the core is compressed to such a high density that the matter becomes degenerate at an extremely high pressure that is capable of withstanding the extreme gravity. The result is a small star, about the size of the Earth, that can live forever but slowly cools and fades: a white dwarf. In high-mass stars, the core suddenly collapses in just a few seconds. This releases so much energy that the outer layers are ejected at a very high speed. This is a supernova explosion. The surviving core remnant is either an extremely dense neutron star or a black hole.

1.5 Summary

1. The Sun is a good example of an average star. It is about a million times larger in volume than the Earth and it has about a million times more mass. It radiates so brightly that, even at a distance of 150 million km, the tiny fraction of its radiation that hits the Earth is enough to sustain life.
2. The age of the Sun is 4.5 billion years. When it was formed, the Sun consisted of 70% H, 28% He, and only 2% of heavier elements.
3. A star is in perfect equilibrium between the inward pull of gravity and the outward force produced by the gradient of the gas pressure and the radiation pressure.
4. The energy source of a star is nuclear fusion, which occurs in its very hot and very dense center.
5. A star can go through various nuclear fusion phases. In each successive phase, the nuclear products of the previous phase are fused into more massive elements.
6. At the end of its evolution, a star ejects its envelope. The remaining core is either a white dwarf, a neutron star, or a black hole. There is no more nuclear fusion and the remnant slowly cools.

Reference

Cox, A. N. 2000, Allen's Astrophysical Quantities (4th ed.; New York: Springer)

Henny J.G.L.M. Lamers and Emily M. Levesque

Chapter 2

Observations of Stellar Parameters

In this section, we briefly review observations of the fundamental parameters of stars. These are mass (M), radius (R), luminosity (L), and effective temperature (T_{eff}). The existence of observed relations between these parameters provides important information about the internal structure of stars. Any reliable theory of stellar structure must be able to explain these observed relations. One of the most valuable methods for studying stellar evolution is the analysis of the relation between L and T_{eff}. This relation can be plotted on the Hertzsprung–Russell diagram (HRD). We describe this diagram and its observational analog, the color–magnitude diagram (CMD), for stars in clusters of different ages. Changes in the distribution of stars in the HRD with time provide empirical evidence and diagnostics of stellar evolution. We also introduce the various types of stars and their names, which are typically related to their location in the HRD.

2.1 The Distance of Stars

The distance to a star can be measured by several methods. The most direct one is the parallax method, shown in Figure 2.1. The position of a target star, as compared to background stars, is observed at two epochs that are half a year apart. A nearby star will show a difference in projected location, compared to background stars, at these two epochs due to the parallax. The epochs are chosen in such a way that the line connecting the two locations of the Earth in its orbit is perpendicular to the direction of the star. In this way, the displacement, called parallax angle or simply parallax, reaches a maximum.

Simple geometry shows that the distance d is related to the parallax angle θ by $d = R/\tan\theta$, where R is the radius of Earth's orbit. For very small angles, $d \approx R/\theta$ if θ is expressed in radians. Since $R = 1.5 \times 10^{13}$ cm and an angle of 1 arcsec corresponds to 4.85×10^{-6} radians, a parallax angle of 1 arcsec corresponds to a distance of 3.09×10^{18} cm = 3.26 lightyears. This distance is called a **parsec** (pc). Thus we find that d (in parsecs) = $1/\theta$ (in arcsec).

doi:10.1088/978-0-7503-1278-3ch2

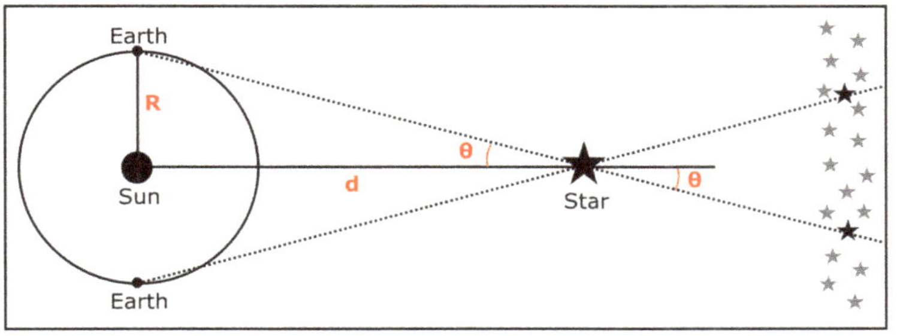

Figure 2.1. Principle of parallax measurements. The parallax angle θ can be measured by comparing the position of the star to background stars. In practice, the angle is very small and on the order of milliarcseconds. The black stars show the projected position of the nearby star at two epochs.

Apart from the yearly parallax motion of a star against the background, most nearby stars also show a continuous motion along the sky due to their space velocity. This is called the proper motion of a star. It can be derived by comparing observations that were made exactly one year apart. The parallax measurement must be corrected for it in order to provide an accurate distance.

Parallax measurements from ground-based observatories reach an accuracy of $\sim 10^{-2}$ arcsec. The Hipparcos satellite reached $\sim 10^{-3}$ arcsec accuracy[1]. The *Gaia* satellite will reach $\sim 5 \times 10^{-4}$ arcsec for stars with $V \approx 20$ mag (10^{-8} times as bright as the second brightest star Vega in the sky) and $\sim 2 \times 10^{-5}$ arcsec for stars with $V \approx 15$ mag (10^{-6} times the brightness of Vega) at the end of its mission in 2021. The distance accuracy will be 2% for a solar type star at 250 pc and 20% for a star that is 10^3 times more luminous than the Sun at the distance of about 10 kpc near the Galactic Center (GAIA Collaboration 2016).

2.2 The Mass of Stars

In principle, the mass of a star can be measured from its radius and surface gravity; however, this is typically not an accurate method because the surface gravity has to be derived from the stellar spectrum and involves detailed knowledge of the star's atmospheric structure. Accurate mass determinations are possible from the orbits of planets (in the case of the Sun) or from the orbits of binary stars.

Figure 2.2 shows a binary system with circular orbits. The orbital period is given by Kepler's third law:

$$(2\pi/P)^2 = G(M_1 + M_2)/a^3 = G(M_1 + M_2)/(a_1 + a_2)^3 \tag{2.1}$$

and

$$a_1/a_2 = M_2/M_1, \tag{2.2}$$

where M_1 and M_2 are the masses of the stars and a_1 and a_2 are the orbital radii. The total separation is $a = a_1 + a_2$. If the orbits of the stars can be measured (either directly for visual binaries or indirectly for eclipsing binaries) and combined with radial velocity measurements (using the Doppler effect), the masses of both stars can be determined.

[1] https://en.wikipedia.org/wiki/Hipparcos

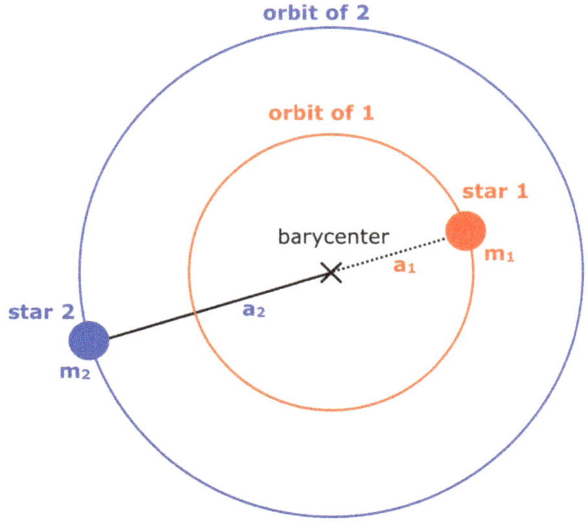

Figure 2.2. Orbits of binary stars in circular orbits.

2.3 The Luminosity of Stars

The **luminosity** L of a star can be derived from its radiative flux received at Earth and its distance using the relation

$$L = 4\pi d^2 f_{\rm bol} = 4\pi R^2 \sigma T_{\rm eff}{}^4 \text{ (in erg s}^{-1}) \tag{2.3a}$$

or

$$L/L_\odot = (R/R_\odot)^2\,(T_{\rm eff}/5777\text{ K})^4 \quad \text{with } L_\odot = 3.85 \times 10^{33}\text{ erg s}^{-1}, \tag{2.3b}$$

where the bolometric flux $f_{\rm bol}$ is the flux integrated over all wavelengths and corrected for attenuation by the interstellar medium and Earth's atmosphere. The Stefan–Boltzmann radiation constant is $\sigma = 5.67 \times 10^{-5}$ erg cm^{-2} K^{-4}. It is named after the Austrian physicist Jozef Stefan (1835–1893), who proposed it in 1879, and his Austrian student Ludwig Boltzmann (1844–1906), who proved it in 1884.

Equation (2.3a) states that L is proportional to the surface area of the star, $4\pi R^2$, and the amount of radiation per unit area, $\sigma T_{\rm eff}{}^4$, which is described by the **effective temperature $T_{\rm eff}$.**

The effective temperature is approximately equal to the gas temperature in the photosphere of the star at optical depth $\tau \approx 1$. This relation connects three basic stellar parameters L, R, and $T_{\rm eff}$. If L and $T_{\rm eff}$ can be determined from observations, then R is known.

2.4 Magnitude, Color, and Temperature

The outer layer of a star from which radiation escapes is called the photosphere. It has a density and temperature structure with both quantities decreasing outward. The temperature structure can be derived from a detailed study of the spectrum of the star. For understanding stellar evolution, we are mainly interested in the effective temperature, defined in Equation (2.3a). As described above, $T_{\rm eff}$ is approximately the gas temperature in the photosphere at optical depth $\tau \approx 1$. The effective temperature can be derived from the spectrum or color of the star, corrected for the influence of Earth's atmosphere and reddening by interstellar dust. The temperature of a star can also be expressed by its spectral type, running from hot to cool with the designations O (~30,000–40,000 K), B, A, F, G, K, and M (~2500–3500 K). The Sun, with $T_{\rm eff}$ = 5777 K, has the spectral type G2. The relationship between spectral type and $T_{\rm eff}$ for stars in the Milky Way is tabulated in Table 2.1.

The **color** of a star is defined as the ratio between the flux in two wavelength bands, expressed as the difference in magnitudes. It is measured from the brightness ratio of the star through two filters. The color of the standard star Vega with $T_{\rm eff} \approx$ 10,000 K is used for reference. There are many possible astronomical filters but the most commonly used are the Johnson U ($310 < \lambda < 390$ nm), B ($390 < \lambda < 480$ nm), and V ($500 < \lambda < 590$ nm) filters. The flux through these filters is expressed in **apparent magnitudes,** m_λ. For historical reasons, this is defined by filters

$$m_\lambda = -2.5 \times \log\left(\frac{f_\lambda}{f_0}\right), \tag{2.4}$$

with the flux f in erg cm^{-2} s^{-1}. Note that the higher the flux is, the smaller the magnitude. The normalization values of f_0 depend on the filter used, with $f_0(U)$ = 4.2×10^{-8} erg cm^{-2} s^{-1} nm^{-1} at $\lambda = 0.36$ μm, $f_0(B) = 6.4 \times 10^{-8}$ erg cm^{-2} s^{-1} nm^{-1} at $\lambda = 0.44$ μm, and $f_0(V) = 3.8 \times 10^{-8}$ erg cm^{-2} s^{-1} nm^{-1} at $\lambda = 0.55$ μm. These normalization fluxes are chosen in such a way that the magnitude of the standard star Vega is zero. The magnitudes through the U, B, and V band filters are indicated by U, B, and V. Because of the logarithmic magnitude scale, colors, which are defined by flux ratios, are expressed in **magnitude differences**. The colors of the standard star Vega are zero per definition. Table 2.1 gives the approximate relation between the U-B and B-V colors and $T_{\rm eff}$. For a full list, see Appendix B.

Table 2.1. The Temperature, Spectral Type, Luminosity, Absolute Bolometric Magnitude, Absolute Visual Magnitude, Bolometric Correction, and Colors of Main-sequence Stars (Cox 2000).

$T_{\rm eff}$ (K)	Type	$\log L/L_\odot$	$M_{\rm bol}$	M_V	BC	U-B	B-V
42,000	O5 V	5.94	−10.1	−5.7	−4.40	−1.19	−0.33
30,000	B0 V	4.78	−7.2	−4.0	−3.16	−1.08	−0.30
20,900	B2 V	2.39	−4.8	−2.45	−2.35	−0.84	−0.24
11,400	B8 V	2.34	−1.1	−0.25	−0.80	−0.34	−0.11
8180	A5 V	1.17	+1.8	+1.95	−0.15	+0.10	+0.15
6650	F5 V	0.54	+3.4	+3.5	−0.14	−0.02	+0.44
5790	G2 V	−0.10	+4.5	+4.7	−0.20	+0.12	+0.63
4830	K2 V	−0.50	+6.0	+6.4	−0.42	+0.64	+0.91
3840	M0 V	−1.06	+7.4	+8.8	−1.38	+1.22	+1.40
3170	M5 V	−1.94	+9.6	+12.3	−2.73	+1.24	+1.64

Q (2.1) *Note that the B-V color hardly changes with $T_{\rm eff}$ for temperatures above about* 30,000 K. *Explain this in physical terms, using the properties of Planck's formula for blackbody radiation.*

The flux integrated over all wavelengths is expressed in **bolometric magnitude**, defined by

$$m_{\rm bol} = -2.5 \times \log\left(\frac{f_{\rm bol}}{f_{\rm ref}}\right), \tag{2.5}$$

with $f_{\rm ref} = 2.5 \times 10^{-5}$ ergs cm^{-2} s^{-1}. If the distance of a star is known, its apparent magnitude can be converted to **absolute magnitude,** indicated by M_λ, which is the apparent magnitude that a star would have if it was at a distance of 10 pc.

$$M_\lambda = m_\lambda + 5 - 5\log(d) \qquad \text{if } d \text{ is in parsecs.} \tag{2.6}$$

The **absolute bolometric magnitude** describes the flux of a star at a distance of 10 pc, integrated over all wavelengths. It is related to the luminosity by

$$M_{\rm bol} = -2.5 \times \log\left(\frac{L}{L_\odot}\right) + 4.74. \tag{2.7}$$

The **bolometric correction (BC)** is the difference between the bolometric magnitude and the visual magnitude. It is defined as

$$BC = M_{\rm bol} - M_v. \tag{2.8}$$

It is a measure of the fraction of the star light that is outside the wavelength range of the V-filter. Hot stars with $T_{\rm eff} > 20{,}000$ K emit most of their radiation in the UV, so the V-filter captures only a small fraction of its light. On the other hand, cool stars with $T_{\rm eff} < 4000$ K emit most of their radiation in the infrared. So both hot and cool stars have a large negative BC (see Table 2.1).

2.5 The Mass–Luminosity Relation

Observations show that the luminosity of a star is related to its mass. This correlation is particularly strong for stars that are fusing H in their center. These are the main-sequence stars. The empirical relation, based on double lined spectroscopic main-sequence binaries, is shown in Figure 2.3.

The figure shows a relation of the type $L \sim M^\alpha$, with a mean slope of $\alpha \approx 3.8$ over the full mass range. However, a closer look shows three distinct mass ranges: $\alpha = 2.4$ for $M < 0.6M_\odot$, $\alpha = 4.0$ at $0.6 < M < 6M_\odot$, and $\alpha = 2.9$ at $M > 6M_\odot$. This relation is only valid for main-sequence stars. Red giants, red supergiants, asymptotic giant branch stars, and degenerate stars do not follow this relation.

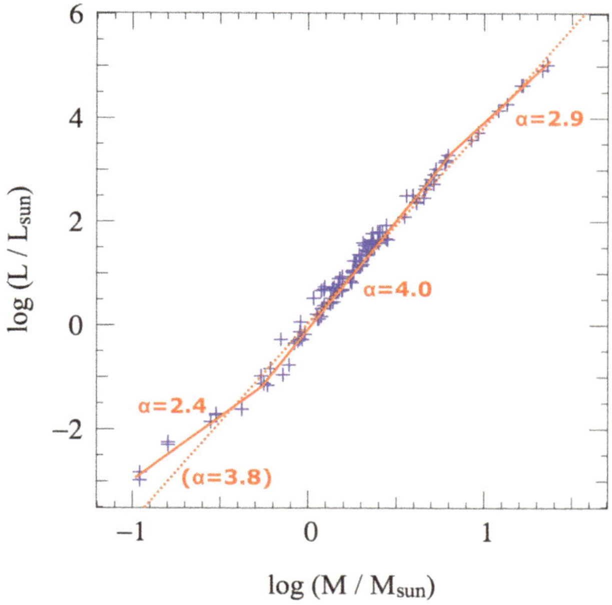

Figure 2.3. Mass–luminosity relation from double-lined spectroscopic main-sequence binaries. The dotted line shows a fit with a slope of $\alpha = 3.8$ over the entire mass range. The full line shows better linear fits in three mass ranges: $M/M_{\odot} = (0.1,0.6)$, (0.6,6) and (6,25), with corresponding slopes. (Adapted from Pols 2011.)

2.6 The Hertzsprung–Russell Diagram and the Color–Magnitude Diagram

The Danish astronomer Ejnar Hertzsprung (1873–1967) in Copenhagen and the US astronomer Henry Norris Russell (1877–1959) at Princeton University independently discovered in 1905 and 1913 that stars occupy specific regions in the **color–magnitude diagram (CMD)**. Since the colors are related to T_{eff} and the magnitudes of stars with known distance are related to L, the CMD can be transformed into a T_{eff}–L diagram. This last version is called the **Hertzsprung–Russell diagram (HRD)** after the discoverers.

Figure 2.4 shows the CMD of stars in the solar neighborhood with accurate distances measured by the *Hipparcos* satellite. This is a mixture of stars of all ages. In this figure, the B-V color is plotted versus M_V, which is a measure for the flux in the V-band corrected for distance. Remember from Equation (2.4) that a higher value of M_V implies a fainter star. The value of the B-V color increases from left to right. This means that T_{eff} decreases from ~30,000 K on the left to 3000 K on the right.

Figure 2.4 shows that the stars are not distributed randomly but that they preferentially occupy specific regions in the CMD: the broad line from upper left to lower right, i.e., from hot and bright to cool and faint, is called **the main sequence (MS)**. The fact that the majority of the stars are on the MS indicates that it represents the longest evolution phase. The wide band of stars that deviate from the MS to the upper right, i.e., to cool and bright, is called the **red giant branch (RGB)**.

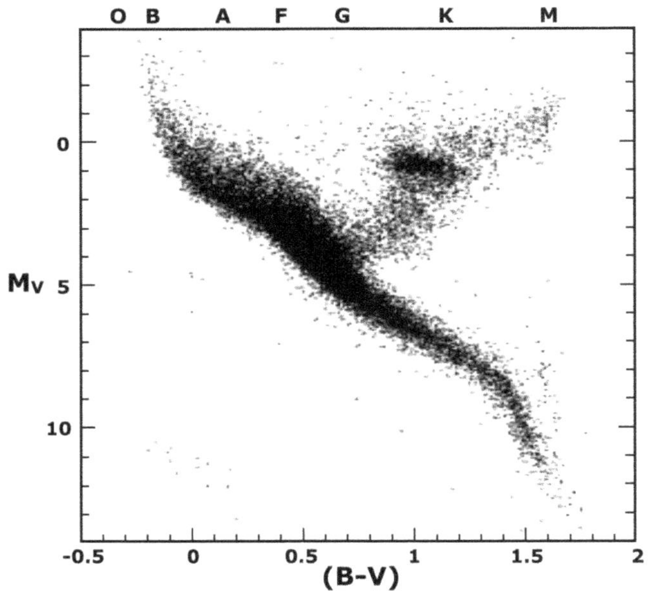

Figure 2.4. Color–magnitude diagram (CMD) of 8784 stars with accurate distances measured by the Hipparcos satellite. (Figure courtesy of ESA 1997.)

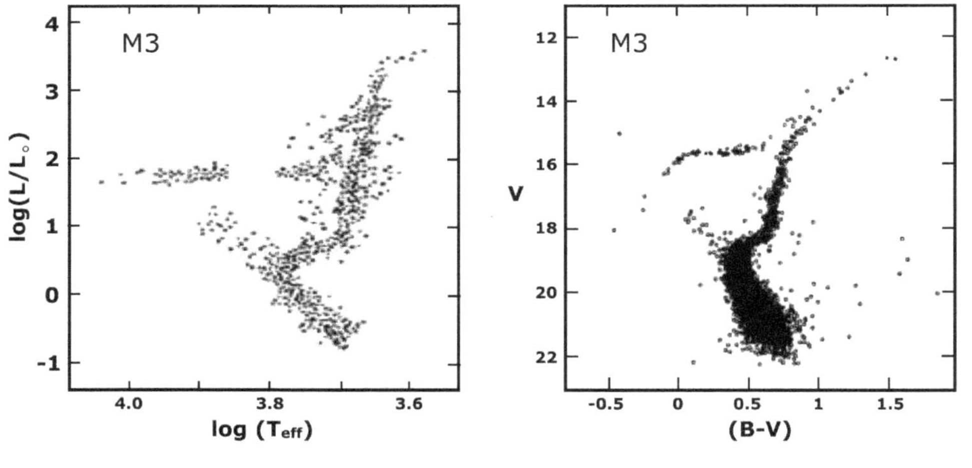

Figure 2.5. The Hertzsprung–Russell diagram (left: in log $L/L_{\odot}$ versus log $T_{\rm eff}$) and the color–magnitude diagram (right: in V versus B-V) of the globular cluster M3. (Left: reproduced from Johnson & Sandage 1956. Right: reproduced from Buonanno et al. 1994, with permission. © ESO.)

The concentration of G- and K-type stars around $M_v \approx 1$ is called the **red clump**. The group of stars in the lower left, i.e., hot and faint, are **white dwarfs.**

Figure 2.5 shows the difference between the HRD and CMD of globular cluster M3 at a distance of 10.4 kpc.

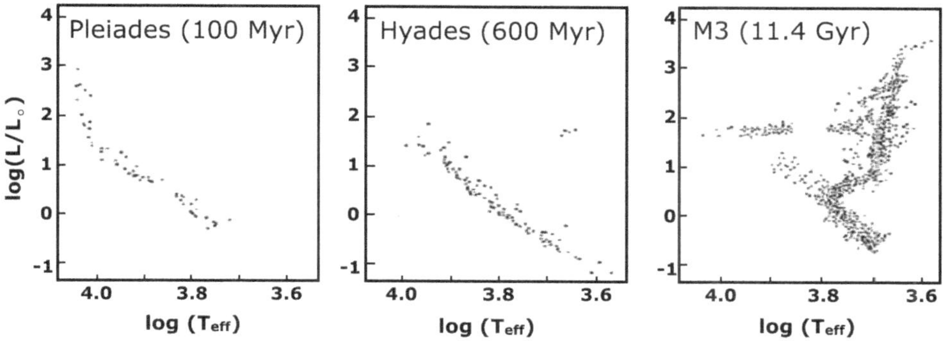

Figure 2.6. HRD of three clusters of increasing age. (Reproduced from Prialnik 2004. © Cambridge University Press. Reprinted with permission.)

Notice three major differences between the two figures.

1. At the hot end (left), the stars in the CMD cover only a small range in B-V but a much larger range in log $T_{\rm eff}$.
2. At the cool end (right), the stars cover a small range in log $T_{\rm eff}$ but a large range in B-V. This difference is due to the highly nonlinear relation between log $T_{\rm eff}$ and B-V as already shown in Table 2.1.
3. The horizontal branch at log $(L/L_{\odot}) \approx 1.7$ in the HRD and at $V \approx 15.5$ in the CMD strongly curves downward in the CMD. This is due to the strong UV flux of hot stars that does not contribute to the flux in the V-band.

Q (2.2) *Explain in physical terms why the red giant branch curves strongly to high values of B-V for a relatively small change in* $T_{\rm eff}$. *(Hint: use the properties of the Planck curve for blackbody radiation)*

Q (2.3) *Explain why the distribution of stars in the CMD of M3 is completely different from Figure (2.4).*

As a star evolves and its temperature, luminosity, and radius change, it moves to a different location in the HRD or CMD. This is clear if we compare HRDs with different ages. Figure 2.6 shows the HRD of three star clusters with different ages. Note that the MS gets shorter (i.e., extends to lower luminosity) with age and the ratio between the number of red giants and the number of main-sequence stars increases with age.

2.7 Nomenclature of Regions in the HRD and CMD

The different regions in the HRD or CMD that are occupied by stars have been given different names to distinguish stellar types. Later, we will show that each of these types refers to a specific evolutionary phase with its corresponding internal structure. Figure 2.7 shows the regions and their names.

The stars are classified according to the region that they occupy in the HRD:

MS stars = Main-sequence stars.

SGB stars = Subgiants (SGB: for subgiant branch).

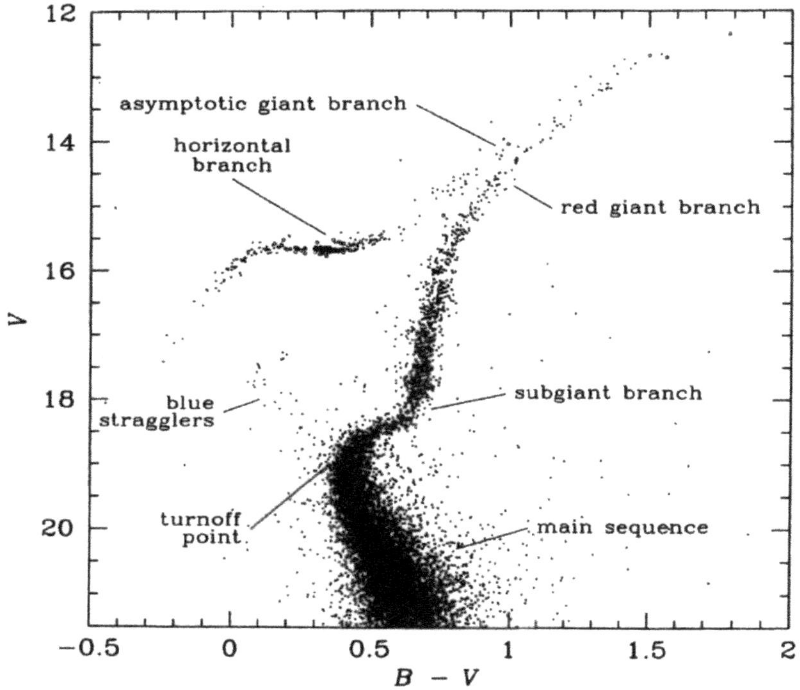

Figure 2.7. Nomenclature of different regions in the CMD of the cluster M3 with an age of 11.4 Gyr. (Reproduced from Renzini & Fusi Pecci 1988.)

RGB stars = Red giants (RGB: for red giant branch).
AGB stars = Asymptotic giant branch stars.
HB stars = Horizontal branch stars.
Blue stragglers = stars on the main sequence but brighter than stars at the turn-off point.

There are other stars that are not shown in the HRD of the old cluster of Figure 2.7.
BSG stars = Blue supergiants = hot stars more luminous than AGB stars.
RSG stars = Red supergiants = cool stars more luminous than AGB stars.
WR stars = Wolf-Rayet stars = luminous hot He-rich stars.
WD stars = White dwarfs = stellar remnants of low-mass stars ($M < 8M_\odot$).
NS = Neutron stars = stellar remnants of high-mass stars ($M > 8M_\odot$).

There are also several types of variable stars not listed here. They will be described later in Chapter 21 when we discuss pulsating stars.

2.8 Summary

1. The observed brightness of a star is expressed in magnitudes and its color is expressed by the difference between magnitudes measured in different filters.

Brighter objects have smaller magnitudes. If the distance of a star is known, the apparent magnitude can be converted into an absolute magnitude.

2. The absolute bolometric magnitude M_{bol} is a measure of the luminosity of a star, with $M_{bol} = 4.74$ corresponding to the luminosity of the Sun and $\log(L/L_\odot) = -0.4 \times (M_{bol} - 4.74)$.
3. The luminosity of a star depends on its effective temperature and radius as $L/L_\odot = (R/R_\odot)^2 \times (T_{eff}/T_{eff\,\odot})^4$ with $T_{eff\odot} = 5777$ K.
4. The parameters of stars can be plotted in the Hertzsprung–Russell diagram (HRD), with $\log(T_{eff})$ on the x-axis and $\log(L/L_\odot)$ on the y-axis, or in the color–magnitude diagram (CMD) with a color on the x-axis and a magnitude (absolute or apparent) on the y-axis.
5. Stars are classified according to their location in the HRD. This classification is related to the evolutionary phase of the stars.
6. Observations show that there is a strict relation of the type $L \sim M^\alpha$ between luminosity and mass for main-sequence stars, with $\alpha = 2.4$ for $M < 0.6M_\odot$, $\alpha = 4.0$ at $0.6 < M < 6M_\odot$, and $\alpha = 2.9$ at $M > 6M_\odot$.

Exercises

2.1 (a) Calculate the radius of an M5 I supergiant with $\log(L/L_\odot) = 5.50$ and $T_{eff} = 2700$ K.
(b) Assume a mass of approximately $20M_\odot$ and calculate the mean density of the star.
(c) Calculate the escape velocity.
(d) Compare these values with those for the Sun.

2.2 (a) Calculate the luminosities of the horizontal branch stars with $B\text{-}V \approx +0.30$ and $BC = -0.11$ in the cluster M3 (NGC 5272), whose CMD is shown in the right panel of Figure 2.5. The distance to M3 is 10.4 kpc and its interstellar extinction is negligible. Compare your result with the HRD that is shown in Figure 2.5 (left panel).
(b) What is the bolometric correction of the two stars at $B\text{-}V \approx -0.25$?

2.3 The star τ Sco has an apparent visual magnitude of $V = +2.8$ and a spectral type of approximately B0V. Parallax measurements indicate a distance of 470 ly.
(a) Calculate the absolute visual magnitude.
(b) Adopt the bolometric correction from Table 2.1 and calculate the luminosity L.
(c) Adopt the value of T_{eff} from Table 2.1 and calculate the radius.
(d) Estimate the mass.
(e) Calculate the acceleration of gravity at the stellar surface and the escape velocity.
(f) Calculate the mean density of the star.
(g) Compare these values with those of the Sun.

2.4 The *Gaia* satellite will measure parallaxes with an accuracy of 2×10^{-5} arcsec for stars with $V < 15$.

(a) What is the distance d of an M5 I supergiant with $\log(L/L_{\odot}) = 5.50$ and $BC = -3.70$ that has $V = 12$, if the effect of interstellar extinction is ignored?
(b) What is the relative distance accuracy $\sigma(d)/d$ of such a star?
(c) Assuming that $T_{\rm eff}$ is known, what is the relative accuracy in radius $\sigma(R)/R$?

References

Buonanno, R., Corsi, C. E., Buzzoni, A., et al. 1994, A&A, **290,** 69

Cox, A. N. 2000, Allen's Astrophysical Quantities (4th ed.; New York: Springer)

ESA 1997, The Hipparcos and Tycho catalogues (ESA-SP 1200; Paris: ESA), http://www.cosmos.esa.int/web/hipparcos/catalogues

GAIA Collaboration 2016, A&A, **595,** 2

Johnson, H. L., & Sandage A. R. 1956, ApJ, **124,** 379

Pols, O. 2011, Stellar Structure and Evolution, Utrecht University Lecture Notes, https://www.astro.ru.nl/~onnop/education/stev_utrecht_notes/

Prialnik, D. 2004, in An Introduction to Stellar Structure and Evolution 2004 (Cambridge: Cambridge Univ. Press), 12

Renzini, A., & Fusi Pecci, F. 1988, ARA&A, **26** 199

AAS | IOP Astronomy

Understanding Stellar Evolution

Henny J.G.L.M. Lamers and Emily M. Levesque

Chapter 3

Hydrostatic Equilibrium and Its Consequences

This chapter deals with the condition of hydrostatic equilibrium, which implies that no part of the star is expanding or contracting. This requires that, at all depths in the star, the inward-directed force from gravity is counterbalanced by the outward-directed force of the pressure gradient. Although stars change during their evolution, the evolutionary timescales are so long compared to the timescale for reaching hydrostatic equilibrium that for most of its evolution a star can be considered to remain in hydrostatic equilibrium. This allows a very significant simplification for computing stellar evolution. We will see that the condition of hydrostatic equilibrium provides a basic property of stars expressed in *the virial theorem.*

3.1 Conservation of Mass: The Mass Continuity Equation

The mass M_r within a sphere of radius r is related to the density structure $\rho(r)$ by

$$dM_r/dr = 4\pi r^2\rho(r) \qquad \text{or} \qquad \rho(r) = (dM_r/dr)/4\pi r^2 \tag{3.1}$$

3.2 Hydrostatic Equilibrium

The equation of hydrostatic equilibrium follows from the equation of motion with the condition that the net force is zero. The equation of motion of one cm^3 gas with density ρ in the shell at distance r and with a thickness dr can be written as

$$\rho\frac{d^2r}{dt^2} = -\rho\frac{GM_r}{r^2} - \frac{dP}{dr}, \tag{3.2}$$

where the first term describes the radial acceleration (positive for outward motion), the second term is the force of gravity, and the last term is the force due to a pressure gradient. Equation (3.2) is the radial variant of the more familiar form of Newton's equation $ma = f$.

doi:10.1088/978-0-7503-1278-3ch3

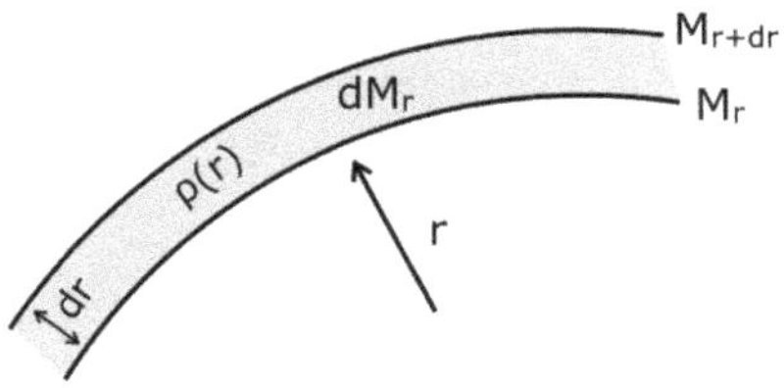

Figure 3.1. Relation between mass, M_r, and density, ρ, in a spherical shell.

In hydrostatic equilibrium, abbreviated as HE, the acceleration d^2r/dt^2 is zero and so

$$\boxed{\frac{dP}{dr} = -\rho\frac{GM_r}{r^2} = -\rho g_r}, \tag{3.3}$$

where g_r is the local acceleration of gravity. This is the **equation for hydrostatic equilibrium (HE)** in a star. If this is combined with Equation (3.1) for ρ, we obtain an alternative expression for the pressure gradient

$$\frac{dP}{dM_r} = -\frac{GM}{4\pi r^4}. \tag{3.4}$$

This expression has an interesting consequence. The integration from $r = 0$ to the radius R of the star provides a lower limit to the pressure at the center of the star

$$P(M) - P_c = -\int_0^M \frac{GM_r dM_r}{4\pi r^4} < -\int_0^M \frac{GM_r dM_r}{4\pi R^4} = -\frac{GM^2}{8\pi R^4}, \tag{3.5}$$

where $P_c = P(r = 0) = P(M_r = 0)$ is the pressure at the center of the star. At the radius R of the star, which encompasses its total mass M, the gas pressure $P(M)$ goes to zero and so

$$P_c > \frac{GM^2}{8\pi R^4}. \tag{3.6}$$

This is a very safe lower limit because, in the integration above, we adopted $r^4 < R^4$, while in fact $r^4 \ll R^4$ near the center. We can obtain a better estimate if we integrate Equation (3.3), adopt the mean density $\bar{\rho}$ in the star, and substitute $M_r = (4\pi/3)r^3\bar{\rho}$.

$$\int_0^R dP = P(R) - P_c = \frac{4\pi}{3}G\bar{\rho}^2\cdot\int_0^R r\,dr = -\frac{2\pi}{3}GR^2\bar{\rho}^2 = -\frac{3}{8\pi}\frac{GM^2}{R^4}. \tag{3.7a}$$

This provides a better estimate for the central pressure of any star in HE

$$P_c \approx \frac{3}{8\pi}\frac{GM^2}{R^4}. \tag{3.7b}$$

Q (3.1) *Argue that, for any realistic density distribution of the star, this will also be a lower limit.*

We see from these arguments that the central pressure in a star will be on the order of

$$\boxed{P_c \sim \frac{GM^2}{R^4},} \tag{3.8}$$

with the proportionality factor depending on the density distribution. The more strongly peaked the density is, the larger the proportionality factor.

Q (3.2) *Could you have guessed that the central pressure is of the order of GM^2/R^4? Hint: consider the weight of a column of gas on the center.*

Having obtained an estimate for the central pressure of a star, we can use the gas law to derive an estimate for the central temperature if the gas pressure is larger than the radiation pressure. We will see later in Section 4.5 that this is a valid assumption except for degenerate stars and main-sequence stars more massive than about $50M_\odot$.

The ideal gas law states that $P = nkT = (\rho/\mu m_\mathrm{H})\,kT$, where n is the number of particles per cm^3 and μm_H is the mean mass per particle. The pressure in the center of the star is $P_c \sim (\bar{\rho}/\mu m_\mathrm{H})kT_c \sim GM^2/R^4$. Adopting $\bar{\rho} \sim M/R^3$, we obtain an order of magnitude estimate of the central temperature of a star in HE

$$\boxed{T_c \sim \frac{\mu m_\mathrm{H}}{k}\frac{GM}{R}}. \tag{3.9}$$

3.2.1 Estimate of the Central Properties of the Sun

Let us test the derived estimates for the Sun. The predicted central pressure in Equation (3.8) is

$$P_c^\odot \simeq \frac{GM^2}{R^4} = \frac{7\times10^{-8}(2\times10^{33})^2}{(7\times10^{10})^4} \simeq 1\times10^{16}\ \mathrm{dyne\ cm^{-2}} \simeq 10^{10}\ \mathrm{atmosphere}.$$

Detailed solar models show that $P_c^\odot \approx 2.3\times10^{17}\ \mathrm{dyne\ cm^{-2}}$ (Guenther et al. 1992). The large discrepancy between this value and our estimate is due to the fact that $\rho_c \ggg \bar{\rho}$.

The predicted central temperature of the Sun according to Equation (3.9) is

$$T_c^\odot \simeq \frac{0.5\times1.7\times10^{-24}}{1.4\times10^{-16}}\times\frac{7\times10^{-8}\times2\times10^{33}}{7\times10^{10}} \simeq 2\times10^7\ \mathrm{K}.$$

Detailed solar models show that $T_c^\odot \approx 1.6\times10^7$ K (Guenther et al. 1992). We see that the estimate of the central temperature is more accurate than the estimate of the central pressure.

3.3 The Virial Theorem: A Consequence of HE

The **virial theorem (VT)** links the total potential energy of a star in HE to the total internal (kinetic) energy of the star. We will see later that this helps with understanding why stars expand or contract during certain evolutionary phases. The word "virial" is derived from the Latin word of "vis" meaning "energy" and was introduced by the Polish–German thermodynamicist Rudolph Clausius (1822–1888) in 1870.

The mathematical derivation of the virial theorem is easy to follow. Since it is a consequence of HE, we start from Equation (3.3), multiply both sides by $4\pi r^3\, dr$, and integrate from $0 \rightarrow R$. This yields

$$\int_0^R 4\pi r^3\, dP = -\int_0^R \frac{GM_r}{r}\rho 4\pi r^2\, dr. \tag{3.10}$$

Now consider both terms separately. The potential energy of a cm^3 of gas with density ρ at distance r is $-(GM_r/r)\,\rho$, where M_r is the mass enclosed within radius r. So the right side of Equation (3.10)

$$-\int_0^R \left(\frac{GM_r}{r}\right)\rho 4\pi r^2\, dr = E_{\mathrm{pot}} \tag{3.11}$$

is the **total potential energy of the star** (in the literature it is sometimes expressed by the symbol Ω). For the left term of Equation (3.10), we use the rule for integration by parts

$$\int_a^b d(uv) = uv|_a^b = \int_a^b u\, dv + \int_a^b v\, du.$$

If we define $u = 4\pi r^3$ and $v = P$, so $dv = dP$ and $du = 12\pi r^2\, dr$, we find that

$$\int_0^R 4\pi r^3\, dP = 4\pi r^3 P|_0^R - \int_0^R 12\pi P r^2\, dr = -3\int_0^R 4\pi P r^2\, dr. \tag{3.12}$$

We used $4\pi r^3 P|_0^R = 0$ because at the lower boundary $r = 0$ and at the upper boundary $P(R) = 0$. For a star whose HE is dominated by gas pressure, we can link P to the internal energy. For an ideal gas $P = nkT$, where n is the number of particles per cm^3 and the internal energy per cm^3 is $u = (3/2)\, nkT$, $P = (2/3)u$. Using this in Equation (3.12), we find that

$$\int_0^R 4\pi r^3\, dP = -3\int_0^R P \times 4\pi r^2\, dr = -2\int_0^R u \times 4\pi r^2\, dr = -2E_{\mathrm{kin}}, \tag{3.13}$$

where E_{kin} is the **total kinetic energy of star** (sometimes expressed by the symbol U). Combining the results of Equations (3.10), (3.11), and (3.13) yields

$$\boxed{-E_{\mathrm{pot}} = 2E_{\mathrm{kin}}} \text{ or } \boxed{E_{\mathrm{kin}} = -\tfrac{1}{2}E_{\mathrm{pot}} > 0}. \tag{3.14}$$

This is the virial theorem (VT). From the way we derived it, it is clear that it only applies to the star as a whole, not to individual layers or regions. It can also be expressed in terms of the **total energy of a star**

$$\boxed{E_{\mathrm{tot}} = E_{\mathrm{pot}} + E_{\mathrm{kin}} = \tfrac{1}{2}E_{\mathrm{pot}} < 0}. \tag{3.15}$$

Q (3.3) *Does the VT apply to massive main-sequence stars, where HE is partly supported by radiation pressure? Why?*

Q (3.4) *Does the Virial Theorem apply to degenerate stars?*
Hints: (a) Did we use the ideal gas law in deriving it? (b) How did we use it?

3.3.1 Consequences of the VT for Contracting Stars

1. If a star is formed from a very cold contracting extended interstellar cloud with a very small total energy, it must have lost half of its final potential energy in the form of radiation.
2. When a star contracts and decreases its potential energy (i.e., E_{pot} becomes more negative), the virial theorem requires that

$$dE_{\text{kin}}/dt = -\tfrac{1}{2}\, dE_{\text{pot}}/dt. \tag{3.16}$$

   So only half of the released energy goes into thermal energy for heating the star; the other half must be radiated away.
3. When a star runs out of nuclear energy and compensates for its radiative energy loss by contraction, it must release twice as much potential energy as it radiates because half of the released potential energy is used for heating the star. So the VT dictates that a contracting star must radiate and increase its temperature.
4. The VT also applies to galaxy clusters in equilibrium, where the internal energy is the total kinetic energy of the motions of the galaxies. In fact, the presence of dark matter in galaxy clusters was discovered from the determination of the total kinetic energy of the observed galaxies. The total potential energy of the observed galaxies, required by the VT, turned out to be about a factor of 10 higher than could be explained by the observed galaxies. This called for the presence of unseen matter that exerts gravity. This is called dark matter.

3.4 Summary

1. The central pressure of a star in HE is on the order of $P_c \sim GM^2/R^4$.
2. If the pressure in a star in HE is dominated by gas pressure, then the central temperature is on the order of $T_c \sim (\mu m_{\text{H}}/k)(GM/R)$.
3. A star in hydrostatic equilibrium must obey the condition expressed in the virial theorem: its (positive) kinetic energy plus half of its (negative) potential energy is zero.
4. The total energy of a star in HE is negative and equal to half of the potential energy.
5. A contracting star uses half of the released potential energy to increase its average internal temperature. The other half is radiated away.

Exercises

3.1 We have derived the VT mathematically, using some tricks such as multiplication of the integral and substitutions of variables for integration in parts. Can you also explain in physical terms that a relation between E_{pot} and E_{kin} is to be expected?

3.2 Why is the estimate of T_c (Equation 3.9) better than that of P_c (Equation 3.8) in comparison with the solar value?
Hint: consider how the density is used in both estimates.

3.3 We will show later that for nonrelativistic degenerate gas $P = 1/3u$. What does that imply for the VT for degenerate stars?

3.4 If $\bar{T}$ and $\bar{\rho}$ are the mean temperature and density of a star, then show by using the VT that $\bar{T}$ can be described by a relation that is similar to Equation (3.9).

3.5 The VT also applies to the orbits of the individual planets around the Sun. Show that Kepler's third law follows directly from the VT.

3.6 **Computer exercise.**
The estimates of P_c and T_c, derived from the HE condition in Equations (3.7a) and (3.9), were obtained by assuming a constant mean density. These estimates can be improved by adopting a more realistic density distribution. It turns out that a density distribution of the type $\rho/\rho_c \approx 3e^{-10x}$ with $x = r/R$ is a reasonable approximation for main-sequence stars in the range of $0.11 < x < 0.98$. Adopt this, plus a constant density core with $\rho/\rho_c = 1$ at $x < 0.11$.
(a) Derive an expression for the mean density by computer.
(b) Use the HE condition to derive an expression for the central pressure.
(c) Derive an expression for the central temperature, using the ideal gas law.
(d) Apply these estimates to a zero-age main-sequence star of $1M_\odot$ and compare the results with the values from stellar models: $\rho_c = 78$ g cm^{-3}, $P_c = 1.2 \times 10^{17}$ dyne cm^{-2}, and $T_c = 1.36 \times 10^7$ K.
(e) Do the same for a zero-age main-sequence star of $60M_\odot$: $\rho_c = 2.2$ g cm^{-3}, $P_c = 1.2 \times 10^{16}$ dyne cm^{-2}, and $T_c = 8.31 \times 10^7$ K.
(f) Comment on the differences.

Reference

Guenther, D. B., Demarque, P., Kim, Y.-C. & Pinsonneault, M. H. 1992, ApJ, **387 372**

AAS | IOP Astronomy

Understanding Stellar Evolution

Henny J.G.L.M. Lamers and Emily M. Levesque

Chapter 4

Gas Physics of Stars

In this chapter, we describe the relation between the major properties of the gas inside stars, i.e., density ρ, temperature T, and pressure P. This relation is called the **equation of state (EoS)**. As the conditions in stars cover a wide range in characteristics, the EoS varies between different regimes in the (T, P, ρ) domain. At low densities, the EoS is described by the well-known ideal gas law. At very high densities, however, quantum mechanical effects play a role and the energy distribution of particles can no longer be described by ideal gas laws.

If that is the case, matter is said to be *degenerate* and the EoS is very different from the ideal gas law. Because of the large ratio in mass between ions and electrons, ions become degenerate at a much higher density than electrons. This implies that in the same layers of a star, a different EoS may have to be applied for the electrons and ions. Apart from the pressure produced by particles, photons can also contribute to the pressure. This is called *radiation pressure*. In the cores of massive main-sequence stars radiation pressure may even dominate over the gas pressure. We will derive expressions for the various EoS and show the conditions where they apply.

4.1 Mean Particle Mass

The mean particle mass depends on the composition and the degree of ionization of the gas. It is convenient to express the mean particle mass in atomic mass units (AMU). An AMU is so close to a proton mass that we will use $m_{\rm H}$ for simplicity (see Table 1.1).

The atomic mass A_j is also expressed in units of $m_{\rm H}$; for instance, the atomic mass of an O atom is $A_{\rm O} \times m_{\rm H}$ with $A_{\rm O} = 16$. We express the chemical composition in terms of the mass fractions X_j of each element (e.g., $X_{\rm H}$, $X_{\rm He}$, $X_{\rm C}$...) for H, He, C, etc. We adopt the definition frequently used in the literature:

$$X \equiv X_{\rm H}, \quad Y \equiv X_{\rm He}, \text{ and } Z \equiv 1 - X - Y. \tag{4.1}$$

doi:10.1088/978-0-7503-1278-3ch4

From this we can see that Z is the mass fraction of all elements together except H and He. The composition of the solar atmosphere is $X = 0.70$, $Y = 0.28$, and $Z = 0.02$.

The number of free electrons of element j is given by the symbol q_j. For instance, $q_C \approx 1$ at solar photosphere and $q_C = 6$ in the fully ionized solar core. For fully ionized gas

$$q_H = 1, \ q_{He} = 2, \text{ and } q_j \approx 0.5A_j \qquad \text{for } j > 2. \tag{4.2}$$

With these definitions, we find that the total number of ions per gram is $\Sigma_j(X_j/A_j m_H)$ and the total number of electrons per gram is $\Sigma_j(X_j q_j/A_j m_H)$. It is useful to define

mean mass per ion: $\mu_i m_H = 1/\Sigma_j(X_j/A_j m_H)$, so $\mu_i = \{\Sigma_j(X_j/A_j)\}^{-1}$, (4.3*a*)

mean mass per electron: $\mu_e m_H = 1/\Sigma_j(X_j q_j/A_j m_H)$; so $\mu_e = \{\Sigma_j(q_j X_j/A_j)\}^{-1}$, (4.3*b*)

mean mass per particle:
$$\mu m_H = \frac{1}{\Sigma_j(X_j/A_j m_H) + \Sigma_j(X_j q_j/A_j m_H)},$$
$$\text{so } \mu = \{\Sigma_j(X_j/A_j)(1 + q_j)\}^{-1}, \text{ and} \tag{4.3c}$$

number density of particles: $n = \rho/\mu m_H = 5.98 \times 10^{23} \rho/\mu$. (4.3*d*)

With these definitions, the number of electrons, ions, and all particles per cm^3 are, respectively, $n_e = \rho/\mu_e m_H$, $n_i = \rho/\mu_i m_H$, and $n = \rho/\mu m_H$. It can be shown that for **a fully ionized gas**

$$\mu_e = \frac{2}{1 + X} \tag{4.4}$$

(see Exercise 4.1).

Note that the mean mass per electron is not the mass of the electron. For instance, if the degree of ionization is so low that a gram of gas would contain only 1 electron, the mean mass per electron would be 1 g.

4.2 A General Expression for the Pressure

The pressure exerted by particles or photons can be derived by considering a small volume, a box of $1 \times 1 \times 1$ cm^3, filled with particles of velocity v. When a particle hits the side of a box at an impact angle θ and bounces off, its momentum $2m\text{v} \cos\theta$ is transferred to the walls. The factor of 2 comes from the recoil during the bounce (see Figure 4.1).

Let us first assume for simplicity that all particles have the same mass m, the same speed v, and the same momentum $p = m\text{v}$. The time between two collisions with the same wall is

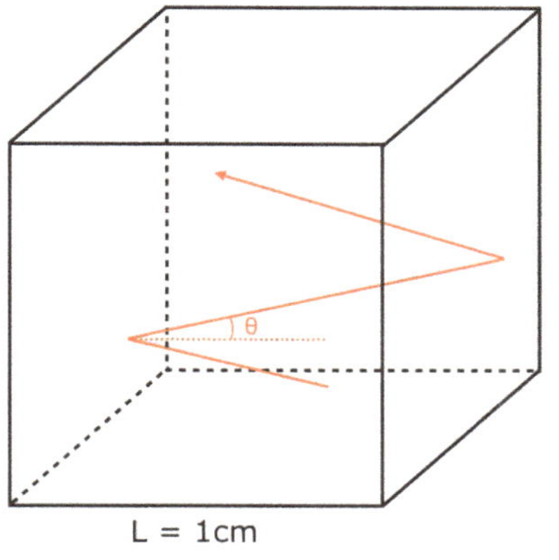

Figure 4.1. Schematic figure showing the calculation of the pressure produced by a particle bouncing in a box of 1 × 1 × 1 cm.

$$\Delta t = 2L/\mathrm{v}\cos\theta = 2/\mathrm{v}\cos\theta, \qquad \text{with } L = 1 \text{ cm for a box of } 1 \text{ cm}^3. \tag{4.5}$$

The momentum transfer per collision is $\Delta p = 2p\cos\theta$. So the momentum transfer per particle per second onto one side of 1 cm^2 is

$$\frac{1}{\Delta t}\Delta p = \mathrm{v}p\cos^2\theta. \tag{4.6}$$

Integrate this over all possible angles for an isotropic velocity distribution, with

$$\int_0^{\pi/2} \sin\theta\, d\theta\, \cos^2\theta = -\frac{1}{3}\cos^3\theta\Big|_0^{\pi/2} = \frac{1}{3}, \tag{4.6a}$$

and multiply it by the total number of particles n per cm^3. The pressure on one wall exerted by one type of particle with velocity v, momentum p, and density n is $P = n\mathrm{v}\,p/3$. In general, the particles will have a distribution in velocity and momentum. If $n(p)$ is the momentum distribution of the particles and $\mathrm{v}(p)$ is the velocity distribution, normalized by the integral of $\mathrm{v}(p)\, n(p)dp = n$, the general expression for the pressure is

$$\boxed{P = 1/3\int_0^\infty \mathrm{v}(p)\, p\, n(p)\, dp}. \tag{4.7}$$

We will use this expression to calculate the pressure under various conditions.

4.3 Radiation Pressure

The velocity of photons is c and their momentum is $p = h\nu/c$. This implies that $n(p)dp = n(\nu)d\nu$ and $dp = \frac{h}{c}d\nu$. For blackbody radiation, $n(\nu)$ is described by the Planck function

$$n(\nu)d\nu = \frac{8\pi\nu^2}{c^3}\frac{d\nu}{e^{h\nu/kT} - 1} \qquad \text{or} \qquad n(p)dp = \frac{8\pi\nu^2}{c^3}\frac{1}{e^{h\nu/kT} - 1}\frac{c}{h}dp. \tag{4.8}$$

The Planck function is named after the German physicist Max Planck (1858–1947). Substituting it into Expression (4.7) yields

$$P_{\text{rad}} = \frac{1}{3}\int cp\, n(p)\, dp = \frac{1}{3}\int_0^\infty \frac{h\nu}{c}\, c\, n(\nu)\, d\nu, \tag{4.9}$$

which results in the final expression for the **radiation pressure**

$$\boxed{P_{\text{rad}} = \tfrac{1}{3}aT^4}, \tag{4.10}$$

with $a \equiv \frac{4\sigma}{c} \equiv \frac{8\pi^5 k^4}{15c^3 h^3} = 7.56 \times 10^{-15}$ erg cm^{-3} K^{-4} $= 7.56 \times 10^{-15}$ dyne cm^{-2} K^{-4}.

The radiation pressure is often expressed as a fraction of the total (gas + radiation) pressure

$$P_{\text{gas}} = \beta P, \quad \text{with} \quad 0 < \beta < 1 \quad \beta \approx 0\text{: } P_{\text{rad}} \text{ dominates} \tag{4.11a}$$

$$P_{\text{rad}} = (1 - \beta)P \qquad \beta \approx 1\text{: } P_{\text{gas}} \text{ dominates.} \tag{4.11b}$$

Almost all stars are dominated by gas pressure. Radiation pressure is only significant in the interiors of massive stars ($M > 30M_\odot$), which have a very high luminosity; in the atmospheres of luminous hot and cool stars, where radiation pressure drives a stellar wind; and during the very high luminosity phase of supernova explosions.

4.4 Pressure of an Ideal Gas

The Maxwell momentum distribution of particles is

$$n(p)dp = n_{\text{tot}} \frac{4\pi p^2\, dp}{(2\pi mkT)^{3/2}}\, e^{-p^2/2mkT}. \tag{4.12}$$

It is named after the British physicist James Clerk Maxwell (1831–1879), who published it in 1859 at the age of 28. Combining this with the expression for the momentum, $v = p/m$, yields

$$P = \frac{1}{3}\int v\ pn(p)\, dp = n_{\text{tot}}kT = \rho kT/m = \frac{\Re \rho T}{\mu}, \tag{4.13}$$

where $\Re = k/m_{\text{H}} = 8.31 \times 10^7$ erg K^{-1} mole is the gas constant. We can express the contribution to the pressure by the different constituents of the gas.

The pressure by ions: $P_i = n_i kT = \dfrac{\Re \rho T}{\mu_i}$ with $1/\mu_i \simeq X + \dfrac{Y}{4} + \dfrac{Z}{20}$, (4.14a)

The pressure by electrons: $P_e = n_e kT = \dfrac{\Re \rho T}{\mu_e}$ with $1/\mu_e \simeq (1 + X)/2,$ (4.14*b*)

The total gas pressure: $P_{\rm gas} = P_i + P_e = n_{\rm tot} kT = \dfrac{\Re \rho T}{\mu}$ with $1/\mu = 1/\mu_e + 1/\mu_i.$ (4.14*c*)

The approximate expressions for μ are valid for a fully ionized gas. Remember that $\mu_e\, m_{\rm H}$ and $\mu_i\, m_{\rm H}$ are not the masses *of* the electrons and ions but the mean mass *per* electron and *per* ion (see Equation (4.3)).

The pressure P at any location in a star is the sum of the contributions by all constituents at that location. If we only consider ions, electrons, and photons and ignore other particles, such as neutrinos, we find that the **total pressure** is

$$P = P_i + P_e + P_{\rm rad} = P_{\rm gas} + P_{\rm rad} \qquad \text{with } P_{\rm gas} = P_i + P_e \tag{4.15}$$

Q (4.1) *Calculate $P_{\rm gas}$ and $P_{\rm rad}$ in the center of the Sun and show that $P_{\rm rad} \ll P_{\rm gas}$ (use data in Appendix C).*

4.5 Electron Degeneracy

At very high densities, quantum mechanical effects become important. This changes the relation between P, T, and ρ, i.e., the equation of state. The velocity distribution of the electrons at very high densities cannot be Maxwellian because of two principles of quantum mechanics:

1. **Heisenberg uncertainty principle:**
   $\Delta x\, \Delta p > h$ in one-dimensional and $\Delta{\rm Vol}\, \Delta^3 p > h^3$ in three-dimensional phase space with the Planck constant $h = 6.626 \times 10^{-27}$ erg s^{-1} and h^3 is the unit of phase space volume. The uncertainty principle is named after the German theoretical physicist Werner Heisenberg (1901–1976), who published it in 1927 at the age of 26.
2. **Pauli exclusion principle:**
   No two identical particles (same quantum state) can exist at the same time and place (i.e., in the same phase-space volume h^3). This implies that there are at most two electrons (spin up and spin down) in a unit of phase space. The exclusion principle is named after the Austrian–Swiss–American theoretical physicist Wolfgang Pauli (1900–1958), who published it in 1924 at the age of 24.

Figure 4.2 shows a schematic demonstration of the changes in the 1D velocity distribution $n(v_x)$ if more and more identical particles are squeezed into a fixed volume. The horizontal upper limit is set by Pauli's exclusion principle. The first three distributions (black lines) are Maxwellian (M). The blue lines illustrate partial degeneracy (PD). The green lines show complete degeneracy (CD) when the velocity

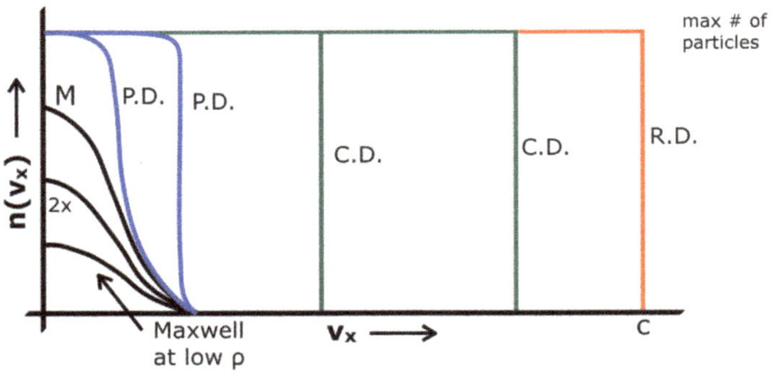

Figure 4.2. One-dimensional velocity or momentum distributions at increasing density. M = Maxwellian (black lines); PD = partial degeneracy (blue lines); CD = complete degeneracy (green lines); RD = relativistic degeneracy (red line).

distribution is rectangular up to the **Fermi momentum** $p_F = mv_F$, which is the highest value of the momentum p for a given density of complete degenerate electron gas. The last distribution (red line), which extends to $p_x = mc$, is for complete relativistic degeneracy (RD). The Fermi momentum, or Fermi energy, is named after the Italian–American nuclear physicist Enrico Fermi (1901–1954), the architect of the nuclear bomb.

The upper limit for a 1D distribution is a constant; however, momentum is a 3D vector, so the upper limit in 3D phase space is a line of the form $f(p) = \int n(p)\ 2\pi p^2\ dp \sim p^3$ if $n(p)$ is constant.

Q (4.2) *What is the role of temperature in the transition from the Maxwellian to the partial degenerate distributions?*
Hint: consider the width of the distributions for different temperatures.

Q (4.3) *What is the role of the particle mass?*

Figure 4.3 shows the momentum distribution for electrons for different temperatures (left) and different densities (right). The left figure is for $n_e = 6 \times 10^{27}$ cm^{-3}, which corresponds to $\rho = 2 \times 10^4$ g cm^{-3} for He and metals, for which $\mu_e = 2$. The right figure is for a very low temperature and $n_e = 2 \times 10^4$ and 4×10^4 g cm^{-3}.

4.5.1 Nonrelativistic Complete Degeneracy (CD)

In the case of complete degeneracy, the $n(p)$ distribution is rectangular for $p < p_F$. So the electron distribution is described by

$$n_e(p)\ d^3p = \frac{2}{h^3}\,(4\pi p^2\ dp) \qquad \text{if } p < p_F \tag{4.16a}$$

$$n_e(p)\ d^3p = 0 \qquad \text{if } p \geqslant p_F. \tag{4.16b}$$

Using these equations, we can derive the Fermi momentum for a total electron density n_e.

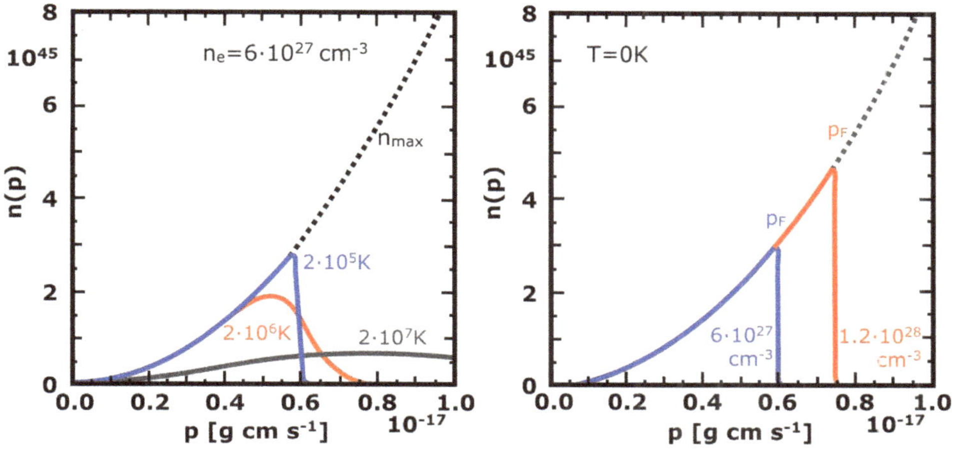

Figure 4.3. Momentum distribution of electrons at several combinations of temperature (left) and densities (right). (© Pols 2011.)

$$n_e = \int_0^{p_F} n_e(p)\ d^3p = \int_0^{p_F} \frac{2}{h^3}\ 4\pi p^2\ dp = \frac{8\pi}{3h^3}\, p_F^3. \tag{4.17}$$

So

$$p_F = \left(\frac{3h^3\, n_e}{8\pi}\right)^{1/3}. \tag{4.18}$$

Now we can find P_e, with $v = p/m_e$

$$P_e = \frac{1}{3}\int p\ v\ n_e(p)\, dp = \frac{1}{3}\int_0^{p_F} \frac{p^2}{m_e}\frac{2}{h^3}\ 4\pi p^2\ dp = \frac{8\pi}{15h^3} p_F^5. \tag{4.19}$$

We have seen above that $p_F \sim n_e^{1/3}$, so the pressure is $P_e\,(CD) \sim n_e^{5/3}$, independent of T!

The **electron pressure for complete degeneracy** is

$$\boxed{P_e(CD) = K_1(\rho/\mu_e)^{5/3}}, \tag{4.20}$$

with $K_1 = 1.00 \times 10^{13}\ \frac{\text{dyne cm}^{-2}}{(\text{g cm}^{-3})^{5/3}}$, if $\mu_e = 2$, i.e., if $X_H = 0$.

4.5.2 Extreme Relativistic Degeneracy (ERD)

If the density is so high that the electrons are relativistic, then $v = c$ and $vp = cp$. So

$$\begin{aligned} P_e &= \frac{1}{3}\int pv\, n_e(p) 4\pi p^2\ dp = \frac{c}{3}\int p n_e(p)\ 4\pi p^2\ dp \\ &= \frac{2c}{3h^3}\int_0^{p_F} 4\pi p^3 dp = \frac{2c}{3h^3}\,\pi p_F^4 \qquad \text{with } p_F \sim n_e^{1/3}. \end{aligned} \tag{4.21}$$

Again, the electron pressure is independent of T, as in the CD case, but now $P_e \sim n_e^{4/3}$.

The **electron pressure for relativistic degeneracy** is

$$\boxed{P_e(RD) = K_2(\rho/\mu_e)^{4/3}}, \tag{4.22}$$

with $K_2 = 1.24 \times 10^{15} \dfrac{\text{dyne cm}^{-2}}{(\text{g cm}^{-3})^{4/3}}$.

4.5.3 Partial Degeneracy (PD)

In the case of partial degeneracy, the momentum distribution of electrons at low values of p is rectangular in 1D, but at higher values it approaches the tail of the Maxwell distribution.

The **electron pressure for partial degeneracy** is

$$\boxed{P_e(PD) = \frac{8\pi}{3h^3}(2m_e\, kT)^{3/2}\, kT\, F_{3/2}(\psi)}, \quad \text{where } F_{3/2}(\psi) \equiv \int_{\infty}^{0} \frac{x^{3/2}}{e^{r-\psi}+1} dx. \tag{4.23}$$

This expression is given here for the sake of completeness. Its derivation can be found in the literature (e.g., Maeder 2009).

4.6 The Equation of State (EoS) for Electron Gas

We have seen that the pressure of CD electron gas has two limiting cases. In the nonrelativistic case, $P_e = K_1 \times n_e^{5/3}$, and in the relativistic case, $P_e = K_2 \times n_e^{4/3}$. We can derive the electron density n_e^{crit} at which the gas goes from CD to RD by requiring

$$\begin{aligned} P_e(CD) = P_e(\text{RD}) &\rightarrow K_1(\rho/\mu_e)^{5/3} = K_2(\rho/\mu_e)^{4/3} \rightarrow \\ &\rightarrow (\rho/\mu_e)_{\text{crit}} = (K_2/K_1)^3 = 1.91 \times 10^6 \text{ g cm}^{-3}. \end{aligned} \tag{4.24}$$

Similarly, the boundary between the ideal gas law and CD follows from

$$n_e\, kT = K_1(\rho/\mu_e)^{5/3} \rightarrow T = \frac{K_1 m_{\text{H}}}{k}(\rho/\mu_e)^{2/3} = 1.21 \times 10^5 (\rho/\mu_e)^{2/3} \text{ K}. \tag{4.25}$$

The boundary between the ideal gas law and RD follows from

$$n_e\, kT = K_2(\rho/\mu_e)^{4/3} \rightarrow T = \frac{K_2 m_{\text{H}}}{k}(\rho/\mu_e)^{1/3} = 1.50 \times 10^7 (\rho/\mu_e)^{1/3} \text{ K}. \tag{4.26}$$

The boundary between radiation pressure and the ideal gas law follows from

$$n_e\, kT = \frac{a}{3}T^4 \rightarrow T = \left(\frac{3k}{am_{\text{H}}}\right)^{1/3}(\rho/\mu_e)^{1/3} = 3.20 \times 10^7 (\rho/\mu_e)^{1/3} \text{ g cm}^{-3}. \tag{4.27}$$

We note that the boundaries are not sharp: the EoS near the boundaries gradually goes from one regime into the other. For instance, near the boundary between nonrelativistic and relativistic degeneracy, given by Equation (4.24), some of the electrons will be relativistic, with $v = c$, while others are still nonrelativistic, with $v \lesssim c$.

Figure (4.4) shows the regimes in the (ρ, T) diagram, where the various expressions for the EoS apply, for an assumed composition of $X = 0.70$ and $Z = 0.02$.

We will see later that the EoS in the core of a main-sequence stars of $0.1\ M_\odot$ is close to the degeneracy limit. In the cores of main-sequence stars of $0.2 < M < 30 M_\odot$, the EoS of ideal gas applies. In the cores of main-sequence stars of $M > 100 M_\odot$, the EoS is at the boundary between the ideal gas and radiation dominated case.

4.7 Neutron Degeneracy

At very high density, the Fermi energy e_F of the degenerate electrons is so high that the electrons can overcome the energy difference between neutrons and protons, $(m_n - m_p)c^2$, which is 1.29 MeV. This happens at a density in excess of about 3×10^7 g cm^{-3}.

Q (4.4) *Why does this require a density limit and not a temperature limit?*

The reaction $p + e \rightarrow n + \nu$ results in a decrease of the number of electrons and protons and an increase in the neutron density. Just like electrons, neutron gas can also become degenerate at high density.

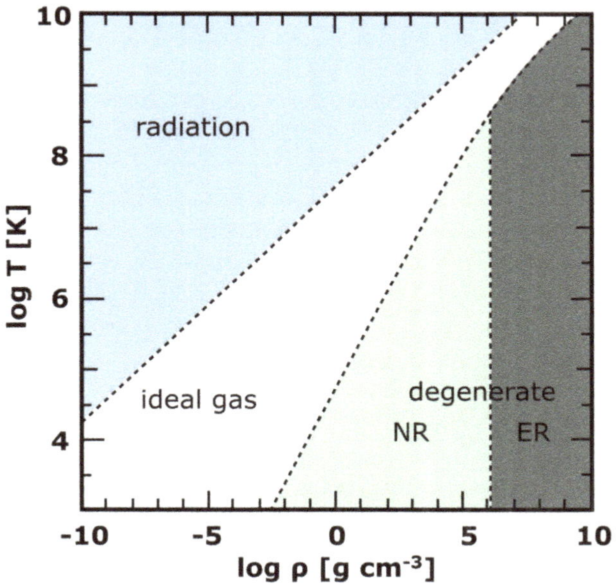

Figure 4.4. Regions in the (ρ, T) diagram, where the different EoS apply for a fully ionized gas with a composition of $X = 0.70$, $Y = 0.28$, and $Z = 0.02$. At this composition, $\mu_e = 2/(1 + X) = 1.18$ and so $n_e = \rho/(m_H \mu_e) = 5.1 \times 10^{23}\ \rho$ cm^{-3}. (Adapted from Pols 2011.)

The electron pressure of nonrelativistic completely degenerate neutron gas is given by a similar expression as that for CD electrons (Equation (4.20))

$$P_n(\mathrm{CD}) = K_{1,n}(\rho/\mu_n)^{5/3} \qquad \text{for } 3 \times 10^7 < \rho < 6 \times 10^{15}\ \mathrm{g\ cm^{-3}}, \tag{4.28}$$

with $\mu_n = 1$ for gas consisting completely of neutrons and $K_{1,n} = 5.38 \times 10^9 \frac{\mathrm{dyne\,.\,cm^{-2}}}{(\mathrm{g\ cm^{-3}})^{5/3}}$. At higher density, the neutron gas becomes relativistically degenerate and the neutron pressure is

$$P_n(RD) = K_{2,n}(\rho/\mu_n)^{4/3} \qquad \text{for } \rho > 6 \times 10^{15}\ \mathrm{g\ cm^{-3}}, \tag{4.29}$$

with $K_{2,\mathrm{n}} = 1.33 \times 10^{15} \frac{\mathrm{dyne\ cm^{-2}}}{(\mathrm{g\ cm^{-3}})^{4/3}}$.

The similarity of Equation (4.28) for neutron stars and Equation (4.20) for the electron degenerate gas of white dwarfs implies that both types of stars will have a similar mass–radius relation (which will be derived in Chapter 20). Moreover, the similarity of Equation (4.29) for neutron stars and Equation (4.22) for white dwarfs implies that for both types of stars there is a maximum mass at which they can be stable.

4.8 Polytropic Gas

We have seen above that in several cases the EoS can be written in the very simple form of

$$\boxed{P \sim \rho^{\gamma}.} \tag{4.30}$$

Such a relation is called a **polytrope** and γ is the **polytropic index**.

The $P(r)$ and $\rho(r)$ structure of polytropic stars can be derived directly from the hydrostatic equilibrium condition because there is no T-dependence!

A historical note: in 1926, Sir Arthur Eddington (1882–1944) "guessed" that stars behave as polytropes, calculated the first model for the solar interior using the polytropic approximation $P \sim \rho^{4/3}$, and found approximately the correct M, R, and even L. His guess was based on the assumption that the ratio between the gas pressure and the radiation pressure might be roughly constant throughout the star. Modern models show that his assumption is incorrect.

Which stars behave as polytropes?

a. Complete electron degenerate stars: because $P \sim n^{5/3} \sim \rho^{5/3}$.
b. Relativistic degenerate stars: because $P \sim n^{4/3} \sim \rho^{4/3}$.
c. Stars with a constant ratio $\beta = P_{\mathrm{gas}}/P_{\mathrm{tot}}$ (Eddingtons' assumption)

$$P_{\mathrm{rad}} = (1 - \beta)P \rightarrow P \sim P_{\mathrm{rad}} \rightarrow P \sim \rho^{4/3}.$$

d. Fully convective stars (we will prove this below) $P \sim \rho^{5/3}$.

In degenerate stars, where $P(r)$ and $\rho(r)$ are independent of T, the condition of HE does not provide any information on the temperature or luminosity of the stars. However, in nondegenerate polytropic stars, the HE condition not only sets $P(r)$ and $\rho(r)$ but also the $T(r)$ structure, because if the gas behaves as a polytrope, $P \sim \rho^\gamma$, *and* if the gas behaves as an ideal gas, $P \sim \rho T$, then $T \sim \rho^{\gamma-1}$ and $T \sim P^{(\gamma-1)/\gamma}$. The temperature structure in a star defines the flow of energy; as a result, the luminosity of such a star is also known.

4.8.1 Proof That a Fully Convective Adiabatic Star Is a Polytrope

The energy transport in some types of stars or some stellar layers is not via the transport of radiation, but by convection: hot rising cells release their energy at the top of the convection zones and cool descending cells gain energy in deeper layers. Let us assume that the convection is (almost) adiabatic, which means that convective cells have no energy loss or gain. We will show later that this is a reasonable approximation in a convective layer.

The first law of thermodynamics states that $du + P\,dV = dQ$, where u is the specific internal energy (in ergs gram^{-1}), $V = 1/\rho$ is the specific volume of a gram of gas, and Q is the heat content. The adiabatic condition implies that $dQ = 0$ so $du = -P\,dV$.

For an ideal gas $u = (3/2)\,NkT$, where $N = nV$ is the number of particles per gram, n is the number of particles per volume, and $P = nkT$, we find $PV = NkT$ and $u = 3/2\;PV$.

For adiabatic expansion or contraction of rising and descending gas cells

$$du = -P\,dV \rightarrow d\left(\frac{3}{2}PV\right) = -P\,dV \rightarrow \frac{3}{2}P\,dV + \frac{3}{2}V\,dP = -P\,dV$$
$$\rightarrow \frac{5}{2}P\,dV = -\frac{3}{2}V\,dP \rightarrow \frac{dP}{P} = -\frac{5}{3}\frac{dV}{V}. \tag{4.31}$$

So

$$P \sim V^{-5/3} \rightarrow \boxed{P \sim \rho^{+5/3}}. \tag{4.32}$$

This shows that a convective region has a polytropic EoS. For ideal gas $P \sim \rho T$, so the temperature structure of a convective region is given by

$$T \sim \rho^{2/3} \quad \text{and} \quad T \sim P^{2/5}. \tag{4.33}$$

Q (4.4) *Can a fully convective star be completely adiabatic?*

4.8.2 The Polytropic Index of Partially Ionized Gas

In the central regions of stars, the gas is fully ionized and acts like an ideal gas (unless it is degenerate at high density); however, in higher layers closer to the

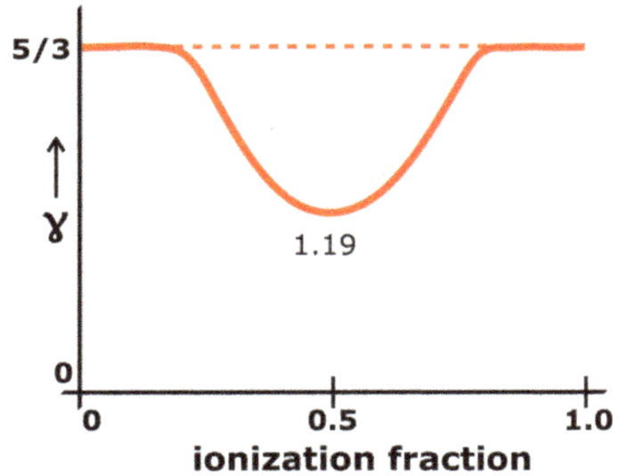

Figure 4.5. Polytropic index γ for partially ionized H-gas.

surface, where the temperature is lower, the gas may be partially ionized. In those layers, the gas is not ideal and $u \neq (3/2)NkT$ because there is also a potential energy from ionization that is released in the case of recombination. For instance, an increase in T will increase the degree of ionization, which will cost energy and vice versa for recombination. The internal energy is therefore higher than that for a fully ionized gas.

The internal energy of a gram of partially ionized H-gas is given by

$$u = 3/2NkT + q_{\rm H}\chi_{\rm H}/m_{\rm H}, \tag{4.34}$$

where $q_{\rm H}= n_+/(n_+ + n_0)$ is the ionization fraction of H and $\chi_{\rm H}$ is the ionization potential of H (13.598 eV). The number of particles N includes the H atoms as well as the free electrons, so the number of particles per gram is $N = (1 + q_{\rm H})/m_{\rm H}$ for pure H gas.

The extra energy term $q_{\rm H}\chi_{\rm H}/m_{\rm H}$ in u affects the relation between P and ρ. We have defined the polytrope index γ by $P = \rho^{\gamma}$. We can also define a temperature index Δ by $T \sim P^{\Delta}$, with $\Delta_{\rm ad} = \frac{\gamma - 1}{\gamma}$ and $\gamma = \frac{1}{1 - \Delta_{\rm ad}}$ for an adiabatic gas.

Figure 4.5 shows the value of γ for H-gas as a function of the ionization fraction of H. For neutral H and fully ionized H, the polytropic index is $\gamma = 5/3$ so $\Delta_{\rm ad} = 0.4$. For partially ionized H-gas, γ shows a minimum of $\gamma \approx 1.19$ and $\Delta_{\rm ad} \approx 0.16$ when H is about half ionized. These are approximate values because the degree of ionization depends on T and ρ (see Pols 2011 for a more complete description).

Q (4.5) *Explain why $\gamma = 5/3$ at both the high- and low-temperature ends.*
Q (4.6) *What is the physical reason that γ drops in a partially ionized region?*

4.9 Summary

1. The equation of state (EoS) describes the relation between P, T, and ρ. At low densities this relation is given by the ideal gas law: $P \sim \rho T$.

2. Radiation pressure is proportional to T^4. It only plays an important role in the cores of massive stars and supernovae.
3. At densities in excess of about 10^3 g cm^{-3} inside stars (but depending on T) the momentum distribution of the electrons becomes degenerate: the electrons are forced to high velocities. Completely nonrelativistic degenerate electron gas has an EoS of the type $P_e \sim {n_e}^{5/3}$.
4. At densities in excess of 10^6 g cm^{-3}, electron gas becomes relativistically degenerate with an EoS of the type $P_e \sim {n_e}^{4/3}$.
5. At densities in excess of 3×10^7 g cm^{-3}, electrons and protons form neutrons.
6. At densities in excess of about 10^{12} g cm^{-3}, neutrons become degenerate. Nonrelativistic and completely relativistic neutron gas has an EoS of the same type as degenerate electron gas.
7. Stars made of gas that have an EoS of the type $P \sim \rho^\gamma$ are called polytropic stars. Their P and ρ structure can be derived from the hydrostatic equilibrium equation, independent of T.
8. If a polytropic star consists of gas with an ideal EoS (i.e., $P \sim \rho T$), the combination of P and ρ also sets the T structure of the stars and hence the energy flow through the star.
9. Examples of polytropic stars are
   - degenerate stars,
   - fully convective stars, and
   - stars with a constant ratio between P_{gas}/P_{rad}.

Exercises

4.1 Show that for a fully ionized gas the number of electrons per gram is $(1 + X)/2\ m_H$ and that $\mu_e = 2/(1 + X)$.

4.2 Calculate the ratio P_{rad}/P_{gas} in the center of zero-age main-sequence stars of 1, 10, 30, 60, and $120 M_\odot$. (Use the models by Ekstrom et al. 2012.)

4.3 Plot the run of temperature versus density in the Sun (Appendix C) in Figure 4.4. What do you conclude about the equation of state?

4.4 Plot the values of the central temperatures versus densities of zero-age main-sequence stars of 1, 4, 12, 60, and $120 M_\odot$ in Figure 4.4. What do you conclude? Use the same models as in Exercise 4.2.

4.5 Compare the densities at which neutrons and electrons become degenerate at the same temperature. Hint: consider two effects.
   1. Energy exchange results in $\langle m_e v_e^2 \rangle = \frac{3}{2}kT = \langle m_n v_n^2 \rangle$. Calculate the ratio of the momenta p_n/p_e of neutrons and electrons.
   2. Degeneracy occurs when $d^3p \sim h^3/\Delta\mathrm{Vol}$, with $\Delta\mathrm{Vol} \sim 1/n_n$. Calculate the ratio n_n/n_e when the values of $(d^3p \times \Delta\mathrm{Vol})$ for electrons and neutrons are equal.
   3. Calculate the corresponding density ratio ρ_n/ρ_e.

4.6 Compare the expected radius of a nonrelativistic complete electron degenerate star with that of a nonrelativistic complete neutron degenerate star of the same mass. Comment on the result.

References

Ekström S., Georgy C., Eggenberger P., et al. 2012, A&A, **537,** 146
Maeder A. 2009, in Physics, Formation and Evolution of Rotating Stars (Berlin: Springer), 165
Pols O. 2011, in Stellar Structure and Evolution, Utrecht University Lecture Notes, 27, https://www.astro.ru.nl/~onnop/education/stev_utrecht_notes/

AAS | IOP Astronomy

Understanding Stellar Evolution

Henny J.G.L.M. Lamers and Emily M. Levesque

Chapter 5

Opacities in Stars

For understanding and calculating stellar structure, particularly the transport of energy by radiation, we have to know the absorption coefficient of the gas as a function of density, temperature, and composition. A variety of particles, each with their own atomic line or continuum transitions, may contribute to the opacity; however, in each temperature and density region only one or two opacity processes dominate. In this chapter, we discuss these different processes and the conditions in the stars where they are important.

Almost all opacity processes are wavelength-dependent; however, for calculating the radiative transfer inside a star, we can use a cleverly defined wavelength-independent mean value, the Rosseland-mean opacity. Below we derive its definition and present approximate expressions for this opacity for the different processes.

5.1 The Rosseland-mean Opacity

The **absorption coefficient per gram** at frequency ν, κ_ν (in cm^2 g^{-1}), is defined as the cross section for absorption or scattering of photons of frequency ν if these photons pass through a gram of gas. We can also define the **absorption coefficient per cm^3** of gas, k_ν (in cm^2 cm^{-3} = cm^{-1}), with the obvious relation $k_\nu = \rho\kappa_\nu$.

In an optically thick medium, such as inside a star, the flux of radiation at frequency ν that passes through a cm^3 at a given radius r depends inversely on the absorption coefficient k_ν and on the local gradient of the Planck function B_ν. This can be understood intuitively: the higher the absorption coefficient, the more difficult it will be for the radiation to pass through. On the other hand, the larger the gradient dB_ν/dr of the radiation density, the higher the flux. We can therefore expect

$$F_\nu \sim \frac{-1}{k_\nu}\frac{dB_\nu}{dr} \,. \tag{5.1}$$

doi:10.1088/978-0-7503-1278-3ch5

The minus sign indicates that a positive (outflowing) flux requires an outward decreasing temperature. If we define the total (frequency integrated) flux as $F = \int_0^\infty F_\nu \, d\nu$, then

$$F \sim -\int \frac{1}{k_\nu} \frac{dB_\nu}{dr} d\nu = -\int \frac{1}{k_\nu} \frac{dB_\nu}{dT} d\nu \times \frac{dT}{dr}, \tag{5.2}$$

where we changed dB_ν/dr into $dB_\nu/dT \times dT/dr$.

For calculating stellar structure, we want to avoid integration over frequency. Instead, we prefer to write the flux in a frequency-independent expression that is similar to Equation (5.1) as $F \sim -k^{-1} \, dB/dr$. This can be done if the frequency-independent k is properly defined. This frequency-independent k is called the Rosseland-mean opacity, indicated by $k_{\rm R}$, after the Norwegian astrophysicist Svein Rosseland (1894–1985), who proposed this in 1924. The definition of $k_{\rm R}$ follows from

$$F \sim \frac{-1}{k_{\rm R}} \frac{dB_\nu}{dr} = \frac{-1}{k_{\rm R}} \frac{dB_\nu}{dT} \frac{dT}{dr} = \int_0^\infty F_\nu \, d\nu = \int \frac{-1}{k_\nu} \frac{dB_\nu}{dT} d\nu \frac{dT}{dr}. \tag{5.3}$$

Comparing the third and the fifth expression for F results in the definition of the **Rosseland-mean opacity per cm^3**

$$\boxed{\frac{1}{k_{\rm R}} \equiv \int_\infty^0 \frac{1}{k_\nu} \frac{dB_\nu}{dT} d\nu \Big/ \int_\infty^0 \frac{dB_\nu}{dT} d\nu}. \tag{5.4}$$

The **Rosseland-mean opacity per gram** is $\kappa_{\rm R} = k_{\rm R}/\rho$. The absorption coefficients described in the sections below are all Rosseland-mean values.

5.2 Electron Scattering: σ_e

Free electrons scatter photons. The absorption cross section for electron scattering is σ_e (in cm^2 g^{-1}). Deep inside stars, the gas is fully ionized and electron scattering is the dominant opacity with

$$\sigma_e = \sigma_{\rm T} N_e, \tag{5.5}$$

where $\sigma_{\rm T} = 6.65 \times 10^{-25}$ cm^2 electron^{-1} is the Thomson cross section for electrons and $N_e = 1/\mu_e m_{\rm H}$ is the number of electrons per gram. In Section 4.1, we have defined $\mu_e m_{\rm H}$ as mean particle mass per electron with $\mu_e = 2/(1 + X)$ for a fully ionized astrophysical gas. This means that the **electron scattering coefficient** is

$$\boxed{\sigma_e = \frac{\sigma_{\rm T}}{m_{\rm H}} \frac{1}{\mu_e} = \frac{0.40}{\mu_e} \simeq 0.40 \frac{1 + X}{2} \approx 0.2(1 + X) \text{ cm}^2 \text{ g}^{-1}}. \tag{5.6}$$

N.B. The cross section (in cm^2 g^{-1}) for *scattering* of photons is indicated by the symbol σ. The cross section (in cm^2 g^{-1}) for *absorption* of photons is indicated by the symbol κ.

5.3 Free–Free Absorption: $\kappa_{\rm ff}$

Free–free absorption is the inverse of bremsstrahlung. In bremsstrahlung, an electron that is deflected by an ion emits a photon. In free–free absorption a photon is absorbed by an electron when it briefly interacts with an ion. If Z_i is the charge of the ion, the interaction depends on Z_i^2.

Because free–free absorption depends on encounters between electron and ions, we expect the cross section to be $\kappa_{\rm ff} \sim Z_i^2 n_i n_e$ (in cm^2 cm^{-3}), where $n_i = \rho/\mu_i m_{\rm H}$ and $n_e = \rho/\mu_e m_{\rm H}$ are the number of ions and electrons per cm^3.

The **absorption coefficient for free–free absorption** is

$$\boxed{\kappa_{\rm ff} = 7.5 \times 10^{22}\left(\frac{1+X}{2}\right)\left\langle\frac{Z_i^2}{A_i}\right\rangle \rho T^{-7/2} \ {\rm cm^2\ g^{-1}}}. \qquad (5.7)$$

where $<Z_i^2/A_i>$ is the mean value of this ratio for all ions and $T^{-7/2}$ is the temperature dependence. The steep temperature dependence implies that the Rosseland-mean value of the free–free absorption is strongest at low temperatures.

Q (5.1) *Why is $\kappa_{\rm ff}$ in cm^2 g^{-1} proportional to ρ?*
Q (5.2) *Show that for a low-metallicity composition of $Z << 1$ the factor $<Z_i^2/A_i> \approx 1$.*

5.4 Bound–Free Absorption: $\kappa_{\rm bf}$

If a photon with sufficient energy hits an atom that is not fully ionized it may kick out an electron. This photoionization results in the bound–free absorption of photons. The Rosseland-mean value of the bound–free absorption is calculated by summing all possible bound–free transitions of many ions. The result is known as **Kramers' bound–free opacity law**, after the Dutch physicist Hendrik Kramers (1894–1952), who derived it in 1923.

$$\boxed{\kappa_{\rm bf} = 4.3 \times 10^{25}\,(1+X) Z \rho T^{-7/2}\ {\rm cm^2\ g^{-1}}}. \qquad (5.8)$$

Note that $\kappa_{\rm bf} >> \kappa_{\rm ff}$ for a metal abundance of $Z > 0.001$.

Q (5.3) *Why is $\kappa_{\rm bf}$ in cm^2 g^{-1} proportional to ρ?*

5.5 Bound–Bound Absorption: $\kappa_{\rm bb}$

If the gas inside a star is not fully ionized, it can absorb photons of many possible wavelengths, resulting in electron transitions from one bound state to another. The calculation of this Rosseland-mean opacity is very difficult due to the numerous possible transitions of different ionization stages for the various elements. The results of such calculations can be found on the web. It turns out that the many line transitions of Fe group elements give a peak in the Rosseland-mean opacity at temperatures around 10^5 to 10^6 K, depending on the density.

5.6 Total Rosseland-mean Opacity: κ_R

Tables of opacities in stellar interiors and atmospheres can be found on the web at

https://opalopacity.llnl.gov/existing.html

Figure 5.1 shows the opacities from the OPAL database for solar composition as a function of temperature for several densities. Data for other compositions are also given in the database.

Note the following features.

- At high T and low density, all matter is ionized, so $\kappa = \sigma_e \simeq 0.3$ cm^2 g^{-1}.
- As T decreases, the bound–free and free–free absorption coefficient increases as $\kappa \sim \rho T^{-7/2}$. This explains the decrease from the peak toward higher T.
- At intermediate temperatures (depending on ρ) the gas is partly ionized. This results in many more possible electron transitions and huge opacities. This produces the so-called **Fe-opacity peak** around 10^5 K for densities 10^{-6} to 10^{-4} g cm^{-3} (i.e., in the envelopes of stars).
- The peak shifts to higher T as the density increases.
- The peak at $10^4 < T < 10^5$ K at very low density ($\sim 10^{-10}$ to 10^{-8} g cm^{-3}) is due to H. We will see later that this peak is responsible for convection in the outer layers of cool stars.

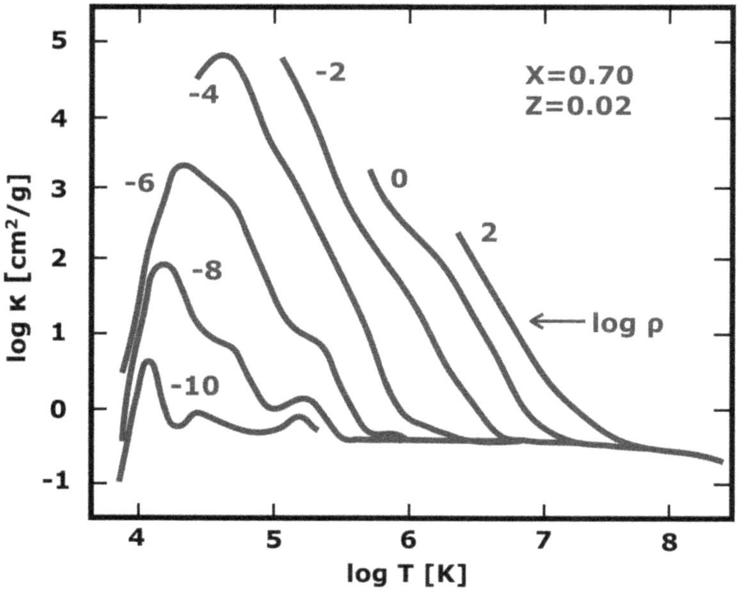

Figure 5.1. Total Rosseland-mean opacity κ_R for solar composition as a function of T and ρ. (Figure is based on data from OPAL opacity tables[1].)

[1] http://opalopacity.llnl.gov/existing.html

- At very low $T < 10^4$ K, the opacity, due to H^-, decreases steeply toward lower T as $\kappa(H^-) \sim T^9$. This explains the steep slope of κ in Figure 5.1 at low T. We will show later that this is important for stars on the Hayashi track, AGB stars, and red supergiants. It explains why the effective temperature of stars has a lower limit.

Q (5.4) *Why does the Fe opacity peak shift to higher T with increasing ρ?*

5.7 The Mean-free Path of Photons: ℓ

The mean-free path of a photon is $\ell \simeq \frac{1}{\kappa\rho}$ in $\left[\frac{\text{cm}^2}{\text{g}}\right]^{-1}\left[\frac{\text{g}}{\text{cm}^3}\right]^{-1} = [\text{cm}]$.

Inside main-sequence stars at $T > 10^6$ K:

$$\left.\begin{array}{l}\rho \sim 1 \text{ g cm}^{-3} \\ \kappa \sim 1 \text{ cm}^2 \text{ g}^{-1}\end{array}\right\}, \quad \text{so } \ell \sim 1 \text{ cm} \tag{5.9}$$

This implies that when a photon travels about 1 cm it will be absorbed and re-emitted or scattered. Photons are scattered so frequently over a short distance that they quickly lose information about their direction. This means that the radiation inside a star must be (almost) isotropic.

Q (5.5) *If radiation inside a star was isotropic, how could there be a radiative outward moving flux?*

5.8 Summary

1. The Rosseland-mean opacity is defined in such a way that the radiative transfer in a star can be calculated in a frequency-independent way.
2. At high temperatures the absorption coefficient σ_e for electron scattering is independent of density and temperature. For highly ionized gas, it is about $0.3 \text{ cm}^2 \text{ g}^{-1}$.
3. The absorption coefficients κ_{ff} and κ_{bf} depend on density and temperature as $\sim\rho T^{-7/2}$.
4. The total absorption coefficient has a strong peak that reaches up to $\kappa_R \approx 10^5 \text{ cm}^2 \text{ g}^{-1}$ in the temperature range of 10^5 to 10^6 K, depending on density. It is mainly due to the huge number of spectral lines of partially ionized metals.
5. At $T < 10^4$ K the opacity, mainly due to H^-, decreases very steeply to lower temperature as $\kappa \sim T^9$.

Exercises

5.1 (a) Identify in Figure 5.1 the ranges in T and ρ, where electron scattering dominates, and where the bound–free and free–free opacity dominates.
(b) Check the dependence of κ on T and ρ in these regions.

5.2 The average *radial* distance traveled by a photon in a random walk is $r \sim \ell\sqrt{N}$, where N is the number of random steps and ℓ is the step length (i.e., the mean-free path length).

(a) Use $\ell \approx 1$ cm and $r = R_\odot$ to estimate the number of scatterings a photon will undergo before reaching the surface of the star.

(b) Calculate the total path length L and the time $\tau_\mathrm{L} = L/c$ it takes a photon created in the center to leave the Sun.

(c) Is it still the same photon? (Same λ, ν?)

5.3 We have argued that radiation inside a star must be almost isotropic. This is tested in this exercise.

Estimate the fraction of the radiation that is nonisotropic in the Sun at $r = 0.5R_\odot$.

Hint: split the flow of radiation into two parts, an outward and inward directed part, (as shown below) and calculate the difference. Compare this net flow with the radiation density at that location. Adopt this distribution of radiation density: the length of the arrows is proportional to the radiative intensity in that direction.

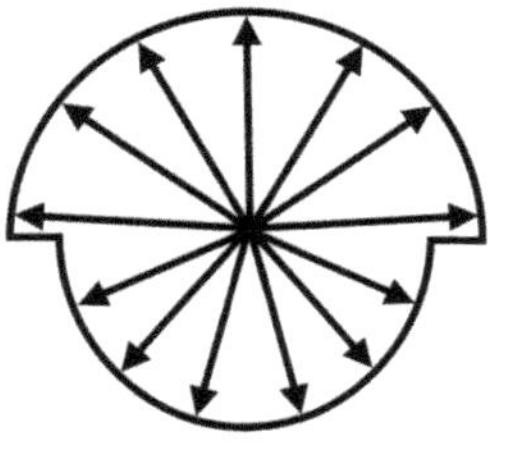

5.4 The internal structure of the Sun is listed in Appendix C.

(a) Sketch the internal structure of the Sun in Figure 5.1.

(b) Identify the three regions in the Sun, where the opacity is dominated by (1) electron scattering, (2) free–free and bound–free absorption, and (3) H^- opacity.

Reference

Opal opacity tables: http://opalopacity.llnl.gov/existing.html

IOP Astronomy

Understanding Stellar Evolution

Henny J.G.L.M. Lamers and Emily M. Levesque

Chapter 6

Radiative Energy Transport

The energy of a star is generated in its central regions and transported outward by radiation, convection, or diffusion. If the energy is transported by radiation, the star is in radiative equilibrium. Radiative energy transport requires a specific temperature gradient throughout the star that depends on the amount of flux that must be transported as well as the opacity. This temperature gradient is described by the Eddington equation. We will show that the combination of hydrostatic equilibrium and radiative equilibrium results in a relation between the mass and the luminosity of a star. This also leads to an upper limit for the luminosity of stars.

6.1 Eddington's Equation for Radiative Equilibrium

In 1926, the British astrophysicist Sir Arthur Eddington (1882–1944) derived an expression for the energy transport in stars by means of radiation. This is one of the fundamental equations for understanding the structure and evolution of stars. We start by giving this expression and then it will be derived intuitively in three steps that are easy to remember.

The **Eddington equation** for radiative transfer through an optically thick medium is

$$\boxed{\frac{dT}{dr} = -\frac{3}{4}\frac{1}{ac}\frac{\kappa\rho}{T^3}\frac{L_r}{4\pi r^2}}, \tag{6.1}$$

where a is the radiation density constant (see Section 4.3), and $ac = 4\,\sigma$, where σ is the Stefan–Boltzmann constant (see Section 2.3).

Eddington's equation is also called the **equation for radiative equilibrium (RE).** This is because it describes the temperature structure of a star in which the energy generated in the center is in equilibrium with the energy that is transported outward by radiation. NB: Radiative equilibrium does *not* mean that gravity is balanced by radiation pressure.

doi:10.1088/978-0-7503-1278-3ch6

The following is an intuitive derivation of the Eddington equation (see Figure 6.1): (A). Consider a cm^3 of gas in a layer of 1 cm thickness, so the surface area is 1 cm^2. Now consider the amount of radiation that enters and leaves this volume from the top and the bottom. The medium inside the stars is optically thick, so each cm^2 radiates an amount of σT^4, where T is the local temperature. If T_2 and T_1 are the temperatures at the bottom and the top of the cm^3, then the net flux that passes from the bottom to the top is expected to be

$$F \sim \sigma\left(T_2^4 - T_1^4\right) = -d(\sigma T^4)/dr. \tag{6.2a}$$

(B). The flux will also depend on the transparency of the volume. So we can expect $F \sim$ transparency of layer $\sim$ 1/opacity per cm. So

$$F \sim 1/k \sim 1/\kappa\rho. \tag{6.2b}$$

(C). The flux is related to the luminosity that passes through the star at radius r

$$F \sim L_r/4\pi r^2. \tag{6.2c}$$

Combining conditions, (A), (B), and (C), yields

$$\frac{L_r}{4\pi r^2} \sim \frac{-1}{\kappa\rho}\frac{d}{dr}\sigma T^4 = -\frac{1}{\kappa\rho}\frac{ac}{4}4T^3\frac{dT}{dr}. \tag{6.2d}$$

So from this simple intuitive derivation, we expect

$$\frac{dT}{dr} \sim \frac{-1}{ac}\frac{\kappa\rho}{T^3}\frac{L_r}{4\pi r^2}. \tag{6.2e}$$

In simple terms: the higher the opacity, the higher the temperature gradient must be in order to transport sufficient radiation through each layer.

Equation (6.2e) is very similar to the Eddington Equation (6.1) for energy transport by radiation. The difference of a factor of 3/4 comes from proper integration of the radiation intensity over all angles. The simplest way to remember and write the Eddington equation is therefore

$$\frac{L_r}{4\pi r^2} = -\frac{4}{3}\frac{1}{\kappa\rho}\frac{d\sigma T^4}{dr}, \tag{6.3}$$

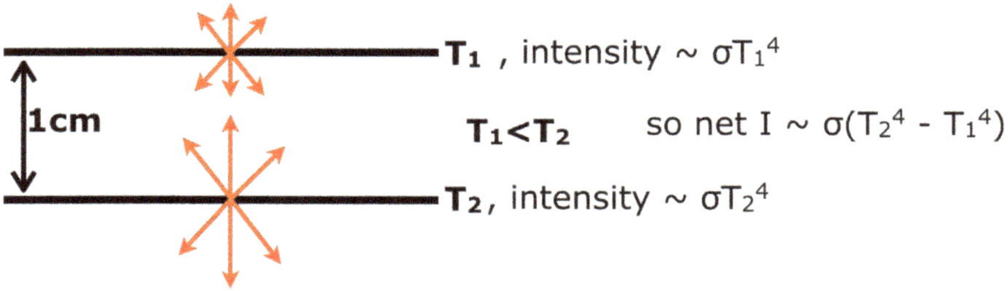

Figure 6.1. Schematic description of radiative transport.

where one only has to remember the factor of 4/3; the other factors are trivial.

Comments:
- Radiative equilibrium means that the energy is transported in a star by radiation. It does **not** mean that gravity is balanced by radiation pressure.
- The full derivation of Eddington's equation is given in several textbooks in terms of radiative diffusion (e.g., Kippenhahn & Weigert 1990).
- The Eddington equation is sometimes derived using the radiation pressure (e.g., Prialnik 2004). However, this might be confusing since radiative transfer is related not to the pressure balance but to the diffusion of energy.

Q (6.1) *Free–free absorption and bound–free absorption are real absorption: photons disappear and are re-emitted at another wavelength. But electron scattering does not "absorb" photons; it just sends them in another direction (with a very small mean-free path). With this in mind, why does electron scattering play a role at all in radiative transfer and in the structure of stars?*

6.2 Mass–Luminosity Relation for Stars in HE and RE

Stars in HE that transport energy by radiation obey a mass–luminosity relation. This can be derived rigorously for stars with a constant ratio $P_{\rm rad}/P_{\rm gas}$ (i.e., for polytropic stars with $\gamma = 4/3$ in RE), but we will derive it intuitively.

We have seen that, for stars in HE,

$$T_c \simeq \frac{\mu}{\mathcal{R}}\frac{GM}{R}. \tag{6.4a}$$

We have also seen that, for stars in RE,

$$\frac{L_r}{4\pi r^2} = -\frac{4}{3}\frac{1}{\kappa\rho}\frac{d\sigma T^4}{dr}. \tag{6.4b}$$

Let us make the following order-of-magnitude approximations

$$-\frac{dT^4}{dr} \simeq \frac{T_c^4}{R}, \quad \rho \simeq (4\pi/3)M/R^3 \text{ and } L_r/4\pi r^2 \simeq L/4\pi R^2. \tag{6.5}$$

This yields

$$L \simeq 4\pi R^2 \frac{4}{3}\frac{\sigma}{\kappa}\frac{4\pi R^3}{3M}\frac{1}{R}\frac{\mu^4}{\mathcal{R}^4}\frac{G^4M^4}{R^4} \sim \frac{\mu^4}{\kappa}M^3. \tag{6.6}$$

So, for stars in HE and RE, we expect the following **mass–luminosity relation**

$$\boxed{L \approx -\frac{\sigma G^4}{\mathcal{R}^4}\frac{\mu^4}{\kappa}M^3 \sim \frac{\mu^4}{\kappa}M^3.} \tag{6.7}$$

The most surprising result is that we made no assumption about the energy production process! This shows that a star can only be in HE and RE if it has some fixed luminosity, independent of the energy source. This implies that the energy generation process must adjust itself to the required value; otherwise the star is not in equilibrium.

Considering the simplicity of our derivation, it agrees surprisingly well with the observed mean *M-L* relation for main-sequence stars of $L \sim M^{3.8}$ (Section 2.5).

Q (6.2) *Why is this relation not valid for lower main-sequence stars, red giants, and red supergiants?*

Q (6.3) *Is the relation valid for degenerate stars? Why?*

6.2.1 Consequences of the *M-L* Relation

1. The dominant opacity in massive main-sequence stars is electron scattering, so $\kappa = \sigma_e$ with $\sigma_e \simeq 0.20(1 + X)$ cm^2 g^{-1}. The values of σ_e and μ for different compositions are
   - $\sigma_e = 0.34$ and $\mu = 0.61$ if $X = 0.70$ and $Y = 0.30$ and
   - $\sigma_e = 0.20$ and $\mu = 1.33$ if $X=0$ and $Y=1$.

   Because $L \sim \mu^4/\sigma_e$, a He star of $1 M_\odot$ will be about 40 times more luminous than a star of $1 M_\odot$ with normal composition. This explains why hot horizontal branch stars are much more luminous than the main-sequence stars at the turn-off point, although they have about the same mass (Figure 2.5).
2. The strong dependence of L on μ (as $L \sim \mu^4$) and the weaker dependence of L on $1/\sigma_e$ explains why main-sequence stars increase in luminosity during the H-fusion phase when their He abundance increases.
3. It also explains why the post-MS evolution tracks of massive stars are approximately horizontal: their composition hardly changes during the short phase when they move to the right in the HRD. The red giant and supergiant phases are an exception because these stars are largely convective.

6.3 The Eddington Limit: The Maximum Luminosity and the Maximum Mass

For any star in HE, the inward force due to gravity should be larger than the outward force by radiation pressure, because HE requires

$$\left|\frac{dP_{\rm rad}}{dr}\right| + \left|\frac{dP_{\rm gas}}{dr}\right| = \frac{GM(r)}{r^2}\rho_r \tag{6.8}$$

and so

$$\left|\frac{dP_{\rm rad}}{dr}\right| < \frac{GM(r)}{r^2}\rho_r \ \text{with}\ P_{\rm rad} = \frac{a}{3}T^4. \tag{6.9}$$

Radiative equilibrium requires

$$\frac{dP_{\rm rad}}{dr} = \frac{d}{dr}\left(\frac{aT^4}{3}\right) = \frac{4}{3c}\frac{d}{dr}(\sigma T^4) = -\frac{\kappa\rho_r}{c}\frac{L(r)}{4\pi r^2}. \tag{6.10}$$

Combining these two equilibrium conditions, with r^2 and ρ_r canceling, gives

$$L(r) < \frac{4\pi cGM(r)}{\kappa(r)}. \tag{6.11}$$

NB: Layers in a star where Condition (6.11) is violated cannot be in radiative equilibrium. Such layers may exist in a star, but only if they are convective.

The most luminous main-sequence stars are the massive stars. In these stars, electron scattering is the dominant opacity, with σ_e being about constant throughout the star. This gives an upper limit for the luminosity of massive stars, which is called the **Eddington limit.**

$$\boxed{L < L_{\rm E} = 4\pi cGM/\sigma_e}. \tag{6.12}$$

If a star's luminosity exceeded $L_{\rm E}$, it would be blown up by its radiation pressure. Substitution of the electron scattering coefficient $\sigma_e = 0.20\ (1 + X)\ {\rm cm^2\ g^{-1}}$ gives

$$\boxed{\frac{L_{\rm E}}{L_\odot} = \frac{1.3\times10^4}{\sigma_e}\frac{M}{M_\odot} \simeq 3.8\times10^4\frac{M}{M_\odot},} \quad \text{if } X = 0.70. \tag{6.13}$$

The influence of the radiation pressure is often expressed in terms of the **Eddington factor $\Gamma_{\rm E}$**

$$\Gamma_{\rm E} \equiv \frac{L}{L_{\rm E}} = \frac{\kappa L}{4\pi cGM}, \quad \text{with } \Gamma_{\rm E} < 1 \text{ for stable stars.} \tag{6.14}$$

We have seen in Figure 2.3 that for MS stars with $10\ M > 6\ M_\odot$ the empirical mass–luminosity relation is $L/L_\odot \approx 12\ (M/M_\odot)^{2.9}$. If we extrapolate this relation to higher masses and combine it with the Equation (6.13) for the Eddington limit, we find that

$$M_{\rm max} \approx 70M_\odot \text{ and } L_{\rm max} \approx 3\times10^6 L_\odot. \tag{6.15}$$

The empirically derived maximum mass is $M_{\rm max} \simeq 160$–$300M_\odot$ and $L_{\rm max} \simeq 3\times10^6\ L_\odot$ for stars in the LMC cluster R136 (Crowther et al. 2010). The difference between the predicted and the empirically derived values of $M_{\rm max}$ and $L_{\rm max}$ suggest that the mass–luminosity relation for the most massive stars is flatter than the extrapolation that we adopted here. We will show later in Chapter 13 that this is indeed the case.

6.4 Summary

1. If the energy inside a star is transported outward by radiation, a specific T-gradient is required that is described by the Eddington Equation (6.1).

2. For a star in which the energy transport is by radiation, the combination of the hydrostatic and radiative equilibrium results in a relation between luminosity and mass of the type $L \sim (\mu^4/\kappa)M^3$. This simple prediction agrees reasonably well with the observed M-L relation for main-sequence stars of $L \sim M^{3.8}$.
3. The strong dependence of $L \sim \mu^4$ in the relation predicts that He-stars are about a factor of 40 brighter than main-sequence stars of the same mass. This explains the brightness of stars on the horizontal branch. The same μ-dependence also explains the brightening of stars during the main-sequence phase.
4. A star can only be in hydrostatic equilibrium if the outward force due to the radiation pressure gradient is smaller than the inward force due to gravity. This sets an upper limit to the luminosity of stars, known as the Eddington limit.
5. Application of the Eddington limit to massive stars, in which electron scattering is the dominant opacity, results in an estimate for the maximum stellar luminosity and mass of $L_{\max} \approx 3 \times 10^6\ L_\odot$ and $M_{\max} \approx 10^2 M_\odot$, in reasonable agreement with the observed value.

Exercises

6.1 The predicted mass–luminosity relation implies that $L \sim \mu^4/\kappa$.
   (a) Compare this with the results of the stellar models for zero-age main-sequence stars with $Z = 0.014$ and $Z = 0.002$, listed in Appendix D,
   (b) Explain the trend that the luminosity of massive stars is about the same for both compositions, but that the luminosity of the metal-poor low-mass stars is higher than those of $Z = 0.014$ stars.

6.2 The Eddington limit was derived in Section 6.3 by extrapolating the empirical mass–luminosity relation of $L/L_\odot \approx 12(M/M_\odot)^{2.9}$ to higher mass. This can be improved by adopting the predicted M-L relation for zero-age main-sequence stars, listed in Appendix D.
   (a) Derive the approximate M-L relation in the mass range of 30 to $60 M_\odot$.
   (b) Extrapolate this relation and derive the Eddington limit, in mass and luminosity.

References

Crowther, P., Schnurr, O., Hirschi, R., et al. 2010, MNRAS, **408, 731**

Kippenhahn, R., & Weigert, A. 1990, Stellar Structure and Evolution (Berlin: Springer)

Prialnik, D. 2004, in An introduction to Stellar Structure and Evolution 2004 (Cambridge: Cambridge Univ. Press), 215

Henny J.G.L.M. Lamers and Emily M. Levesque

Chapter 7

Convective Energy Transport

In most stars, energy is transported outward by radiation. In those layers where the transport of energy by radiation is not efficient enough, convection will take over. Hot bubbles rise from deeper layers, deliver their heat in higher layers, and descend again as cooler bubbles. A similar process occurs in the Earth's atmosphere when the surface is heated by the Sun to such a degree that it cannot get rid of the heat by conduction or radiation. In that case, the Earth's atmosphere becomes convective, enabling birds of prey to soar high in the sky with minimum effort.

In this chapter, we derive the conditions necessary for layers to become convective if they are chemically homogeneous and if there is a chemical gradient. Subsequently, we discuss the efficiency of convective energy transport. This results in expressions for the average velocity of the convective cells and for the temperature difference between the hot rising cells and the cool descending cells. We will show that both this temperature difference and the average velocity of the cells are surprisingly small and explain the underlying reason. We discuss which layers in stars are convective and how this affects mixing of their chemical composition.

7.1 The Schwarzschild Criterion for Convection

We start with the qualitative picture that was introduced by the German astrophysicist Karl Schwarzschild (1873–1916) in 1906 (Schwarzschild 1906). Consider a gas layer in a star and assume that a blob of gas in that layer accidentally starts moving upward. If it falls back immediately, then that layer is obviously stable. However, if the blob keeps rising, that layer is unstable for rising blobs and the layer will be convective.

Consider the following picture of a blob of gas at its initial location r_1, which is in equilibrium with its surroundings. The surrounding medium has a temperature, pressure, and density $T^s(r_1)$, $P^s(r_1)$, and $\rho^s(r_1)$. Suppose the blob moves upward to location r_2, where the ambient medium has $T^s(r_2)$, $P^s(r_2)$, and $\rho^s(r_2)$. The conditions

doi:10.1088/978-0-7503-1278-3ch7

inside the blob at r_2 are $T^b(r_2)$, $P^b(r_2)$ and $\rho^b(r_2)$. These may be different from the surroundings at r_2 (see Figure 7.1).

When the blob rises, it will remain in pressure equilibrium with the surroundings; otherwise, the blob would expand or contract to equalize the interior pressure with the outside pressure. This implies that $P^b(r_2) = P^s(r_2)$, and because of the ideal gas law, $P \sim \rho T$,

$$T^b(r_2)\,\rho^b(r_2) \;=\; T^s(r_2)\,\rho^s(r_2). \tag{7.1}$$

Will the blob keep rising or will it fall back to its original position? The blob will keep rising (due to buoyancy force) if its density at r_2 is lower than that of the surrounding medium, i.e., if

$$\rho^b(r_2) < \rho^s(r_2). \tag{7.2}$$

Combining (7.1) and (7.2) shows that the blob will keep rising if $T^b(r_2) > T^s(r_2)$. This will be the case if the temperature inside the blob decreases upward more slowly than in the surrounding medium. We can express this condition in terms of the T-gradients in the blob and in the surrounding medium:

$$\left|\frac{dT}{dr}\right|_b < \left|\frac{dT}{dr}\right|_s. \tag{7.3}$$

Q (7.1): *What is the reason for the absolute value signs?*
Q (7.2): *Show that a descending blob will keep descending under the same condition.*

Assume that a layer is in radiative equilibrium: $(dT/dr) = (dT/dr)_{\rm rad}$. If the blob of gas rises adiabatically, i.e., without losing energy, its T-gradient is $(dT/dr)_{\rm ad}$. In that case, Condition (7.3) can be written as the **Schwarzschild criterion for convection:**

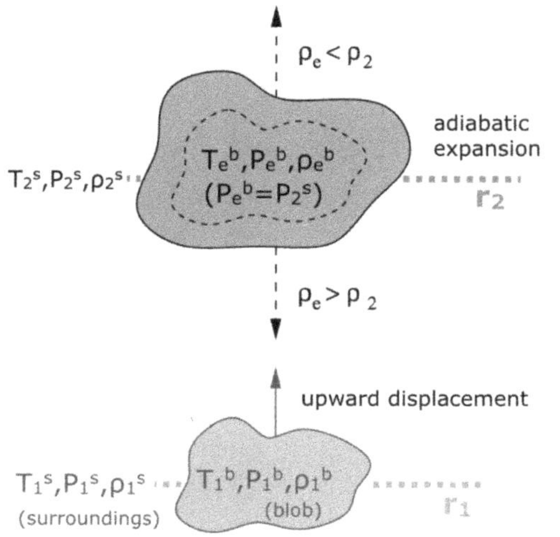

Figure 7.1. Schematic representation of convective blobs. (Adapted from Pols 2011.)

$$\boxed{|dT/dr|_{\text{ad}} < |dT/dr|_{\text{rad}}}. \tag{7.4}$$

This criterion is easy to remember: the star always adopts the least steep temperature gradient. It shows that if a layer is in radiative equilibrium, it will automatically become convective when criterion (7.4) is satisfied.

We will show below that the mean temperature gradient in a convective zone is very close to the adiabatic gradient.

7.1.1 The Schwarzschild Criterion in Terms of the Polytropic Index

We can express the Schwarzschild criterion for convection in terms of the *local* adiabatic polytropic index $\gamma \equiv d\ell n\, P/d\ell n\, \rho$.

The Schwarzschild criterion states that convection occurs if

$$\left(\frac{dT}{dr}\right)_{\text{ad}} > \left(\frac{dT}{dr}\right)_{\text{rad}} \rightarrow \left(\frac{d\ell n\, T}{dr}\right)_{\text{ad}} > \left(\frac{d\ell n\, T}{dr}\right)_{\text{rad}}. \tag{7.5}$$

Combining $T \sim P/\rho$ for an ideal gas and the definition of the polytropic index, $P \sim \rho^{\gamma}$, yields $T \sim P^{\,(1-1/\gamma)}$ and thus

$$\frac{d\ell n\, T}{dr} = \left(1 - \frac{1}{\gamma}\right)\frac{d\ell n\, P}{dr} \quad \text{with} \quad \frac{d\ell n\, P}{dr} \text{ from HE.} \tag{7.6}$$

So the Schwarzschild criterion can also be written as

$$\left|\frac{dT}{dr}\right|_{\text{rad}} < \left(\frac{\gamma_{\text{ad}} - 1}{\gamma_{\text{ad}}}\right)\frac{T}{P}\left|\frac{dP}{dr}\right|. \tag{7.7}$$

A layer is convective if

$$\left(1 - \frac{1}{\gamma_{\text{ad}}}\right)\frac{d\ell n\, P}{dr} > \left(1 - \frac{1}{\gamma_{\text{rad}}}\right)\frac{d\ell n\, P}{dr} \quad \rightarrow \quad \boxed{\gamma_{\text{ad}} < \gamma_{\text{rad}}}. \tag{7.8}$$

We have seen in Section (4.7.2) that $\gamma_{\text{ad}} = 5/3$ for a fully ionized or fully neutral ideal gas, but that it drops to lower values of $\gamma_{\text{ad}} \approx 1.2$ in H-ionization zones. This makes these zones more sensitive to convection.

7.2 Convection in a Layer with a μ-gradient: Ledoux Criterion

We have derived the Schwarzschild criterion for convection by considering the rise of bubbles in a medium of constant chemical composition. We now consider the case of a chemically stratified star with a mean particle mass μ decreasing radially outward.

Q (7.3) *Why would the mean particle mass decrease with the distance from the center?*

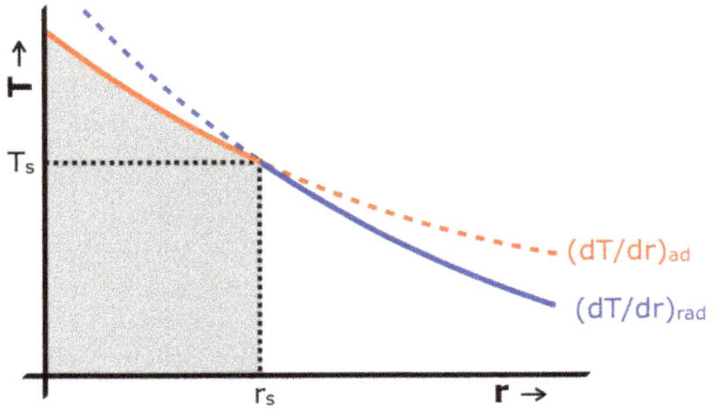

Figure 7.2. Schematic picture of the Schwarzschild criterion for convection. Dashed line: radiative (blue) and adiabatic (red) T-gradients. Solid lines: actual T structure. The region below r_s is convective.

If a bubble (b) rises adiabatically and the surroundings are in radiative equilibrium, then the condition for convection to occur is

$$\left|\frac{d\rho}{dr}\right|_b > \left|\frac{d\rho}{dr}\right|_s \text{ or} \left|\frac{d\ln\rho}{dr}\right|_b > \left|\frac{d\ln\rho}{dr}\right|_s. \tag{7.9}$$

With $P = \Re\rho T/\mu$ and $\rho = \mu P/\Re T$, we find that convection occurs if

$$\left(\frac{d\ln P}{dr}\right)_b + \left(\frac{d\ln\mu}{dr}\right)_b - \left(\frac{d\ln T}{dr}\right)_b > \left(\frac{d\ln P}{dr}\right)_s + \left(\frac{d\ln\mu}{dr}\right)_s - \left(\frac{d\ln T}{dr}\right)_s. \tag{7.10}$$

Here, we have allowed μ to be distance-dependent, which causes a rising blob to have a different μ than its surroundings. Since the bubble will remain in pressure equilibrium with its surroundings, and the chemical composition of the bubble will not change when it rises, we find the condition for convection:

$$-\left(\frac{d\ln T}{dr}\right)_b > -\left(\frac{d\ln T}{dr}\right)_s + \left(\frac{d\ln\mu}{dr}\right)_s. \tag{7.11}$$

If convective cells rise adiabatically in a medium that is in radiative equilibrium and has a μ-gradient, the condition for convection is therefore

$$\boxed{\left|\frac{d\ln T}{dr}\right|_{\text{rad}} > \left|\frac{d\ln T}{dr}\right|_{\text{ad}} + \left|\frac{d\ln\mu}{dr}\right|.} \tag{7.12}$$

This is the **Ledoux criterion for convection**, after the Belgian astrophysicist Paul Ledoux (1914–1988), who introduced it in 1947 (Ledoux 1947).

Note the following.

1. In a homogeneous layer, the Ledoux criterion is the same as the Schwarzschild criterion.
2. For given adiabatic and radiative temperature gradients in a star, a chemically stratified zone is more stable against convection than a chemically homogeneous zone.

3. Convection is very efficient in chemical mixing, so the chemical stratification will disappear in a convective zone and the zone will adopt the average of μ.

7.2.1 Semi-convection

We have seen that a layer with a μ-gradient is more stable against convection than a chemically homogeneous layer. However, detailed calculations show that the zone between the Schwarzschild criterion and the Ledoux criterion is not completely stable if there is a μ-gradient. Some type of pulsational instability may develop that results in chemical mixing. This is called **semi-convection**. In a semi-convective layer the energy is transported by radiation but the gas is mixed and the μ-gradient is erased (see Maeder 2009).

7.3 The Mixing Length: How Far Does a Convective Cell Rise before It Dissolves?

A rising convective cell will dissolve into its surroundings (i.e., it will lose its identity as a distinct cell), as the temperature of the gas inside the cell gradually adjusts to the temperature of its surroundings by the loss of radiation or heat at its boundary. The distance a hot cell rises or a cold cell descends before it dissolves in its surroundings is called the **mixing length, ℓ_m**. The proper calculation of the mixing length is complicated because it involves 3D hydrodynamics (Robinson et al. 2003).

As a reasonable guess, we can assume that ℓ_m will be on the order of the pressure scale height inside the star and make a rough estimate of the pressure scale height. HE requires

$$\frac{dP}{dr} = -\rho\frac{GM_r}{r^2} \quad \text{with } \rho = \frac{P\mu}{\mathcal{R}T} \quad \rightarrow \quad \frac{d\ln P}{dr} = -\frac{GM_r}{r^2}\frac{\mu}{\mathcal{R}T}. \tag{7.13}$$

If T and M_r/r^2 do not vary too strongly with distance around r, we can approximate

$$P(r+h) \simeq P(r)e^{-h/H_p}, \tag{7.14}$$

where the pressure scale height is

$$\boxed{H_p = \frac{\mathcal{R}T}{\mu}\frac{r^2}{GM_r} = \frac{\mathcal{R}T}{\mu g}}. \tag{7.15}$$

(The scale height of a physical quantity is defined as the length in the radial direction over which the value of that quantity decreases by a factor of e^{-1}.)

The **mixing length for convection** is usually written as

$$\ell_m = \alpha \times H_p \quad \text{with } \alpha \simeq 1. \tag{7.16}$$

The assumption of near-constant T is a rather bad one because there must be a temperature gradient in a star in order to transfer energy outward. The proper value of the pressure scale height can only be derived when the structure of the star has

been calculated; however, for our purposes of estimating the properties of convection, this approximation is sufficient.
NB: Expression (7.15) also applies to the atmospheres of planets.

Q (7.4) *Estimate the pressure scale height in the Earth's atmosphere. Do you need an oxygen mask when you climb Mount Everest* ($\sim$10 *km*)?

7.4 The Efficiency of Convective Energy Transport

In convective layers, convection transports the stellar luminosity outward. We will provide a simple estimate to get a "feeling" for this physical process. Consider the energy transport through a sphere at a radius r inside a convective zone. Hot cells rise and cooler cells descend through this layer. Due to the conservation of mass, the amount of mass that moves up at r is equal to the amount that descends (see Figure 7.3). Let us assume that

- half of the matter moves up and the other half moves down;
- because of mass conservation, the upward velocity ($+\mathrm{v_c}$) is equal to the downward velocity ($-\mathrm{v_c}$); and
- the difference in temperature between upward- and downward-moving cells is $T_{\mathrm{up}} - T_{\mathrm{down}} = \Delta T$.

In that case, the difference in thermal energy content per cm^3 between a rising cell and a descending cell is $\Delta e_{\mathrm{th}} = (3/2)nk\Delta T$

The energy transport, $L_c(r)$, by convection through the shell at r is the difference between the thermal energy of the rising flow and the descending flow:

$$\begin{aligned} L_c(r) &= 4\pi r^2 \mathrm{v_c}\,\frac{3}{2}nkT_{\mathrm{up}} - 4\pi r^2 \mathrm{v_c}\,\frac{3}{2}nkT_{\mathrm{down}} \\ &= 6\pi r^2\, nk\Delta T\ \mathrm{v_c} = 6\pi r^2 \frac{\rho}{\mu m_{\mathrm{H}}} k\,(\Delta T\cdot \mathrm{v_c}) = \frac{6\pi r^2 \rho \mathcal{R}}{\mu}(\Delta T\cdot \mathrm{v_c}). \end{aligned} \tag{7.17}$$

We will show later that in convective layers almost all energy is transported by convection and only a minor fraction is transported by radiation. This implies that

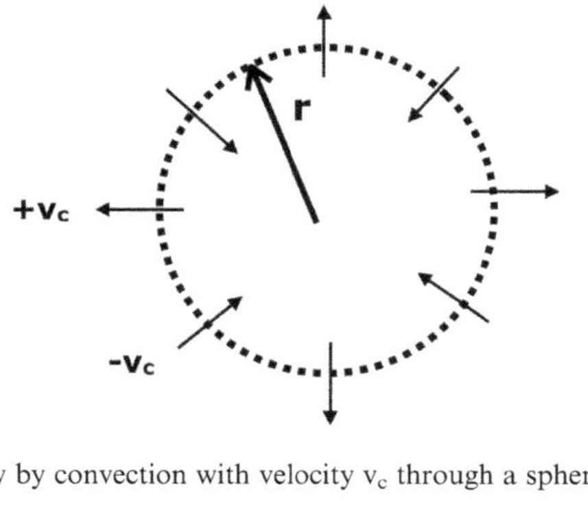

Figure 7.3. Transport of energy by convection with velocity $\mathrm{v_c}$ through a sphere with radius r inside the star.

the convective luminosity is almost equal to the total luminosity at distance r: $L_c(r) \approx L(r) \equiv L_r$. Equation (7.17) results in a condition for the value of $\Delta T \cdot v_c$

$$\boxed{\Delta T \cdot v_c \simeq (L_r/6\pi r^2)\,(\rho\mathcal{R}/\mu)^{-1}}. \tag{7.18}$$

For an order of magnitude estimate, we can substitute characteristic values for a typical star, e.g., the Sun.

$$\left.\begin{array}{l} r \simeq 0.5R_\odot = 3.5\times 10^{10}\,\text{cm} \\ \rho \simeq 1\ \text{g cm}^{-3} \\ \mu \simeq 0.5 \\ \mathcal{R} = 8.3\times 10^{7}\ \text{erg}\ (\text{K}^{-1}\,\text{mole}) \\ L_\odot = 4\times 10^{33}\ \text{erg s}^{-1} \end{array}\right\} \Delta T \cdot v_c \simeq 10^{4}\ \text{cm s}^{-1}\ \text{K} \tag{7.19}$$

This value of $\Delta T \cdot v_c$ is extremely small—much smaller than we could have expected intuitively!

(a). We might have expected that ΔT is a "reasonable fraction" of the mean T at r, e.g., $\Delta T \simeq 0.1\bar{T} \simeq 10^5$ K. In that case, we find that v_c is only 0.1 cm s^{-1}! This is much smaller than the speed of sound, $v_{sound} \simeq 10^7$ cm s^{-1}, which is surprising because we might have expected that the flow speed would be of the order of the sound speed.

(b). On the other hand, if we had assumed that the speed of the cells would be approximately equal to the sound speed, $v_c \simeq v_{sound} \simeq 10^7$ cm s^{-1}, then we would have found that ΔT is only 10^{-3} K! This is much smaller than the average $T \approx 10^6$ K at r.

(c). From (a) and (b), we can conclude that either the velocity v_c of the convective cells is surprisingly small, the temperature difference between rising and descending cells is surprisingly small, or both. This last option is what happens in reality, as we will show below.

Q (7.5) *Explain in words the physical reason for $\Delta T \cdot v_c$ being very small.*
Hint: think in terms of gas-energy content and compare it with a typical stellar luminosity.

7.5 The Convective Velocity

Above we derived an estimate for the product $\Delta T \cdot v_c$. Thus, if we can estimate v_c, we know ΔT. We can estimate the convection velocity by considering a toy model in which we treat a convective cell as a balloon. When a helium-filled balloon is released, it accelerates upward but quickly reaches a constant velocity when the upward buoyancy force is equal to the downward drag due to atmospheric friction.

Figure 7.4 shows the model. Assume that a convective cell is cylindrical. The density in the balloon is ρ_b and that of the surroundings is ρ_s. The upward buoyancy force on the cell is the product of the volume × density difference × gravity = weight difference × gravity:

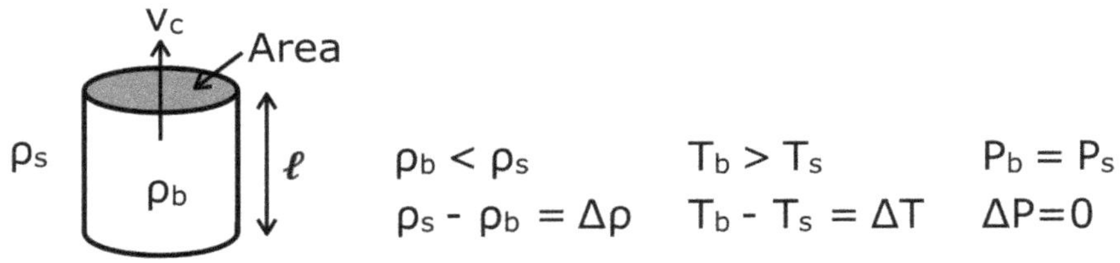

Figure 7.4. Balloon analogy of a convection cell for estimating the convection velocity. The cylinder has a top area A, a length ℓ, and a velocity v_c. The properties of the balloon (b) and the surroundings (s) are indicated.

$$F\uparrow \;=\; A\,\ell\,\Delta\rho\,\frac{GM(r)}{r^2}\ \text{in}\ [\text{cm}^2]\ [\text{cm}]\ [\text{g cm}^{-3}]\ [\text{cm s}^{-2}] = [\text{g cm s}^{-2}]. \tag{7.20}$$

The downward force is the drag experienced by the cell when it rises through the medium. This force, called the ram force, is equal to the momentum transfer of gas that is pushed away at the top.

$$F\downarrow \;=\; \rho_s v_c \times v_c A = \rho_s v_c^2 A\ \text{in}\ [\text{g cm}^{-3}]\ [\text{cm s}^{-1}]^2\ [\text{cm}^2] = [\text{g cm s}^{-2}], \tag{7.21}$$

where $\rho_s v_c$ is the momentum and $v_c A$ is the volume of the medium that is displaced per second.

The cell reaches a constant upward velocity when $F\uparrow = F\downarrow$,

$$\rho_s v_c^2 A \;=\; A\,\ell\,\Delta\rho\, g(r), \tag{7.22}$$

so the speed of the convective cell is approximately

$$v_c^2 \;=\; g(r)\,\ell\,(\Delta\rho/\rho_s). \tag{7.23}$$

We have argued above that the cell will be in pressure equilibrium with its surroundings, so $|\Delta\rho_s/\rho_s| \;=\; |\Delta T/T|$. This results in the expression

$$\boxed{\frac{v_c^2}{\Delta T} \approx \frac{g(r)\,\ell}{T}}. \tag{7.24}$$

Remember that the mixing length ℓ is approximately the pressure scale height $H_p \;=\; \Re T/\mu g$. This implies that $v_c^2/\Delta T \approx \Re/\mu \approx 1.4 \times 10^8\ \text{cm}^2\ \text{s}^{-2}\ \text{K}$.

This equation of motion that specifies the value of $v_c^2/\Delta T$ in combination with the equation for energy transport (7.18), which specifies the value of $\Delta T \cdot v_c$, describes the values of v_c and ΔT separately.

7.6 Typical Values of Convective Velocity and the Timescale

To get a "feeling" for the properties of a convective flow, let us estimate the values of v_c and ΔT at a characteristic radius of $r = 0.5R_\odot$ (although the Sun is not convective at that radius).

$$\left.\begin{array}{l} g(0.5R_\odot) \simeq 10^5 \text{ cm s}^{-2} \\ \ell_p \approx 0.1R_\odot = 7\times10^9 \text{ cm} \\ T \approx 2\times10^6 \text{ K} \end{array}\right\} \frac{v_c^2}{\Delta T} \simeq 3.5\times10^8 \text{ cm}^2\text{ s}^{-2}\text{K}. \tag{7.25}$$

Combining this with the estimate in Equation (7.19) yields

$$\left.\begin{array}{r} \Delta T\cdot v_c \simeq 10^4 \text{ cm Ks}^{-1} \\ v_c^2/\Delta T \simeq 3.5\times10^8 \text{ cm}^2\text{ s}^{-2}\text{K} \end{array}\right\} \boxed{v_c \simeq 10^4 \text{ cm s}^{-1}} \lll v_{\text{sound}} \sim 10^7 \text{ cm s}^{-1}. \tag{7.26}$$

Substitution of this value in Equation (7.23) results in an estimate of ΔT

$$\Delta T\cdot v_c = 10^4 \text{ cm Ks}^{-1} \rightarrow \boxed{\Delta T \simeq 1K} \rightarrow \Delta T/T \simeq 10^{-6}. \tag{7.27}$$

We see that the temperature difference between rising and descending cells is extremely small!

Using the estimate of v_c and the typical distance that a convective cell rises, i.e., a pressure scale height, ℓ_p, we can estimate the typical rise time of convective cells

$$t_{\text{rise}} \simeq \frac{\ell_p}{v_c} \simeq \frac{7\times10^9 \text{ cm}}{10^4 \text{ cm s}^{-1}} \simeq 7\times10^5 \text{ s} \rightarrow \boxed{t_{\text{rise}} \sim \text{week}}. \tag{7.28}$$

These derivations showed that the velocity of convective cells is only a small fraction of the local speed of sound and that the time for rising and descending cells is on the order of weeks. This timescale is much smaller than the evolution (Kelvin–Helmholtz or nuclear) timescale, but significantly longer than the dynamical timescale, which for the Sun is on the order of an hour (see Section 9).

This implies that, although a convection zone is locally unstable, it is globally in hydrostatic equilibrium. As a result, we can apply the HE equation throughout a star, even in the convective layers, when calculating stellar structure.

7.7 The Super-adiabatic Temperature Gradient in Convection Zones

The mean T-gradient of the surroundings in a convection zone must be steeper than $(dT/dr)_{\text{ad}}$ of the convective cells; otherwise, the convection would stop.

Q (7.6) *Explain why this is required.*

If (dT/dr) describes the mean temperature gradient in a convection zone, then the Schwarzschild criterion requires $|dT/dr| > |dT/dr|_{\text{ad}}$. The difference between (dT/dr) and $(dT/dr)_{\text{ad}}$ is called the **super-adiabatic T-gradient.** So

$$\left|\frac{dT}{dr}\right| = \left|\frac{dT}{dr}\right|_{\text{ad}} + \left|\frac{dT}{dr}\right|_{\text{sa}}. \tag{7.29}$$

We have shown above that the super-adiabatic T-gradient produces a difference of only $\sim$1 K over a distance of a pressure scale height. Thus,

$$\left|\frac{dT}{dr}\right|_{sa} \lll \left|\frac{dT}{dr}\right|_{ad} \quad \text{or} \quad \left|\frac{dT}{dr}\right| = \left|\frac{dT}{dr}\right|_{ad}(1+\epsilon), \quad \text{with } \epsilon \lll 1. \tag{7.30}$$

When computing stellar models, we can therefore safely adopt $|dT/dr| = |dT/dr|_{ad}$ *in a convection zone.*

7.8 Convective Overshooting

The Schwarzschild criterion is derived under the assumption that the buoyancy force provides the upward force. At the top of the convection zone, the upward force vanishes. The same applies at the bottom of the convection zone. However, the moving cells do not suddenly halt there. Because of their inertia, they will overshoot at the convection boundaries. This is analogous to the jump of a rubber duck bath toy, filled with air, that is held underwater and then released. If a star has a convective core, the cells will rise above the boundary set by the Schwarzschild criterion. If a star has a convective outer layer, the cells will descend lower than the predicted boundary. The **convective overshooting length** can be expressed in terms of the local pressure scale height as

$$\ell_{os} = \alpha_{os}\ell_{P}. \tag{7.31}$$

Comparisons between observations and theory in different types of stars suggest that the **overshooting parameter** $\alpha_{os} \approx 0.2$ (Maeder & Meynet 1989).

Q (7.7) *What sorts of observations can be used to estimate the value of* α_{os}*?*

[Historical note: The first observational indications for the occurrence of overshooting in the cores of massive stars appeared in about 1985. It was discovered that nuclear products (especially He and N) appear at the surface of massive MS stars before mass loss could have peeled off enough of the outer layers to reach the deeper layers, where the initially convective core could have mixed these products upward. The conclusion was that there must have been some mixing reaching layers higher than the predicted boundary of the convective core (Lamers 2008)].

There is a fundamental difference between convection and overshooting:

- Convection: energy transport by convection plus chemical mixing.
- Overshooting: only mixing; the main energy transport is by radiation.

7.9 Convection: Where and Why?

The Schwarzschild criterion for convection is $(dT/dr)_{rad} > (dT/dr)_{ad}$, with

$$\left|\frac{dT}{dr}\right|_{rad} \sim (\kappa\rho/T^3)\,(L_r/4\pi r^2). \tag{7.32}$$

From this, we can see that there are **two main causes for convection** and two types of layers that easily become convective: a high value of the opacity κ or a high-energy flux $L_r/4\pi r^2$.

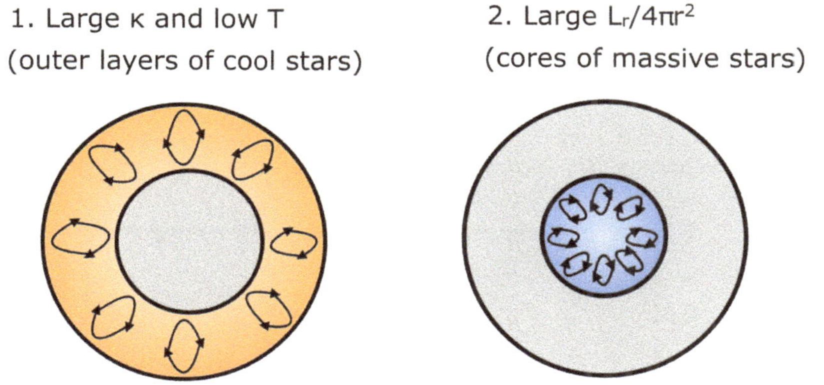

Figure 7.5. Schematic drawing of envelope convection in cool stars and core convection in massive stars. The gray regions are in radiative equilibrium.

1. Layers where γ_{ad} is small or κ is very large (e.g., layers where H is partly ionized; see Figures 4.5 and 5.1) are found in the outer layers of cool stars.
   - MS stars with a spectral type later than about F0:
     - at type G0, the stars have a thin outer convection layer;
     - at type M, the stars are almost fully convective.
   - Red giants and supergiants: almost fully convective with the exception of the core.
2. Layers where the energy flux $L_r/4\pi r^2$ is very large are found in the central regions of massive stars because these stars have a high luminosity created by nuclear fusion in a small core.
   - MS stars of $M \gtrsim 1.2M_\odot$, because H-fusion via the CNO-cycle has a strong T dependence and therefore occurs in a small central region.

This is shown schematically in Figure 7.5.

Figure 7.6 shows the location (in terms of mass fraction) of convective regions in MS stars. The figure shows the convective outer layers of stars of $M < 1.2M_\odot$ and the convective cores of stars of $M > 1.2M_\odot$. The decrease of the convective envelope mass with the increasing star mass in low-mass stars is due to the fact that the ionization zone moves closer to the surface as the mass and the temperature of the star increases. The increase of the mass fraction of the convective cores with stellar mass at the beginning of the core H-fusion phase (zero-age main sequence: ZAMS) is due to the fact that the luminosity increases steeply with M, roughly as $L \sim M^4$ on the MS. This rapid increase in L implies that the Schwarzschild criterion for convection is satisfied in a larger fraction of the mass.

During the main-sequence phase of massive stars the changing chemical composition affects the extent of the convective core. Model calculations show that the mass of the convective core decreases during the core-H fusion phase between the zero-age main sequence, ZAMS, and the end of the core H-fusion phase (terminal-age main sequence, TAMS), even though the luminosity is slightly increasing (see, e.g., Maeder & Meynet 1989). This is because the opacity in the cores of a massive star decreases as the H-abundance decreases and the He abundance increases. We have seen in section 5.2 that the main opacity in the centers of hot stars is electron scattering with $\sigma_e \sim (1 + X)$; σ_e decreases in the core due to core H-fusion. A decrease of σ_e implies that the Schwarzschild criterion is satisfied in a smaller fraction of the mass. This is schematically shown in Figure 7.7.

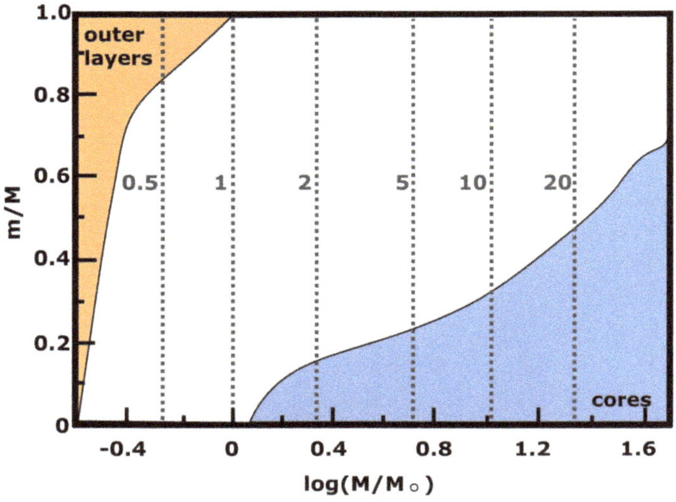

Figure 7.6. Occurrence of convection in stars at the beginning of the core H-fusion phase (ZAMS). The mass of convective envelopes (orange) and convective cores (blue) is expressed as a fraction of the stellar mass, from $m/M = 0$ in the core to $m/M = 1$ at the surface. The vertical lines indicate the stellar mass. (Reproduced from Kippenhahn & Weigert 1990. © Springer-Verlag Berlin Heidelberg 2012.)

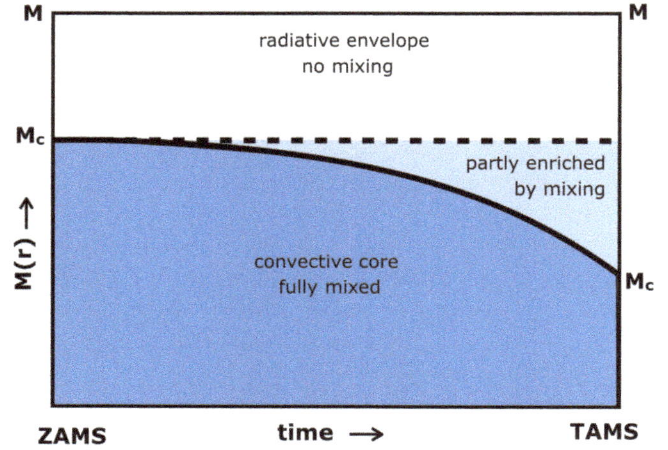

Figure 7.7. Schematic picture showing the decrease of the mass of the convective core during the main-sequence phase of a massive star. The boundary of the convective core is where $|dT/dr|_{\rm ad} = |dT/dr|_{\rm rad} \sim \sigma_e \sim (1 + X)$, with X decreasing to 0 when the core runs out of H (i.e., at the TAMS).

7.10 Chemical Mixing by Convection and Its Consequences

We have seen that the timescale for the rise and descent of convection cells is much shorter than the nuclear timescale. This results in fast mixing throughout convection zones; as a result, a convective region is chemically homogeneous. Figure 7.8 shows this in a schematic way for stars with convective envelopes and with convective cores. In the absence of convection, nuclear fusion would create a chemical profile with the mean atomic mass decreasing outward, e.g., from He ($\mu = 4/3$) to H ($\mu = ½$). However, the presence of convection results in mixing, which erases the μ-gradient in the convection zone and in the overshooting regions.

Q (7.4) *The predicted vertical jumps of the μ-profile shown in figure 7.8 will in reality be slightly smoothed. Which physical effects could contribute to this smoothing?*

We have shown in Section 7.2 that the convection criterion of Ledoux in layers, which have a μ-gradient, is stricter than the Schwarzschild criterion: a star with a chemical gradient is more stable against convection. However, as soon as the Ledoux criterion is fulfilled and convection starts, the chemical gradient is erased and the star develops a fully convective zone according to the Schwarzschild criterion.

Convection and overshooting can prolong the lives of stars. Massive stars on the main sequence have convective cores. Although the nuclear fusion happens in the very center, the nuclear products are mixed throughout the convective core. One consequence is that H is brought into the center from all over the convective core. This results in an increase in the mass fraction of the star that takes part in H-fusion from about 0.14 for solar mass stars to 0.75 for stars with $M > 60 M_{\odot}$. *This considerably extends the expected MS lifetimes of massive stars.*

7.11 Summary

1. Convection occurs in the layers where energy transport by radiation would require the T-gradient to be steeper than the adiabatic one: $|dT/dr|_{\rm rad} > |dT/dr|_{\rm ad}$. This is the Schwarzschild criterion. In those layers, energy is transported outward by hotter rising cells and cooler descending cells. Convective regions are chemically mixed.

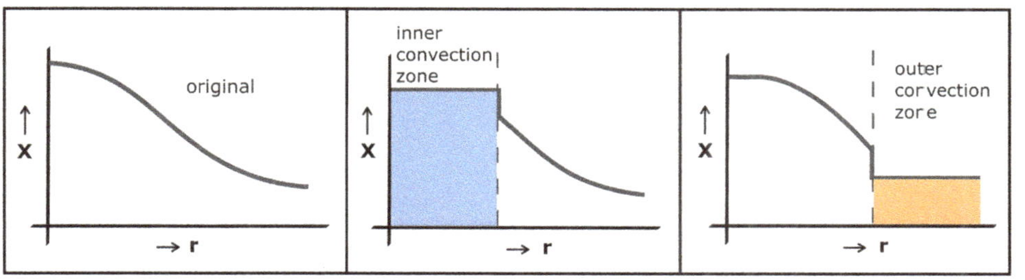

Figure 7.8. Changes in the chemical profile (indicated by X) of a star due to mixing by convection and overshooting. The profile is shown in the case of no convection (left), convection in the core (center), and convection in the envelopes (right).

2. The existence of a μ-gradient suppresses convection, and the Schwarzschild criterion is replaced by the Ledoux criterion. The region between boundaries of the Schwarzschild criterion and the Ledoux criterion is unstable to semi-convection and is also mixed.
3. The temperature structure of a convective region is described by the adiabatic T-gradient.
4. The velocity of the convective cells is very small compared to the local speed of sound and the temperature difference between rising and descending cells is very small compared to the average local temperature.
5. The characteristic time for the rise and fall of convective cells is on the order of weeks to months. Although convective layers are locally unstable, the convective region as a whole is in hydrostatic equilibrium.
6. At the top or bottom of convective zones the moving cells may overshoot beyond the Schwarzschild boundary. This produces extra mixing in the layers on top of convective cores or below convective envelopes.
7. Convection occurs when either of the following criteria is met:
   - High absorption coefficient. This produces convective envelopes in cool stars because κ is highest in cool layers that are not fully ionized.
   - High value of the flux $L_r/4\pi r^2$. This produces convective cores in stars with $M > 1.2M_\odot$ on the main sequence.
8. The mass of the convective core of a massive star decreases during its main-sequence phase.

Exercises

7.1 Calculate the properties of the convective envelope of the Sun.
   (a) Calculate the thermal energy content of a cm^3 of gas in the Sun at $r = 0.9R_\odot$, using data from Appendix C.
   (b) Calculate the speed of sound at that location.
   (c) Calculate g and the pressure scale height
   (d) Assume that the temperature difference between the ascending and descending bubbles is δ_T of the mean local value, and that the velocity of the bubbles is δ_V of the local sound speed. Calculate the luminosity that would be transported by convection in this case.
   (e) Compare this with the true luminosity, L_r, at $0.9\mathrm{R}_\odot$.
   (f) Derive the values of δ_V and δ_T, and of ΔT and v_c.
   Hint: use Equation (7.24).

7.2 Consider the Ledoux criterion.
   (a) Sketch a diagram of outward decreasing μ as a function of r/R. Adopt some lower limit and some upper limit for a zone that would be convective according to the Schwarzschild criterion.
   (b) Show schematically where the convection zone would be according to the Ledoux criterion for convection in a medium with a μ-gradient.

(c) Show how the μ-profile would be changed by convection in these two cases.

7.3 Estimate the duration of the main-sequence phase for stars with convective cores. During the main-sequence phase of the Sun, which lasts 8×10^9 yr, the inner 20% of the solar mass takes part in the H-fusion.
   (a) Use the mass of the convective cores, shown in Figure 7.6, to estimate the duration of the main-sequence phase of stars of 4, 12, 20, and $40 M_\odot$. Use the luminosities from Appendix D.
   (b) Compare the results with the lifetime computed from evolutionary models.

7.4 Sketch the chemical profile of He in a massive star in which the mass of the convective core decreases during the H-fusion phase. Sketch it at three epochs: at the beginning of the main-sequence phase (ZAMS), halfway through the main-sequence phase, and at the end of the main-sequence phase (TAMS).

References

Kippenhahn, R., & Weigert, A. 1990, Stellar Structure and Evolution (Berlin: Springer)

Lamers, H. J. G. L. M. 2008, in ASP Conf. Ser. 388, Mass Loss from Stars and the Evolution of Stellar Clusters, ed. A. de Koter, L. J. Smith, & L. B. F. M. Waters (San Francisco, CA: ASPC), 443

Ledoux, P. 1947, ApJ, **105, 305**

Maeder, A. 2009, Physics, Formation and Evolution of Rotating Stars (Berlin: Springer)

Maeder, A., & Meynet, G. 1989, A&A, **210, 155**

Pols, O. 2011, Stellar Structure and Evolution, Utrecht University Lecture Notes, https://www.astro.ru.nl/~onnop/education/stev_utrecht_notes/

Robinson, F. J., Demarque, P., & Li, L. H., et al. 2003, MNRAS, **340, 923**

Schwarzschild, K. 1906, WisGo, 195, 41 reprinted in Menzel, D. H. 1962, Selected Papers on Physical Processes in Ionized Plasmas (New York: Dover)

Henny J.G.L.M. Lamers and Emily M. Levesque

Chapter 8

Nuclear Fusion

Throughout most of their lives stars radiate energy that is produced by nuclear fusion. There are several subsequent fusion processes that a star can use to generate energy, starting with the fusion of H into He and ending with Si into Fe. We will describe the reaction chains for these processes and their energy production. Each subsequent fusion process requires a higher temperature because of the increasing charge of the nuclei involved. Not all stars reach the required temperatures because low-mass stars may not be able to release sufficient potential energy. There is therefore a minimum stellar mass for every nuclear fusion step. During late stellar evolution phases, massive isotopes are formed by neutron capture. This produces specific sequences of stable isotopes that are observed in stellar spectra. We describe two such sequences: slow neutron capture and rapid neutron capture.

8.1 Reaction Rates and Energy Production

During nuclear fusion two particles (i and j) react, which results in one or two other particles (k and l). The particles involved have a charge Z and a mass A.

So the reaction is $i + j \rightarrow k + l$,
with nuclear charge conservation $Z_i + Z_j = Z_k + Z_l$,
and baryon number conservation: $A_i + A_j = A_k + A_l$.

The **reaction rate,** r_{ijkl} is expressed by the number of reactions per second per gram. The **energy generation rate** by nuclear fusion per second per gram is

$$\epsilon_{ijkl} = r_{ijkl} \cdot Q_{ijkl}, \tag{8.1}$$

where

$$Q_{ijkl} = (m_i + m_j - m_k - m_l) \cdot c^2 \tag{8.2}$$

is the amount of energy produced in one reaction. The term in parentheses is the mass defect of the reaction, i.e., the mass that has been converted into energy.

doi:10.1088/978-0-7503-1278-3ch8

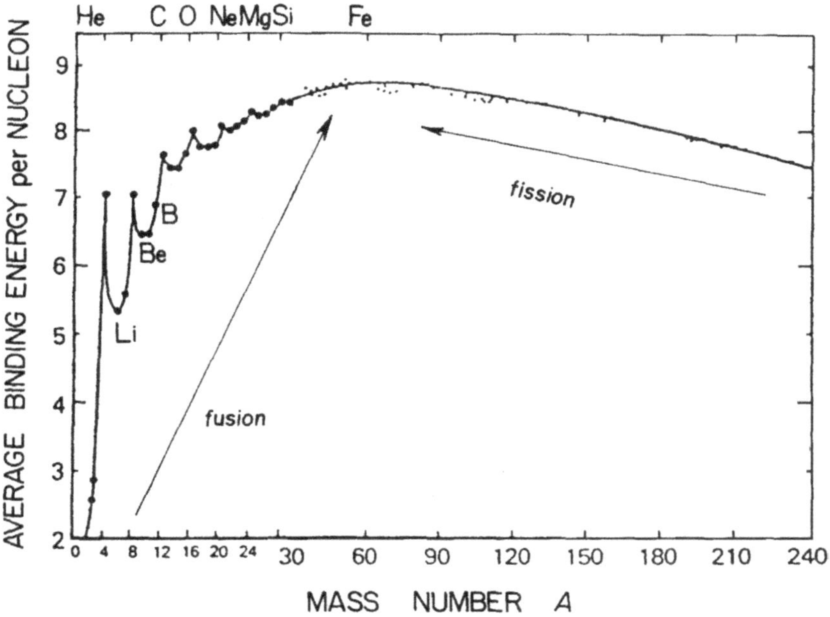

Figure 8.1. Average binding energy in MeV per nucleon (proton or neutron) as a function of atomic mass number. (Reproduced with permission from Rolfs & Rodney 1988. University of Chicago Press © 1988.)

Q corresponds to the difference in binding energy of the nuclei involved in this reaction and is usually expressed in MeV, with 1 MeV $= 1.602 \times 10^{-6}$ ergs.

Figure 8.1 shows the average binding energy per nucleon as a function of atomic mass number. The maximum is reached at $A \approx 56$ (near Fe). In the rising part of the curve, *fusion produces energy* because the total binding energy of the product is larger than the sum of the binding energies of the fused nuclei. In the descending part of the curve, *fusion requires energy* because the total binding energy of the products is smaller than that of the fused nuclei. In that case, *energy can be produced by fission.*

Q (8.1) *Explain why the cosmic abundance of Li and Be is low.*

The mass defect can conveniently be expressed as a fraction of the mass that goes into the fusion process. This is the **mass defect fraction: $\Delta m/m$.**
For instance,

$$\text{for H} \rightarrow \text{He fusion: } \Delta m/m = 0.00712; \tag{8.3}$$

$$\text{for He} \rightarrow \text{C fusion: } \Delta m/m = 0.00065. \tag{8.4}$$

The most important isotopes involved in nuclear reactions in stars are listed in Table 8.1.

Table 8.1. Masses of the Most Important Isotopes Involved in Nuclear Fusion Reactions in Stars, Expressed in Atomic Mass Units: $m_u = m_C/12$

Element	Z	A	M/m_u	Element	Z	A	M/m_u	Element	Z	A	M/m_u
n	0	1	1.0087	N	7	13	13.0057	Ne	10	22	19.9924
H	1	1	1.0078		7	14	14.0031	Mg	14	24	23.9850
	1	2	2.0141		7	15	15.0001	Si	14	28	27.9769
He	2	3	3.0160	O	8	15	15.0031	S	16	32	31.9721
	2	4	4.0026		8	16	15.9949	Fe	26	56	55.9349
C	6	12	12.0000		8	17	16.9991	Ni	28	58	57.9353
	6	13	13.0034		8	18	17.9992	e	−1		0.0005

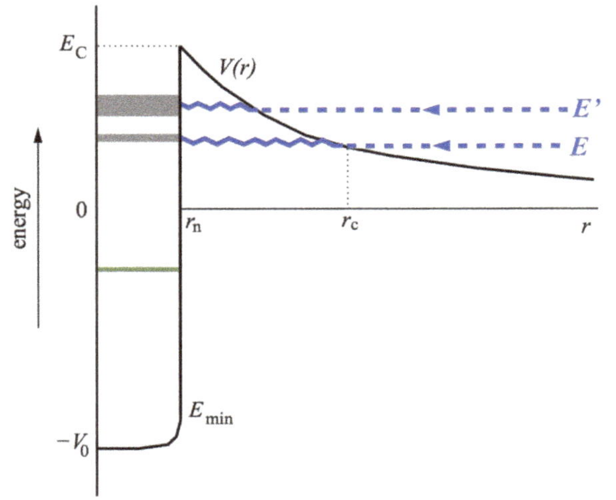

Figure 8.2. Schematic picture of the effect of tunneling through the Coulomb barrier. The dashed blue lines indicate the approach of two particles with energy E' and E. The blue wiggling lines indicate the tunneling from a distance r_c with a certain energy (gray) into the potential well at distance r_n. The potential energy at $r = 0$ is $-V_0$. The green line indicates the presence of a possible resonance level. (Adapted from Pols 2011).

8.2 Thermonuclear Reaction Rates and the Gamow Peak

Ions have a positive charge, so they will repulse one another by electric Coulomb forces. To enable fusion, the particles must overcome this Coulomb barrier. Figure 8.2 shows the potential due to the repulsive Coulomb force that increases toward smaller distance at $r > r_n$. The maximum potential is $E_c = E(r_n)$. At $r < r_n$, the nuclear forces produce an attraction resulting in a negative potential. An incoming particle with an energy $E < E_c$ can only reach $r < r_n$ by quantum mechanical tunneling through the Coulomb barrier.

The velocity of the particles follows the Maxwell distribution. Only the fastest particles have a probability of overcoming the Coulomb barrier; however, their number decreases rapidly with increasing velocity or energy E as

$$N(E) \sim \exp(-E/kT). \tag{8.5}$$

Quantum mechanical tunneling allows particles with an energy less than that of the Coulomb barrier to overcome this barrier. Without this effect, the fusion process in stars would require a much higher T than in reality. The tunneling probability P_t increases with energy as

$$P_t(E) \sim \exp\left(-\frac{b}{\sqrt{E}}\right), \tag{8.6}$$

where b is a constant that depends on the reaction. The net result is that the reaction rate scales with the product of the two functions $N(E)$ and $P_t(E)$ and shows a peak called the **Gamow peak**

$$r(E) \sim N(E)\, P_t(E) \sim \exp\left(-\frac{E}{\kappa T} - \frac{b}{\sqrt{E}}\right). \tag{8.7}$$

It is named after the Russian–American physicist and cosmologist George Gamow (1904–1968), who described it in 1937 (and predicted the cosmic background radiation with a temperature of 7 K; Gamow 1937).

Figure 8.3 shows the Gamow peak for the reaction $^{12}C + p \rightarrow {}^{13}N + \gamma$. This reaction has a peak near $kT = 30$ MeV, corresponding to a temperature of $T \approx 3.5 \times 10^8$ K. The left figure shows the two functions $N(E)$ and $P_t(E)$ and the product $r(E)$. The right figure shows $r(E)$ on a linear scale for three temperatures. Note the Gaussian-like shape of $R(E)$ and the extreme sensitivity to temperature.

Because of the strong dependence of the Coulomb barrier on the charge of the fusing ions, and on the combination of the Maxwell energy distribution and the tunneling effect, the reaction rate and energy production rate depend very strongly on temperature. For instance, the temperatures of the fusion reactions in stars are

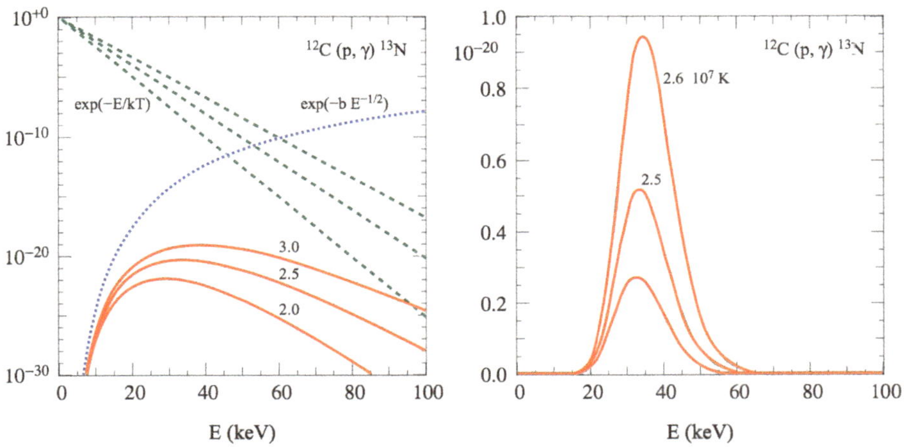

Figure 8.3. Example of the Gamow peak of $r(E)$ for the reaction $^{12}C(p, \gamma)\ ^{13}N$. The peak is the result of the product of the two functions $N(E)$ and $P_t(E)$. Left: the function $N(E)$ is shown for $T = 2.0 \times 10^7$, 2.5×10^7, and 3.0×10^7 K by green dashed lines. The function $P_t(E)$ is shown as a blue dotted line. The product $r(E)$ is shown as solid red lines for these same three temperatures. Right: the function $r(E)$ plotted linearly for three values of $T = 2.4$, 2.5, and 2.6×10^7 K. (Adapted from Pols 2011.)

$$\text{H} \rightarrow \text{He (pp-chain)} \quad \text{at} \quad T \approx 1 \times 10^7\ \text{K with}\ \varepsilon \sim T^4, \tag{8.8}$$

$$\text{H} \rightarrow \text{He (CNO-cycle)} \quad \text{at}\ T \approx 2 \times 10^7\ \text{K with}\ \varepsilon \sim T^{12}, \tag{8.9}$$

$$\text{He} \rightarrow \text{C (triple-}\alpha\text{ process)} \quad \text{at}\ T \approx 1 \times 10^8\ \text{K with}\ \varepsilon \sim T^{40}. \tag{8.10}$$

NB: the reaction rates are not really power laws of T, but can be approximated by partial power laws. The exponents of the T-dependence given above are the exponents near the T-range of the fusion reactions in stars (Lang 1980).

8.3 Abundance Changes

Before discussing the various nuclear reactions in stars in detail, we discuss the changes in abundances due to these reactions. The change in abundance per second of element i due to the reaction $i + j \rightarrow k + l$ can be expressed as

$$dX_i/dt = -r_{ijkl}\, A_i m_u. \tag{8.11}$$

If element i is involved in more fusion reactions, some destroying and others creating i, then the change in abundance should be written as the sum of all possible destruction reactions (− sign) and all formation reactions (+ sign).

$$\frac{dX_i}{dt} = A_i m_u \left\{ \sum\nolimits_x - r_{i \rightarrow x} + \sum\nolimits_y + r_{y \rightarrow i} \right\}. \tag{8.12}$$

8.4 H → He Fusion

There are two major routes for converting H into He in stars:

- The proton–proton chain, which has three branches.
- The CNO cycle, which actually consists of four cycles: one CN-cycle and three ON-cycles.

Although the net reactions of these two routes are the same, they have very different effects on the abundance evolution of stars.

8.4.1 The Proton–Proton Chain: pp Chain

At $5 < T < 15 \times 10^6$ K, H is predominantly fused in stars via the pp chain. This chain has three sub-chains: ppI, ppII, and ppIII. These are shown in Figure 8.4. It starts with the formation of ^{3}He, which fuses into ^{4}He directly (ppI); however, there is a small probability that ^{3}He is converted into ^{4}He via ^{7}Be (ppII) and an even smaller probability that ^{7}Be is converted into ^{4}He via ^{8}B (ppIII). All three chains have the same net reaction (see the bottom of Figure 8.4).

The ppI chain is the most important one for stellar evolution. For instance, it is responsible for about 90% of the nuclear energy production in the Sun. The positrons created in the first step annihilate with free electrons and produce γ photons: $e^+ + e^- \rightarrow 2\gamma$. The neutrinos carry off about 2%, 4%, and 28% of the energy for the ppI, ppII, and ppIII chains, respectively.

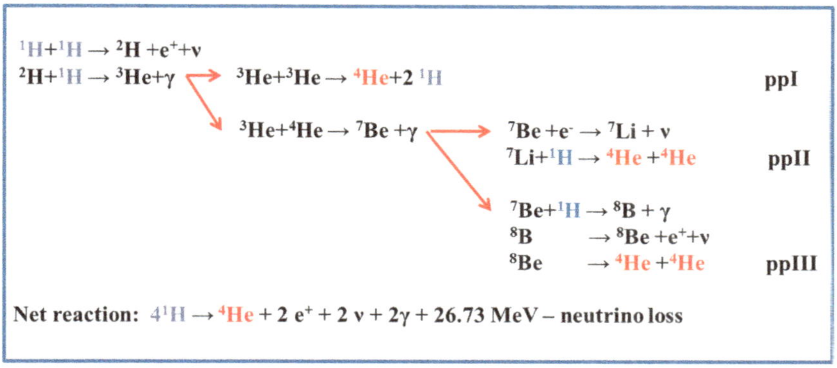

Figure 8.4. Reactions of the ppI, ppII, and ppIII chains for fusing H into He.

The **energy production rate of the pp chain** is

$$\epsilon_{\rm pp} = \epsilon_0^{\rm pp} X^2 \rho (T/10^7{\rm K})^4 \text{ with } \epsilon_0^{\rm pp} = 1.1 \times 10^{-1}\,{\rm erg\ \ cm^3\,g^{-2}\,s^{-1}} \text{at } {\rm T} \approx 10^7\,{\rm K} \quad (8.13)$$

and the **mass defect of the pp chain is $\Delta m/m$** $= 0.00712$

Q (8.2) *Why is $\epsilon_{\rm pp} \sim \rho^1$ and why is $\epsilon_{\rm pp} \sim X^2$?*

8.4.2 The CNO Cycle

At $T > 15$ MK, hydrogen is predominantly fused into helium via the multiple cycles that make up the CNO cycle. These are shown in Figure 8.5. The chain of reactions of the CN-cycle starts with $^{12}C + H \rightarrow ^{13}N + \gamma$, and after a number of steps, ends in $^{12}C + ^4He$, thereby releasing ^{12}C again. Along the way ^{15}N nuclei are formed, from which the ON-cycle can deviate. ^{17}O is the starting point of ON-cycle II and ^{18}O is the starting point of ON-cycle III. All of these cycles produce a ^4He nucleus and return to an earlier reaction in the chain.

The CN-cycle and the ON-cycle I are the two dominant cycles for energy production, with ON-cycle II and ON-cycle III playing only a minor role. This is because the efficiency of each cycle depends on the production rate and the abundance of the starting point of that cycle. The abundance of ^{12}C, which is the starting point of the CN-cycle, is much higher than that of ^{15}O, which is the starting point of the ON-cycle I. The same arguments apply a fortiori to the starting points of ON-cycles II (^{17}O) and III (^{18}O).

Q (8.3) *What is the net reaction of these four cycles?*

The **energy production rate** of the CNO cycle is

$$\begin{aligned} &\epsilon_{\rm CNO} = \epsilon_0^{\rm CNO} X_{\rm H} X_{\rm CN}\, \rho (T/10^7 K)^{18} \\ &\text{with } \epsilon_0^{\rm CNO} = 6.4 \times 10^{-4}\,{\rm erg\,cm^3 g^{-2} s^{-1}} \text{ around } T = 3 \times 10^7{\rm K} \end{aligned} \quad (8.14)$$

where $X_{\rm H}$ is the mass fraction of H and $X_{\rm CN}$ is the mass fraction of $^{12}C + ^{14}N$.

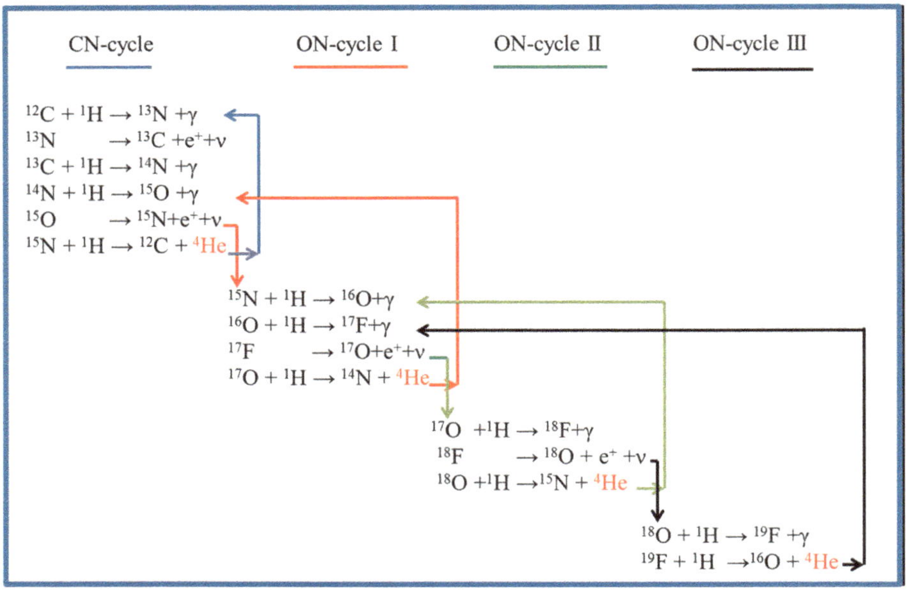

Figure 8.5. Reaction chains of the CN-cycle and the ON-cycles. The positrons annihilate with free electrons to give $e^+ + e^- \rightarrow 2\gamma$. The neutrinos carry off a fraction of the energy.

The **mass defect of the CNO cycle, $\Delta m/m$ = 0.00712,** is the same as that of the pp chain because the net reaction is the same.

Figure 8.6 shows production rates of the pp chain and the CNO cycle as a function of T for a composition of $X_{\rm H} = 0.99$ and $X_{\rm CNO} = 0.01$ and a density of 1 g cm^{-3}. The contributions of the pp chain and the CNO cycle are about equal at $T \approx 15 \times 10^6$ K.

In the center of the Sun at $\sim 14 \times 10^6$ K, the pp chain produces more than 98% of the energy while the CNO cycle produces less than 2%. Note the steeper temperature dependence of the CNO cycle ($\varepsilon \sim T^{18}$) compared to the pp chain ($\varepsilon \sim T^4$).

We have seen in Section 3.2, Equation (3.9), that the central temperature of stars increases toward higher masses: massive MS stars have a higher T_c than low-mass stars. In stars of $M \lesssim 1.3 M_\odot$, the H-fusion occurs mainly via the pp chain, whereas in stars of $M \gtrsim 1.3 M_\odot$ the H-fusion occurs mainly via the CNO cycle.

Q (8.4) *Energy production by the CNO cycle has a much steeper dependence on T than that of the pp chain. What does this imply for the mass of the region where H-fusion occurs in massive stars?*

8.4.3 Equilibrium Abundances of the CNO Cycle

Fusion via the CNO cycle is a cyclic process that quickly reaches equilibrium in the early MS phase. This has two important consequences.

1. The total number of C + N + O ions is conserved because CNO is just a catalyst for the H → He fusion.

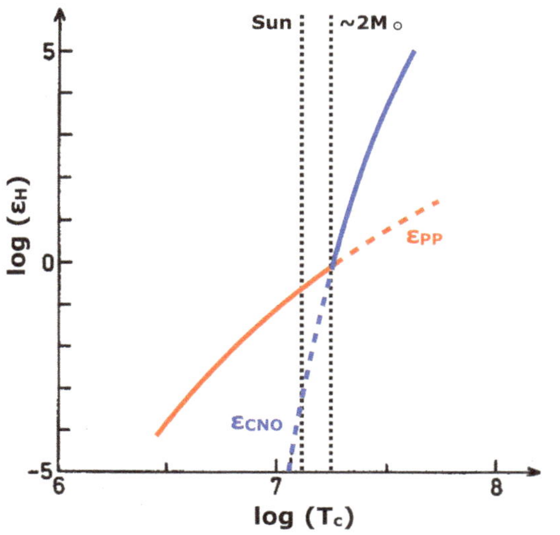

Figure 8.6. Energy production of the pp chain (red) and the CNO cycle (blue) as a function of T at a density of 1 g cm^{-3} for H-gas with $X_{\rm CNO} = 0.01$. The conditions at the centers of the Sun and a star of $2M_\odot$ are indicated. (Reproduced from Kippenhahn & Weigert 1990. © Springer-Verlag Berlin Heidelberg 2012.)

2. In equilibrium, all steps must proceed at the same reaction rate (i.e., the number of reactions per gram per second is the same for each step).

So the rate $r_{i\to j} \sim n_i \times \sigma_{i\to j}$, where σ is the cross-section for the reaction, is the same for all steps; however, some steps have smaller cross-sections so a higher number of those ions is needed to keep the rate the same as those in the other steps.

Let us define the *lifetime* τ *of a nucleus* in this process as

$$\tau_i \equiv n_i/|dn_i/dt| = n_i/r_{ij}, \tag{8.15}$$

with n_i in number per gram and the reaction rate r_{ij} in number g^{-1} s^{-1}. In a cyclic process in equilibrium all reaction rates are equal, so $n_i/n_j = \tau_i/\tau_j$ etc. At $T \simeq 2 \times 10^7$ K, the equilibrium of the reaction cycle implies that the lifetimes of the isotopes are

$$\underset{35\ \rm yr}{\tau(^{15}\rm N)} \ll \underset{1600\ \rm yr}{\tau(^{13}\rm C)} < \underset{6600\ \rm yr}{\tau(^{12}\rm C)} \ll \underset{9\times10^5\ \rm yr}{\tau(^{14}\rm N)} \ll \underset{\sim10^9\ \rm yr.}{\tau_{\rm nuc}} \tag{8.16}$$

We see that in equilibrium the ratio $n(^{14}\rm N)/n(^{12}\rm C) \simeq 140$, whereas the initial 'cosmic' composition has a ratio of $n(^{14}\rm N)/n(^{12}\rm C) \simeq 0.27$. *So the* ^{14}N/^{12}C *ratio increases drastically inside the star due to the* CNO *cycle.*

Q (8.5) *Approximately how long does it take for the* CN *cycle to reach equilibrium?*

The two main abundance effects of H → He via the CNO cycle are

1. H decreases and He increases and
2. ^{12}C decreases and ^{14}N increases.

Q (8.6) *Which one of these two changes occurs faster?*

8.4.4. The NeNa and the MgAl Cycles.

There are many more reactions and reaction chains that occur simultaneously during the H → He fusion. Most of these are not important for generating energy; however, some of them are important for changing abundance ratios. The most important ones, apart from the CN cycle discussed above, are the NeNa cycle and the MgAl cycle. They are responsible for the NeNa and MgAl anti-correlations that are observed in a fraction of the stars in globular clusters (Bastian & Lardo 2018) (Section 17.5).

The **NeNa cycle** can be expressed as

$$\begin{aligned} &^{20}\mathrm{Ne}\ \ (\mathbf{p},\gamma)\ \ ^{21}\mathrm{Na}\ \ (\mathrm{e}^+,\nu)\ \ ^{21}\mathrm{Ne}\ \ (\mathbf{p},\gamma)\ \ ^{22}\mathrm{Na}\ \ (\mathrm{e}^+,\nu) \\ &^{22}\mathrm{Ne}\ \ (\mathbf{p},\gamma)\ \ ^{23}\mathrm{Na}\ \ (\mathbf{p},\mathbf{He^4})\ \ ^{20}\mathrm{Ne}. \end{aligned} \tag{8.17a}$$

The terms in parentheses (in, out) indicate what is added to the original nucleus and what comes out in addition to the resulting nucleus. At $T < 3\times10^7$ K, the abundance of ^{22}Ne decreases and the abundance of ^{23}Na increases.

The **MgAl** cycle proceeds as

$$\begin{aligned} &^{24}\mathrm{Mg}\ \ (\mathbf{p},\gamma)\ \ ^{25}\mathrm{Al}\ \ (\mathrm{e}^+,\nu)\ \ ^{25}\mathrm{Mg}\ \ (\mathbf{p},\gamma)\ \ ^{26}\mathrm{Al}\ \ (\mathrm{e}^+,\nu) \\ &^{26}\mathrm{Mg}\ \ (\mathbf{p},\gamma)\ \ ^{27}\mathrm{Al}\ \ (\mathbf{p},\mathbf{^4He})\ \ ^{24}\mathrm{Mg}. \end{aligned} \tag{8.17b}$$

At $T > 3\times10^7$ K, the abundance of ^{25}Mg decreases and the abundance of ^{26}Al increases.

The bold letters in these reaction rates show that, effectively, four protons are fused into a He core. These reactions are therefore part of the H-fusion phase.

8.5 He→C Fusion: The Triple-α Process

At $T > 10^8$ K, ^4He fuses into ^{12}C via the reactions shown in Table 8.2.

This is called the triple-α process. The **mass defect of the net reaction is $\mathbf{\Delta m/m = 0.00065}$.**

The first reaction is an equilibrium reaction that results in a very small fraction of Be ions. The mean lifetime of ^8Be ions is only 3×10^{-16} s! As a result, the formation of ^{12}C is only possible if the second step occurs on a very short timescale. The second reaction is possible at $T \simeq 10^8$ K because of a resonance in the He + Be reaction: the cross-section as a function of energy has a peak near that temperature.

Table 8.2. He-Fusion by the Triple-α Process

$^4He + ^4He \leftrightarrows ^8Be$
$^4He + ^8Be \rightarrow ^{12}C$
Net: $3\ ^4He \rightarrow ^{12}C + 7.3$ MeV.

Historical note: the British astronomer Fred Hoyle (1915–2001) pointed out in 1954 that the large abundance of C in the Universe could only be explained if such a resonance existed (Hoyle 1954). This resonance was later found experimentally.

The **energy production rate of the triple–α process is**

$$\epsilon_{3\alpha} \sim Y^3\rho^2 T^{40}. \tag{8.18a}$$

Note the very steep T-dependence of this reaction.

Q (8.7) *Why is $\epsilon_{3\alpha} \sim \rho^2$? Why is $\epsilon_{3\alpha} \sim Y^3$? What would the dependence be if He + He→ Be was not an equilibrium reaction?*

Q (8.8) *Would you exist if the ^{12}C nucleus did not have a resonant energy level around* 8 *MeV? Explain.*

Toward the end of the He-fusion phase, when sufficient ^{12}C has been formed, the following reaction occurs:

$$^{12}\mathrm{C} + {}^4\mathrm{He} \rightarrow {}^{16}\mathrm{O} + 7.16\,\mathrm{MeV}. \tag{8.18b}$$

The **mass defect of this process is $\Delta m/m$ = 0.00048**.

8.6 C-fusion, O-fusion, and Ne-photodisintegration

During late evolutionary stages, i.e., after He-fusion by the triple-α process, C-fusion, and later O-fusion can provide nuclear energy to a star. This requires temperatures above 6×10^8 K.

The C+C fusion and the O + O fusion processes have several possible branches. The most common ones for generating energy in stars are listed in Table 8.3. Note that the C-fusion and O-fusion reactions produce a series of elements and isotopes from ^{16}O to ^{24}Mg and from ^{24}Mg to ^{32}S, respectively. The reactions in bold letters are the dominant source of protons, neutrons, and α-particles.

At about the temperature of O-fusion, $T \sim 1.3 \times 10^9$ K, photons are energetic enough to break up Ne nuclei that were formed in large quantities by C-fusion. This **Ne-photodisintegration** process produces He nuclei via $^{20}\mathrm{Ne} + \gamma \rightarrow {}^{16}\mathrm{O} + {}^4\mathrm{He}$. The released helium nuclei can then be captured by existing Ne nuclei and form Mg via $^{20}\mathrm{Ne} + {}^4\mathrm{He} \rightarrow {}^{24}\mathrm{Mg} + \gamma$. This Ne-photodisintegration process is generally considered to be part of C-fusion and O-fusion.

Table 8.3. Reactions of C + C Fusion and O + O Fusion

at $T > 6 \times 10^8$ K	at $T > 1 \times 10^9$ K
$^{12}\mathrm{C} + {}^{12}\mathrm{C} \rightarrow {}^{24}\mathrm{Mg} + \gamma$	$^{16}\mathrm{O} + {}^{16}\mathrm{O} \rightarrow {}^{32}\mathrm{S} + \gamma$
$\rightarrow {}^{23}\mathrm{Mg} + \mathrm{n}$	**$\rightarrow {}^{31}\mathrm{S} + \mathrm{n}$**
$\rightarrow {}^{23}\mathrm{Na} + \mathrm{p}$	**$\rightarrow {}^{31}\mathrm{P} + \mathrm{p}$**
$\rightarrow {}^{20}\mathrm{Ne} + {}^4\mathrm{He}$	**$\rightarrow {}^{28}\mathrm{Si} + {}^4\mathrm{He}$**
$\rightarrow {}^{16}\mathrm{O} + 2{}^4\mathrm{He}$	$\rightarrow {}^{24}\mathrm{Mg} + 2{}^4\mathrm{He}$

8.7 Photodisintegration and the Formation of Heavy Elements

Above $T \gtrsim 1.5 \times 10^9$ K, heavy nuclei can be destroyed by photons. At this T, the average λ of photons, according to Wien's law ($\lambda \simeq 0.4/T$ cm K^{-1}), is $\lambda \approx 3 \times 10^{-10}$ cm and the average energy of photons is $hc/\lambda \simeq 10^{-8}$ erg $\simeq 0.4$ MeV. This destruction of heavy nuclei creates a mixture of protons, neutrons, He-nuclei (α-particles), and isotopes of ions such as Si, Mg, Ne, and C.

An example of such a destruction process is

$$^{28}\text{Si}(\gamma,\ \alpha)\ \ ^{24}\text{Mg}(\gamma,\ \alpha)\ \ ^{20}\text{Ne}(\gamma,\ \alpha)\ \ ^{16}\text{O}(\gamma,\ \alpha)\ \ ^{12}\text{C}(\gamma,\ \alpha)\ \ ^{8}\text{Be}(\gamma,\ \alpha)\ \ 2\alpha. \quad (8.19)$$

This process releases a large number of α-particles, which are subsequently captured. This results in the formation of heavy elements with stable nuclei, of which ^{56}Fe is the most stable one, by consecutive α-captures. An example of the formation of heavy elements by α-captures at $T \sim 3 \times 10^9$ K is

$$^{28}\text{Si}(\alpha,\ \gamma)\ \ ^{32}\text{S}(\alpha,\ \gamma)\ \ ^{36}\text{Ar}(\alpha,\ \gamma)\ \ ^{40}\text{Ca}(\alpha,\ \gamma)\ \ ^{44}\text{Ti}(\alpha,\ \gamma) \ldots {}^{56}\text{Ni}. \quad (8.20)$$

A ^{56}Ni nucleus is unstable and decays into ^{56}Fe via $^{56}\text{Ni} \rightarrow {}^{56}\text{Co} \rightarrow {}^{56}\text{Fe}$ by two steps of β^+ decay, in which a proton is converted into a neutron and emits a positron and a neutrino ($p \rightarrow n + e^+ + \nu_e$) (Maeder 2009). Most of the energy that is created in these reactions is lost from the star because it is carried off by neutrinos.

8.8 Summary of Major Nuclear Reactions in Stars

We have discussed the various nuclear reactions that play a role in stellar evolution; however, not every star goes through all of these reactions because each reaction requires a minimum temperature to be efficient.

Table 8.4 shows the major nuclear reactions in stars, together with the threshold temperature T_{thresh}, that is required. The table also lists the products and the amount of net energy produced per nucleon, corrected for the energy loss by neutrinos, in

Table 8.4. Summary of the Most Important Reaction Rates in Stars

Fuel (1)	Process (2)	T_{thresh} 10^6 K (3)	Product (4)	E_{net} MeV/nucl (5)	T_c 10^6 K (6)	L_{net}/L (7)	Duration yr (8)
H	p-p chain	4	He	6.55	–	–	–
H	CNO cycle	15	He	6.25	35	0.94	1.1×10^7
He	3-α fusion	100	C,O	0.61	180	0.96	2.0×10^6
C	C-fusion	600	Ne,Mg,Na,O	0.54	810	0.16	2.0×10^3
Ne	Ne photdis	900	O,Mg,Si		1600	5.3×10^{-4}	0.7
O	O-fusion	1000	S,Si,P,Mg	0.30	1900	8.2×10^{-5}	2.6
Si	Si nucl equil.	3000	Fe,Ni,Cr,Ti	<0.18	3300	5.8×10^{-7}	0.05

Notes. photdis = photodisintegration. **nucl equil** = nuclear equilibrium = photodisintegration + capture of p, n, and He. Column (5) = energy generated per nucleon (He has 4 nucleons, C has 12 nucleons etc.). Columns (6), (7), and (8) refer to the evolution of a star of $15M_\odot$ (Based on Woosley & Janka 2005, and Maeder 2009).

units of MeV per nucleon. For instance, the pp chain involves four nucleons (protons). The net reaction produces 26.73 MeV, which implies 6.68 MeV per nucleon, of which about 2% is lost by neutrinos, leaving a net energy production of 6.55 MeV per nucleon.

The last three columns give the properties of these processes in a star of $15M_{\odot}$. Column (6) gives the central temperature during each central fusion phase. Column (7) gives the net fraction of the energy that remains in the star. The rest is lost by neutrinos leaving the star. Note that during the later phases (i.e., after C-fusion) the neutrino loss is so large that only a very small fraction of the energy produced by the reaction is used for the luminosity of the star. Column (8) gives the duration of the phase for a star of $15M_{\odot}$. Also note the extremely short duration of the last phases. This is a result of the neutrino loss. The star must produce a certain radiative luminosity $L_{\rm rad}$ to remain in thermal and hydrostatic equilibrium, but because only a very small fraction, $L_{\rm net}/L$, of the energy can be used for this the total amount of energy that the fusion must produce is higher than $L_{\rm rad}$ by a factor $(L_{\rm net}/L)^{-1}$. This shortens the duration of the phases drastically. The duration would have been a factor of $(L_{\rm net}/L)^{-1}$ longer if there had been no neutrino losses.

The reaction rates and energy production for stellar nucleosynthesis have been compiled by the Lawrence Livermore National Laboratory and can be found at https://nuclear.llnl.gov.
Formulae for the reaction rates can be found in Lang (1980).

8.9 Formation of Heavy Elements by Neutron Capture

During the Si-fusion phase, photodisintegration creates a mixture of protons, neutrons, α-particles, and ions. Because neutrons have no charge, they can easily penetrate and be captured by ions, thus creating neutron-rich isotopes. The net result depends on the density of the neutrons, on the time between successive neutron captures by an ion, and on the decay time of unstable nuclei.

8.9.1 Slow Neutron Capture: The s-process

If the neutron capture rate is "relatively slow", a nucleus can capture neutrons until an unstable isotope is formed. This isotope will then suffer β-decay ($n \rightarrow p + e^+$) until it reaches a stable isotope. This stable isotope can again capture neutrons until another unstable isotope is formed, which decays, and so on. This process is shown in Figure 8.7 for a series of neutron captures, starting with an ^{56}Fe nucleus. The isotope ^{59}Fe is unstable and will decay to ^{59}Co. This nucleus can then start capturing neutrons again, until another unstable isotope is formed. In this way, a whole series of stable isotopes of heavy elements can be formed, depending on the stable isotope of the element that started the process.

Typical s-process (s = slow) elements are zirconium (Zr), strontium (Sr), barium (Ba), and lead (Pb). Their enhanced abundances in the photospheres of certain AGB

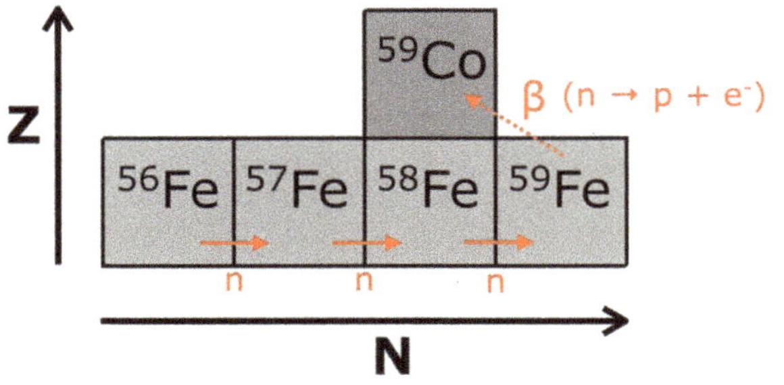

Figure 8.7. Schematic picture of the s-process of slow neutron capture from ^{56}Fe to ^{59}Co. N is the number of nucleons and Z is the charge of the nuclei. Isotope ^{59}Fe is unstable. Red symbols indicate the process: "n" for neutron capture and "β" for beta-decay.

stars shows that these stars must have gone through a phase that produced a large neutron flux.

8.9.2 Rapid Neutron Capture in Supernova: The r-process

If the neutron density is very high, the capture rate of neutrons is so fast that unstable isotopes that are formed have no time to decay between neutron captures. In this way, super-neutron-rich isotopes will be formed. When the neutron flux stops (e.g., because the matter is expelled in a supernova explosion), these neutron-rich isotopes will suffer a series of β-decays until a stable isotope is reached. These final stable isotopes are called r-process elements.

Typical r-process (r = rapid) elements are europium (Eu), gold (Au), xenon (Xe), and platinum (Pt). They can only be formed in supernova explosions.

Figure 8.8 shows both the s-process and the r-process schematically in a part of the isotope diagram. The vertical axis shows the charge (Z) of the isotopes, while the horizontal axis shows their number of nucleons (N). Stable isotopes are depicted by dark gray or colored squares, unstable isotopes are depicted by light gray squares.

Starting with ^{141}Pr (praseodymium) in the lower left of the diagram, the path of slow neutron capture is shown in red. The full part of the line goes through stable isotopes and the dashed part shows β-decay of an unstable isotope. The blue lines show the path of rapid neutron capture starting with the same isotope. The full part of the line shows neutron capture through a series of unstable isotopes. When the neutron flux stops, a series of rapid β-decays (shown by dotted blue lines) starting from very neutron-rich isotopes in the lower right of the diagram ends at a stable isotope. We see that nuclei in this diagram that are shielded in the diagonal direction by a stable isotope can only be formed by the s-process (e.g., ^{148}Sm, ^{154}Gd). We also see that nuclei in this diagram that have an unstable isotope on their left and are *not* shielded by a stable isotope in the diagonal direction can only be formed by the r-process (e.g., ^{148}Nd and ^{150}Nd). Some nuclei can be formed by both processes (e.g., ^{151}Sm and ^{152}Sm).

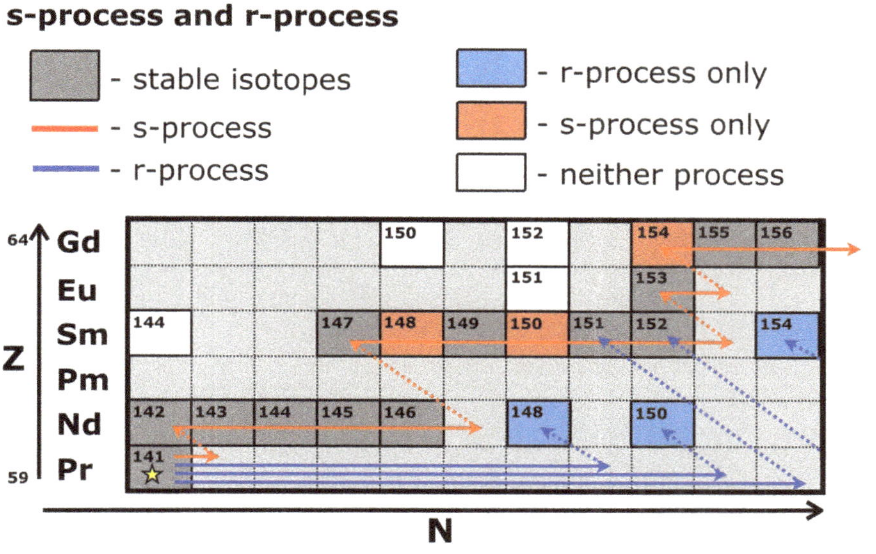

Figure 8.8. Schematic representation of the s- and r-processes of neutron capture and the formation of neutron-rich isotopes. Unstable isotopes are indicated by light gray squares and stable isotopes are indicated by dark (gray, red, or blue) squares. Full lines indicate paths of neutron capture and dashed lines indicate paths of β-decay. Depending on the location of an isotope in this diagram and the stability of its neighbors, a stable isotope can be formed via the r-process, the s-process, or both. Some isotopes cannot be produced by either process.

8.9.3. Proton-rich Elements: p-elements

Some heavy elements have an overabundance of proton-rich isotopes. These are also formed during supernova explosions. They are not formed by proton capture but are instead the result of s-process elements that have lost neutrons by the process ${}^{N}_{Z}A + \gamma \rightarrow {}^{N-1}_{Z}A + n$ during supernova explosions. **Typical p-isotopes are ^{92}Mo (Molybdenum) and ^{144}Sm (Samarium).**

8.10 The Minimum Core Mass for Igniting Fusion Reactions

We have seen in Table 8.4 that each reaction requires a minimum temperature to be ignited. The central temperature of a star can rise if the star, or rather its core, contracts. We have derived before (from HE and the ideal gas law, in Section 3.2) that we can estimate the central temperature of a star as

$$T_c \simeq \frac{m_{\rm H}\mu}{k}\frac{GM}{R}. \tag{8.21}$$

In the case of a contracting core that we are considering here (e.g., after H-fusion), most of the central pressure is due to the mass of the core, as the layers outside the core have a much lower density and contribute little to the central pressure. We can therefore estimate the central temperature in this case as

$$T_c \simeq \frac{m_{\rm H}\mu_c}{k}\frac{GM_c}{R_c} \text{ with } R_c \simeq \left(\frac{M_c}{\rho_c}\right)^{\frac{1}{3}} \rightarrow T_c \approx \frac{m_{\rm H}\mu_c}{k}GM_c^{2/3}\rho_c^{1/3}. \tag{8.22}$$

This shows that when the core of a star with mass M_c contracts, its central temperature will increase with its central density as $T_c \sim M_c^{2/3}\rho_c^{1/3}$. Equation (8.22) might suggest that a core can reach any high T_c by contracting to a small enough radius and high enough density; however, this is not the case. The core may become degenerate before it reaches the required ignition temperature $T_{\rm ign}$. If that happens, the contraction stops because degenerate stars do not contract, as will be shown later in Chapter 11.

In order to reach the ignition temperature of the next fusion reaction, the core must avoid degeneracy.

Degeneracy is prevented if $P_{\rm ideal\ gas} > P_{\rm el.\ degen.}$, which implies

$$P_c = \frac{\mathcal{R}}{\mu_c}\rho_c T_c > K_1\left(\frac{\rho_c}{\mu_e}\right)^{5/3} \rightarrow \boxed{T_c > \frac{K_1}{\mathcal{R}}\left(\frac{\mu_c}{\mu_e}\right)\left(\frac{\rho_c}{\mu_e}\right)^{2/3}}. \tag{8.23}$$

Figure 8.9 shows this condition schematically in a log ρ_c versus log T_c diagram. The arrows indicate the changes in the (ρ_c, T_c) diagram when cores of different mass contract with $T_c \sim \rho_c^{1/3}$ according to Equation (8.22). Lines for different core masses are parallel but displaced by $T_c \sim M_c^{2/3}$. The dashed line shows the relation of $T_c \sim \rho_c^{2/3}$ of Equation (8.23) for a degenerate core. The area to the right of the dashed line indicates electron degenerate gas. The red circles indicate the maximum temperature that a contracting core of a given mass can reach.

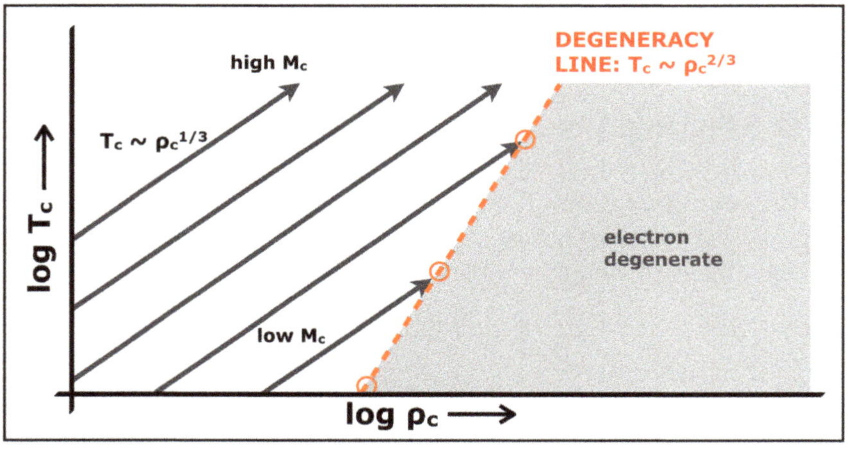

Figure 8.9. Schematic evolution of the central T_c and ρ_c of contracting cores of stars of various core masses. In a contracting core $T_c \sim \rho_c^{1/3}$. The degeneracy limit is given by $T_c \sim \rho_c^{2/3}$. The contraction stops when the core becomes degenerate (red circles).

The figure shows that a star must have a *minimum contracting core mass* to avoid degeneracy before $T_{\rm ign}$ of the next fusion reaction is reached. Combining Equations (8.22) and (8.23), we find that this minimum core mass is given by

$$\boxed{M_{\rm core} > M_{\rm crit} \approx \left\{\frac{\Re K_1'}{\mu_c \mu_e^{\frac{5}{3}} G^2}\right\}^{\frac{3}{4}} T_{\rm ign}^{\frac{3}{4}}}. \tag{8.24}$$

Substituting the constants, we find that $M_{\rm crit}/M_\odot \approx 1 \times (T_{\rm ign}/10^9\ {\rm K})^{3/4}$ if $\mu_e = 2$ and $\mu_c = 2$. These are the characteristic values of μ_e and μ_c of cores during all phases after H-fusion.

8.11 Fusion Phases of Stars in the (ρ_c, T_c) Plane

At the end of each nuclear fusion phase, the core of a star contracts to release potential energy, compensating for the outward transport of energy. When the core density increases, its temperature also increases. The core contraction stops when it becomes degenerate. From then on, ρ_c hardly increases. As there is no more fusion, the central temperature of degenerate stars slowly decreases due to the loss of energy by radiation or conduction.

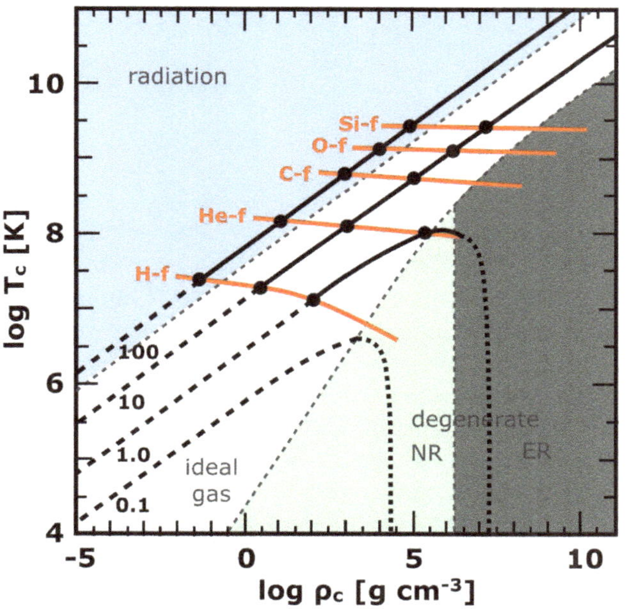

Figure 8.10. Schematic evolutionary tracks of the cores of stars of different mass in the (ρ_c, T_c) diagram. The background colors indicate the regions of the different equations of state (as in Figure 4.4). The red lines indicate the parameters during central fusion phases. The black lines represent the evolution of the central parameters for stars of 0.1 to $100 M_\odot$: dashed = pre-main-sequence phase; full = core contraction; big dots = fusion in the core; dotted = cooling of the degenerate core. (Adapted from Pols 2011.)

The evolution of stars in the (ρ_c, T_c) diagram is shown schematically in Figure 8.10. The different colors indicate regions of the equation of state: white = ideal gas, light blue = radiation pressure dominates, light green = nonrelativistic degeneracy, and gray = relativistic degeneracy. The red nearly horizontal lines indicate the condition where fusion happens in the core. The black lines are the approximate evolutionary tracks in the (ρ_c,T_c) diagram for stars of 100, 10, 1, and $0.1M_\odot$. The dashed portion is the pre-main-sequence evolution (i.e., before H-fusion starts), while the full lines show the (fast) contraction phases of the cores. This contraction occurs on a Kelvin–Helmholtz timescale. The black dots indicate the (slow) phases of fusion in the core, which occur on a nuclear timescale. The dotted part of the tracks, in the degenerate region, shows cooling after fusion has stopped.

Once the center of a star has become degenerate it can no longer contract, so the density reaches its approximate upper limit (the density can, in fact, still increase slightly when the degenerate core becomes more massive due to fusion in the shell surrounding the degenerate core, as we will see later). The downward slope indicates the cooling of the degenerate core near the end of evolution. These tracks are only schematic; in reality, the tracks show wiggles near the fusion phases. This is due to the fact that once a fusion phase starts, stars settle into a new equilibrium configuration (see Figures 23.3 and 26.3).

Note that

- stars of $0.1M_\odot$ do not reach H-fusion,
- stars of $1M_\odot$ barely reach He-fusion, and
- stars of $M \geqslant 8M_\odot$ reach all fusion phases.

8.12 Summary

1. The most important stellar nuclear fusion reactions are listed in Table 8.4
2. Nuclear fusion in stars is possible because of the quantum mechanical tunneling effect that allows charged nuclei to overcome the Coulomb barrier. The combination of the energy distribution of the nuclei and the tunneling effect produces the Gamov-peak at a temperature where fusion is efficient.
3. In stars of $M < 1.2M_\odot$, the H-fusion occurs via the pp chain. In stars of $M > 1.2M_\odot$, the H-fusion occurs via the CNO cycle. During this phase, the N/C ratio increases from its initial number ratio of N/C ≈ 0.3 to 140.
4. Each successive reaction has a higher Coulomb barrier, due to a higher charge of the nuclei, so a higher T_c is needed. The core of the star has to contract to reach this higher central temperature.
5. Each successive reaction has a steeper T-dependence (e.g., $\varepsilon \sim T^4$ and T^{18} for the pp chain and the CNO cycle of H-fusion and $\varepsilon \sim T^{40}$ for He-fusion). This implies that fusion will become more and more concentrated in the core of the star, i.e., in a region that has less mass. This results in the chemical stratification of an "onion skin model" with the most massive products (the latest formed) in the center, surrounded by concentric layers of less massive elements.

6. Each successive reaction has a smaller mass defect $\Delta m/m$ (i.e., it produces less energy), so the reaction rate has to be higher and the fusion faster (more reactions per sec per gram) in order to generate the luminosity L needed to keep the star in hydrostatic and thermal equilibrium. As a result, each fusion phase has a shorter duration than the previous one.
7. At $T > 10^9$ K, neutrinos carry away a large fraction of the energy. This reduces the net energy production of the star and speeds up the evolution very drastically.
8. At $T > 1.5 \times 10^9$ K, heavy elements are destroyed by photodisintegration. In the resulting mixture of heavy and low-mass nuclei, the Fe-group elements (Fe, Cr, Ni, Ti) are formed by the capture of α-particles.
9. Heavy elements beyond the Fe-group are formed by neutron capture. The products depend on the neutron density. Zr, Sr, Ba, and Pb are formed by slow neutron capture in supernovae and in some AGB stars. Eu, Au, Xe, and Pt are formed by rapid neutron capture in supernovae.
10. The fusion phase that a star can reach depends on its core mass. The core mass must be high enough to reach the ignition temperature before degeneracy sets in. Stars with initial masses less than $0.1M_\odot$ do not reach H-fusion. Stars with initial masses greater than $8M_\odot$ go through all evolutionary phases.

Exercises

8.1 (a) Calculate the mass defect fractions of the following fusion reactions

$4\,{}^1\mathrm{H} \rightarrow {}^4\mathrm{He}$

$3\,{}^4\mathrm{He} \rightarrow {}^{12}\mathrm{C}$

${}^{12}\mathrm{C} + {}^4\mathrm{He} \rightarrow {}^{16}\mathrm{O}$

$2\,{}^{16}\mathrm{O} \rightarrow {}^{28}\mathrm{Si} + {}^4\mathrm{He}$

$2\,{}^{28}\mathrm{Si} \rightarrow {}^{56}\mathrm{Fe}$

(b) Describe the trend and discuss what this trend implies for stellar evolution.

8.2 The efficiency of He-fusion has a much steeper T-dependence than H-fusion.

(a) What does it imply for the mass of the He-fusing core compared to that of the H-fusing core of a star with a given initial mass?

(b) What is the consequence for the chemical distribution of the star at the end of the He-fusion phase.

8.3 The lifetimes of the major isotopes involved in the CNO cycle are given in Equation (8.16).

(a) Calculate the equilibrium ratios of these isotopes in the core of the star at the start and at the end of the H-fusion phase.

(b) Sketch the variation of the logarithmic isotope number ratios ${}^4\mathrm{He}/{}^1\mathrm{H}$, ${}^{12}\mathrm{C}/{}^1\mathrm{H}$, ${}^{14}\mathrm{N}/{}^1\mathrm{H}$, ${}^{12}\mathrm{C}/{}^4\mathrm{He}$, ${}^{14}\mathrm{N}/{}^4\mathrm{He}$, ${}^{13}\mathrm{C}/{}^{12}\mathrm{C}$, and ${}^{15}\mathrm{N}/{}^{14}\mathrm{N}$ as a function of time in a single diagram, for a star of about $2M_\odot$ fusing H $\rightarrow$ He at $T = 2 \times 10^7$ K. Start with the cosmic logarithmic abundance ratios by number of (^{1}H, ^{4}He, ^{12}C, ^{13}C, ^{14}N, ^{15}N) = (0.0, −1.0, −3.4, −5.4, −4.0, −6.4). Discuss the trends.

8.4 Suppose that the He-fusion requires a minimum core mass of about $0.3M_{\odot}$. What would be the minimum core mass for the next fusion phases?

References

Bastian N. & Lardo C. 2018, ARA&A, **56**

Gamow, G. 1937, Structure of Atomic Nucleus and Nuclear Transformations (Oxford: Univ. of Oxford Press)

Hoyle, F. 1954, ApJS, **1,** 121

Kippenhahn R. & Weigert A. 1990, Stellar Structure and Evolution (Berlin: Springer)

Lang, K.R. 2008, in Astrophysical Formulae (Berlin: Springer), 418

Maeder, A. 2009, Physics, Formation and Evolution of Rotating Stars (Berlin: Springer)

Pols, O. 2011, in Stellar Structure and Evolution, Utrecht University Lecture Notes, 66 https://www.astro.ru.nl/~onnop/education/stev_utrecht_notes/

Rolfs, C. E. & Rodney, W. S. 1988, Cauldrons in the Cosmos: Nuclear Astrophysics (Chicago, IL: Univ. of Chicago Press)

Woosley S. & Janka T. 2005, NatPh, **1,** 147

Henny J.G.L.M. Lamers and Emily M. Levesque

Chapter 9

Stellar Timescales

In this chapter, we discuss the different timescales on which a star can change its structure if the internal conditions are changing. We will show that in almost all cases a star can regain hydrostatic equilibrium much faster than almost any other change that may occur (except, of course, in the case of explosions). This is crucial for our understanding of stellar evolution because it implies that we can describe stellar evolution by a sequence of hydrostatic equilibrium models with slowly varying chemical structure and energy production.

9.1 The Dynamical Timescale

Suppose that the gas pressure in part of the star or in the whole star suddenly vanishes. This is the most drastic possible disturbance of hydrostatic equilibrium (HE), apart from an explosion. How much time would it take for the star to restore HE? If the pressure vanishes in all layers simultaneously the star would go into free-fall due to the force of gravity. In that case, gravity would be the restoring force for HE.

The free-fall motion of a layer is described by $\Delta r \sim gt^2$, where g is the local acceleration of gravity. This implies that the change in radius occurs on a timescale $t \sim \sqrt{\Delta r/g}$. For an order of magnitude estimate, we adopt $\Delta r \simeq R$ and $g \simeq GM/R^2$. So the **free-fall timescale** is

$$t_{\rm ff} \sim \sqrt{R/(GM/R^2)} = \sqrt{\frac{R^3}{GM}} \text{ so } t_{\rm ff} \simeq \frac{1}{\sqrt{G\bar{\rho}}}. \tag{9.1}$$

Now we consider what happens if the pressure in a layer of the star is suddenly changed: either increased or decreased. Such a disturbance will result in a sound wave that travels with the sound speed c_s. The sound speed is given by $c_s^2 \sim kT/\bar{m}$, where $\bar{m}$ is the mean mass of the particles. Remember from the Virial Theorem (Section 3.3) that the total kinetic energy $E_{\rm kin} \simeq (3/2)k\bar{T}\ M/\bar{m} = -1/2\ E_{\rm pot} \simeq GM^2/R$. This implies

doi:10.1088/978-0-7503-1278-3ch9

that $c_s^2 \sim k\,\bar{T}/\bar{m} \simeq GM/R$. The time it takes for a sound wave to travel a distance equal to the radius R of the star is $t_{\mathrm{sound}} = R/\bar{c}_s$, where $\bar{c}_s$ is the mean sound speed. So the **sound speed travel time** of a star is on the order of

$$t_{\mathrm{sound}} \simeq \frac{R}{\sqrt{\frac{GM}{R}}} \simeq \sqrt{\frac{R^3}{GM}} \simeq \frac{1}{\sqrt{G\bar{\rho}}}. \tag{9.2}$$

Note that the sound speed travel time is of the same order of magnitude as the free-fall time. The free-fall and the sound speed travel timescales together are called **the dynamical timescale**

$$\boxed{\tau_{\mathrm{dyn}} = \frac{1}{\sqrt{G\bar{\rho}}}} \tag{9.3}$$

because it describes how quickly a star can restore hydrodynamical equilibrium. We will see later (Section 21) that the pulsation periods of radially pulsating stars also scale with the dynamical time.

Q (9.1) *Note that the free-fall time is about the same as the sound speed crossing time. Explain this in physical terms.*

Q (9.2) *Show that the dynamical timescale of the Sun is about an hour.*

9.2 The Thermal Timescale or Kelvin–Helmholtz Timescale

When a star ceases nuclear fusion, it keeps radiating because the inside of the star is hotter than the outside, so energy is leaking out. The question is, how long can a star keep up its luminosity if nuclear fusion stops and thermal energy is the only energy source left? This timescale is called the **thermal timescale** or the **Kelvin–Helmholtz timescale** after the discoverers. Lord Kelvin (1824–1907), whose original name was William Thomson, was an Irish–British mathematical physicist and engineer. He estimated the age of the Earth based on the cooling time of the Sun in 1864. Hermann von Helmholtz (1821–1894) was a German physician and physicist who not only studied the conservation of energy in the 1850s and 1860s but also contributed to psychology and the theory of vision.

The thermal timescale is defined by $\tau_{\mathrm{th}} \simeq E_{\mathrm{th}}/L$, where E_{th} is the total thermal energy of the star and L is the luminosity. Remember that $E_{\mathrm{th}} = -\tfrac{1}{2}E_{\mathrm{pot}}$ according to the virial theorem, which implies that $\tau_{\mathrm{th}} \simeq -\frac{E_{\mathrm{pot}}}{L} \simeq \frac{GM^2}{R}/L$.

The **thermal timescale** or the **Kelvin–Helmholtz timescale**, τ_{KH}, is

$$\boxed{\tau_{\mathrm{KH}} = \frac{GM^2}{RL}}. \tag{9.4}$$

Q (9.3) *Show that the Kelvin–Helmholtz timescale of the Sun is about* 3×10^7 yr.

9.3 The Nuclear Timescale

The nuclear timescale is defined as the time a star can radiate by nuclear fusion. It depends on the nuclear energy available and on the luminosity of the star. The nuclear energy available at a given evolutionary phase is set by the total mass M of the star and the fraction f_M of the mass that partakes in the nuclear fusion. The amount of energy that can be created by nuclear fusion of one gram of gas is ε_n. The efficiency factor $\varepsilon_n = c^2 \times (\Delta m/m)$ is in ergs/gram and $\Delta m/m$ is the mass defect that was discussed in Section 8.1, so the amount of energy generated during a nuclear fusion phase is $E_{\rm nucl} \sim f_M M\varepsilon_n$. The **nuclear timescale** is therefore

$$\tau_{\rm nucl} \sim \frac{E_{\rm nucl}}{L} \sim f_{\rm M} M\varepsilon_n/L. \qquad (9.5a)$$

In Section (8.1) we showed that the mass defect is $\Delta m/m = 0.0071$ for H-fusion and $\Delta m/m = 0.00065$ for He-fusion. For main-sequence stars, which have core H-fusion, the mass fraction is about $f_M \sim 0.10$. For the later phase of core He-fusion, the mass fraction f_M is only slightly smaller. This implies that the nuclear timescales are on the order of

$$\tau_{\rm nucl} \sim 10^{-3}\, Mc^2/L \quad \text{for core H-fusion,} \qquad (9.5b)$$

$$\tau_{\rm nucl} \sim 10^{-4}\, Mc^2/L \quad \text{for core He-fusion.} \qquad (9.5c)$$

Let us take the Sun as an example:

$$\left.\begin{array}{l} M = 2\times 10^{33}\ \text{gr},\ L = 3.8\times 10^{33}\ \text{erg s}^{-1} \\ f_M \approx 0.10 \end{array}\right\} \text{ so } \tau^{\odot}_{\rm nucl} \simeq 3\times 10^{17}\ \text{sec} = 10^{10}\ \text{yr}.$$

For other stars, we can estimate the nuclear timescale of H-fusion as

$$\tau_{\rm nucl} \approx \frac{M/M_\odot}{L/L_\odot} \times 10^{10}\ \text{yr} \qquad (9.5d)$$

Q (9.4) *The actual MS phase of massive stars with $M > 30M_\odot$ is longer than $(M/M_\odot)/(L/L_\odot) \times 10^{10}$ yr by a factor of ~3 to 5. Can you think of a reason for this?*

9.4 The Convection Timescale

We have shown in Section 7.6 that the timescale for rising and descending cells in the convection zone of the Sun is on the order of a week. The longest convective timescale is reached in AGB stars, which have a very deep, convective, low-density, and relatively cool envelope. If we adopt $L = 10^3 L_\odot$, $R = 100R_\odot$, $M = 5M_\odot$, a mean density of 6×10^{-5} g cm^{-3}, and a mean temperature of the convective outer layers of 10^5 K at $r = R/2$, we find with Equations (7.23), (7.24), and (7.26) a scale height of $H_p = 3.5R_\odot$ and a convective velocity of 7×10^4 cm s^{-1}. This implies a rise time of the convective cells in an AGB star of about 40 days.

9.5 Comparison of Timescales

The estimates given above show that the dynamical timescale, the convection timescale, the Kelvin–Helmholtz timescale, and the nuclear timescales are very different, and that

$$\boxed{\tau_{\rm dyn} \ll \tau_{\rm conv} \ll \tau_{\rm KH} \ll \tau_{\rm nucl}}. \tag{9.6}$$

The values for the Sun are 1 hr $\ll$ weeks $\ll 3.10^7$ yr $\ll 10^{10}$ yr. For a typical AGB star with H-shell fusion the values are 17 days < 40 days $\ll 8 \times 10^3$ yr $\ll 5 \times 10^7$ yr.

We see that

- except for explosive phases, stars are always in quasi-hydrostatic equilibrium,
- even the convective zones are in global hydrostatic equilibrium, although they are locally in turmoil, and
- the contraction phases last about 1% of nuclear phases.

Q (9.5) *If the Sun suddenly stopped its nuclear fusion, how long would it take an outside observer to notice it (if he does not have a neutrino detector available)?*

9.6 Summary

1. Stars can restore perturbations of hydrostatic equilibrium on the dynamical timescale. This timescale is so short compared to other timescales that stars can be assumed to remain in hydrostatic equilibrium throughout their lives.
2. The Kelvin–Helmholtz timescale describes the duration of the contraction phase of a star when it (temporarily) runs out of nuclear fusion.
3. The nuclear timescale describes the duration of the nuclear fusion phase.
4. Typical timescales for the Sun are as follows.

   Dynamical timescale: 1 hour.
   Rising time of convective cells: weeks.
   Kelvin–Helmholtz timescale: 3×10^7 yr.
   Nuclear timescale for H-fusion: 10^{10} yr.

Exercises

9.1 Calculate dynamical, thermal, and nuclear timescales and their ratios for different stars.
   - MS star of 1 $M_{\rm sun}$.
   - MS star of 60 $M_{\rm sun}$.
   - Red supergiant of 15 $M_{\rm sun}$.
   - White dwarf of 0.6 $M_{\rm sun}$.

   Comment on the consequences of these results.

9.2 If the nuclear fusion in the Sun were to suddenly stop, how long would it take before an observer would notice? Photon travel time τ_L? Dynamical time? Thermal timescale? Nuclear timescale? Give arguments.

Henny J.G.L.M. Lamers and Emily M. Levesque

Chapter 10

Calculating Stellar Evolution

In the previous chapters, we have discussed the physical effects that occur in stars: the properties of the gas; hydrostatic equilibrium and its consequences; energy generation by nuclear fusion; and energy transport by radiation or convection. Now we can put these ingredients together and show how they can be used to calculate the structure of stable stars. We first summarize the equations of stellar structure and their boundary conditions in the core of a star and at its outer radius. We then describe how this set of equations can be solved numerically for a given chemical composition. The evolution of a star is calculated by considering the changes in its chemical structure due to nuclear fusion and convection.

10.1 Assumptions for Computing Stellar Evolution

The stellar models that we describe here are simple and valid for most evolutionary phases. They are based on the following assumptions.

1. The star is **spherically symmetric**, which implies that
   - the physical quantities vary only in the radial direction: $P(r)$, $\rho(r)$, $T(r)$, etc., and
   - the effects of rotation and magnetic fields are ignored. (Rotation will be discussed later in Section 25.)
2. The star is in **hydrostatic equilibrium**: at each depth, the layers are stable.
3. The star is in **thermal equilibrium**: the energy generated in the star equals the energy radiated outward.
4. The energy sources are
   - thermonuclear energy,
   - gravitational energy, which is important for contracting or expanding stars, and
   - internal energy, which is important for cooling white dwarfs.
5. The **energy transport** is by
   - radiation,
   - convection, and

doi:10.1088/978-0-7503-1278-3ch10

– conduction, which is important for degenerate stars.

6. The **chemical composition is changing.**
   – Newly formed stars have a homogenous initial composition described by
     X = mass fraction of H,
     Y = mass fraction of He,
     Z = mass fraction of the rest, mainly C, N, and O.
   – As the star evolves, its chemical composition changes as a function of time and location due to nuclear fusion in the core or in shells.
   – The chemical composition in certain layers may also change due to mixing by convection or convective overshooting.

Under these assumptions, the calculation of stellar evolution consists of
(a) computing a series of subsequent hydrostatic equilibrium models,
(b) each with the chemical structure calculated using the nuclear reaction rates of the previous model,
(c) extrapolating it with a small time step for calculating the new chemical distribution in the star, and
(d) computing the next model in hydrostatic equilibrium with the new chemical structure.

This results in a series of subsequent hydrostatic models of the internal structure of the star as a function of time. The models also give the radius, effective temperature, and luminosity as a function of time. This can be expressed in evolutionary tracks in the Hertzsprung–Russell diagram.

10.2 The Equations of Stellar Structure

	in *r* Euler coordinates	**in *m* = *M*(*r*) Lagrange coordinates**	
Mass continuity	$\frac{dm}{dr} = 4\pi r^2\rho$	$\frac{dr}{dm} = \frac{1}{4\pi r^2\rho}$	(10.1)
Hydrost equilibrium	$\frac{dP}{dr} = -\frac{Gm\rho}{r^2}$	$\frac{dP}{dm} = -\frac{Gm}{4\pi r^4}$	(10.2)
Energy generation*	$\frac{dL}{dr} = 4\pi r^2\rho(\epsilon - \epsilon_v - \frac{Tds}{dt})$	$\frac{dL}{dm} = \epsilon - \epsilon_v - \frac{T\,ds}{dt}$	(10.3)
Energy transport			
by radiation	$\frac{dT}{dr} = -\frac{3}{4ac}\cdot\frac{\kappa\rho}{T^3}\frac{L}{4\pi r^2}$	$\frac{dT}{dm} = -\frac{3}{4ac}\cdot\frac{\kappa}{T^3}\frac{L}{(4\pi r^2)^2}$	(10.4a)
or			
by convection	$\frac{dT}{dr} = \frac{\gamma_{ad}-1}{\gamma_{ad}}\cdot\frac{T}{P}\cdot\frac{dP}{dr}$	$\frac{dT}{dm} = \frac{\gamma_{ad}-1}{\gamma_{ad}}\cdot\frac{T}{P}\cdot\frac{dP}{dm}$	(10.4b)

* The entropy term $-T\,ds/dt$ in the energy equation expresses the energy gained by contraction ($-T\,ds/dt > 0$) or lost by expansion ($-T\,ds/dt < 0$) on a Kelvin–Helmholtz timescale with $T\,ds \equiv du - \frac{P}{\rho}d\rho$. If the star is in thermal equilibrium $T\,ds/dt = 0$.

Equation of state $P = P_{\text{gas}} + P_{\text{rad}} = P_{\text{rad}} + P_e + P_{\text{ion}}$ (10.5)

$P_{\text{rad}} = \frac{a}{3}T^4$

$P_i = \rho \mathcal{R} T / \mu_i$

$P_e = \rho \mathcal{R} T / \mu_e$

or $P_e = K_1 \ (\rho/\mu_e)^{5/3}$ (nonrelativ. electr. degen.)

or $P_e = K_1 \ (\rho/\mu_e)^{5/3}$ (electr. degen.)

or $P_e = \text{equation (4.23)}$ (partial electr. degen.)

Absorption coefficient $\kappa = \kappa_{ff} + \kappa_{bf} + \kappa_e$

$\kappa_{ff} + \kappa_{bf} \sim \rho T^{-7/2}$ (10.6)

$\kappa_e \sim \rho/\mu_e \sim 1 + X$

Nuclear energy production $\epsilon = \epsilon_0 \rho^m T^n - \epsilon_\nu$ (10.7)

ϵ_0, m, n depend on the reaction and the abundance

$-\epsilon_\nu$ is the loss of energy by escaping neutrinos

Composition $X(m)$, $Y(m)$, $Z(m)$ or $X_i(m)$ with $i = 1 \ldots$ all isotopes

These structure equations describe either
T, P, L, ρ, m as a function of r (Euler-coordinates)
or
T, P, L, ρ, r as a function of m (Lagrange-coordinates)

The Lagrangian equations can be derived from the Euler equations using $\frac{dx}{dm} = \frac{dx}{dr} \cdot \frac{dr}{dm}$ with $\frac{dr}{dm} = \frac{1}{4\pi\rho r^2}$.

Q (10.1) *Why is it more practical to use m as the free parameter rather than r?*
Q (10.2) *Which time steps would you use for calculating stellar evolution?*

10.3 Boundary Conditions

The equations of stellar structure in Lagrangian coordinates consist of four differential equations for $r(m)$, $P(m)$, $L(m)$, and $T(m)$. Solving these equations for a star of mass M requires four boundary conditions:

$$r(m = 0) = 0, \tag{10.8a}$$

$$L(m = 0) = 0, \tag{10.8b}$$

$$P(m = M) = 0, \tag{10.8c}$$

$$T(m = M) = T_{\text{eff}} \equiv \{L(M)/4\pi\sigma r(m = M)^2\}^{1/4}. \tag{10.8d}$$

The stellar radius is defined by $P(M) = 0$, which implies $R = r(P = 0)$.

10.4 Solving the Structure Equations

Solving the structure equations numerically is not trivial because there are two inner boundary conditions at $r = 0$ and two outer boundary conditions at $r = R$. As a result, one cannot simply start at the center ($m = 0$) and integrate outward, or start at the outer boundary ($m = M$) and integrate inward. Initially, this problem was solved by starting from the inside with guessed initial values of P_c and T_c, integrating outward to $m=M$, and checking whether $T(M)$ and $P(M)$ satisfy the outer boundary conditions. If not, the process would be repeated with new values of P_c and T_c until the outer boundary conditions were satisfied. Martin Schwarzschild (1912–1997), the son of Karl Schwarzschild, devised a method in 1958 (Schwarzschild 1958), where he integrated the equations from the core with estimated values of P_c and T_c and made a separate integration from the outside with estimated values of L and R. He then tried to find the layer, $r_{\rm fit}$, where the inner model and the outer model would fit (i.e., the values of $T(r_{\rm fit})$, $P(r_{\rm fit})$, $L(r_{\rm fit})$, and $M(r_{\rm fit})$ and their derivatives would be the same for both models). Merging of the two models would then give a full model that agreed with all boundary conditions. Unfortunately, this method was cumbersome because it required many trials for the inner and the outer model before an acceptable model was found.

A more practical method was introduced by Henyey, Forbes, and Gould in 1964 (Henyey et al. 1964). In this implicit Lagrangian method, the structure equations are written as a set of linear equations that describe the relation between each of the parameters at a discrete set of radii r or mass m. Here we describe the principle of the **Henyey method.** For a more extended description, see Maeder (2009).

Introduce the general parameter y^i with $i = 1$ to 4, and define the parameters $y^1 = r$, $y^2 = P$, $y^3 = L$, $y^4 = T$ (Note that the superscript i does not indicate a power but is merely an index).

Divide the star into a large number (N) of spherical shells with radii r_j with $j = 1$ to N. Each parameter at any radius is decribed by y^i_j. For each shell with radius between r_j and r_{j+1} the differential equations that describe the stellar structure in Lagrangian coordinates can be written as

$$\frac{dy^i}{dm_j} = \frac{y^i_{j+1} - y^i_j}{m_{j+1} - m_j} = f\left(y^1_{j+0.5}, \; y^2_{j+0.5}, \; y^3_{j+0.5}, \; y^4_{j+0.5}\right), \tag{10.9}$$

where f describes the dependence of the derivatives on the local parameters at mass $j + ½$. This results in a set of equations of the type

$$E^i_j \equiv \frac{y^i_{j+1} - y^i_j}{m_{j+1} - m_j} - f\left(y^1_{j+0.5}, \; y^2_{j+0.5}, \; y^3_{j+0.5}, \; y^4_{j+0.5}\right) = 0. \tag{10.10}$$

Suppose that a reasonable first-order model is available. The equations of the next step can then be written as a small correction to the first model. For instance, in Lagrangian coordinates (with parameters r, P, L, and T as function of m) the value of T in the next iteration ($T_2 = T_1 + \Delta T$) at a given value of m can be written as a correction

$$\Delta T = (dT/dr)\Delta r + (dT/dP)\Delta P + (dT/dL)\Delta L, \tag{10.11}$$

with all the terms (dT/dP, etc.) given by the partial derivatives of the right sides of the structure equations. For instance, in the case of the ideal gas law, $dT/dP = \mu m_H / k\rho$ if radiation pressure can be ignored. The equation can be written as

$$\Delta T - (dT/dr)\Delta r + (dT/dP)\Delta P + (dT/dL)\Delta L = 0. \tag{10.12}$$

Similar expressions exist for the other parameters. A model agrees with the four structure differential equations at all values of m if all differences Δ are zero at all values of m (i.e., throughout the star) and for all parameters (r, P, L, T). This set of linear equations of the type shown in Equation (10.10) can be written as a large matrix that can be solved with standard mathematical techniques.

The solution of these equations yields the values of y_j^i, which are r, P, L, and T, as a function of m. This defines the model of a star with a given mass and chemical structure in HE and TE. Using these values of r, P, L, and T one can calculate the rates of the nuclear processes in each layer and the change in chemical composition per unit of time, taking a time step Δt and predicting the abundances in each layer at time $t + \Delta t$. These new abundances result in an updated value of the functions f, which contain, for example, the EoS, the opacity, and the reaction rates at any m. By starting the process over again, calculating an equilibrium model at the next time step, and continuing in this manner, this process results in a time series of models that describe the evolution of a star.

10.4.1 The MESA Code

To facilitate the calculation of stellar evolution models, an open source program called MESA ("Modules for Experiments in Stellar Astrophysics") is available to the community (Paxton et al. 2011). It is a suite of robust and efficient libraries for a wide range of applications in computational stellar astrophysics. The modules calculate opacities, nuclear reaction rates, and the equation of state. The one-dimensional stellar evolution module "MESA star" combines numerical and physics modules and solves the combined structure and composition equations simultaneously. It includes effects such as convection and overshooting, mixing, diffusion, and gravitational settling, mass loss, and accretion. It allows the calculation of stellar evolution for both low- and high-mass stars. MESA can be downloaded from the project's website at http://mesa.sourceforge.net/.

10.5 Principles of Stellar Evolution Calculations

The predicted stellar evolution of a star is based on the computation of a series of models in hydrostatic equilibrium and thermal equilibrium with varying chemical structure. This has two interesting consequences.

1. **The structure of a star in HE and TE depends only on the chemical profile as a function of mass**. It is independent of the previous evolution that has led to this chemical distribution. Stars of the same mass that have evolved in different ways but happen to end up with the same chemical structure will

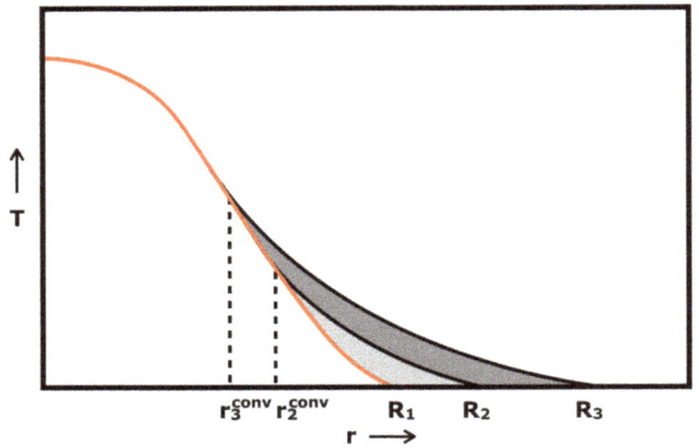

Figure 10.1. Schematic picture of the increase in radius of a star with a convective envelope (gray). The red line indicates a model in radiative equilibrium with radius R_1. The black lines show models 2 and 3 with different depths of the convective envelope, $r^{\rm conv}$, and radii R_2 and R_3.

from then on evolve in exactly the same way (this does not hold if magnetic fields or rotation play a role, as these effects might influence chemical mixing and hydrostatic equilibrium).

2. **Although a star may expand during its evolution, it is always in quasi-hydrostatic equilibrium**. This is because the timescale to reach hydrostatic equilibrium is much shorter than the evolutionary timescale (except during explosive phases). In a stellar model, the forces balance one another. If evolution calculations show that a star is expanding/contacting, one should not ask "which forces make the star expand/contract?" but "why can the star, with this changed chemical profile, only be in equilibrium if it is larger/smaller than before?".

As an example, we can consider the expansion of stars with a convective envelope. If changes in the chemical profile imply that the radius of the star must increase in order to remain in equilibrium, the convection in the envelope will reach deeper layers. The smaller adiabatic T-gradient, compared to the radiative T-gradient, will result in an increase in radius. This is shown in Figure 10.1.

10.6 Summary

1. The structure of a star with a given mass and chemical structure in HE and TE is described by a set of four differential equations. These can be expressed in Lagrangian coordinates, with $m = M(r)$ as the running parameter.
2. There are four boundary conditions: $r = 0$ and $L = 0$ in the center, where $m = 0$, and $P = 0$ and $L = 4\pi R^2 \sigma T^4$ at the surface, where $m = M$.
3. Starting with an initial guess model, the corrections to this model can be described by a large number of linear differential equations. These can be solved by standard matrix inversion techniques, e.g., the Henyey method, to yield $r(m)$, $P(m)$, $L(m)$, and $T(m)$.

4. The resulting model also describes the nuclear reaction rates as a function of m and the change in chemical structure.
5. Taking small time steps and calculating the structure at each time step results in a series of time-dependent models in HE and TE. This series describes the evolution of a star.

Exercises

10.1 How do we know which one of the two energy transport equations to use?

10.2 How can we calculate the entropy term due to expansion or contraction if the evolution is calculated by a series of models in hydrostatic equilibrium?

10.3 **Computer exercise**
Familiarize yourself with the MESA code and calculate a model for the Sun at $t = 4.5$ Gyr with the cosmic abundances.

References

Henyey, L. G., Forbes, J. E. & Gould, N. L. 1964, ApJ, **139,** 306

Maeder, A. 2009, Physics, Formation and Evolution of Rotating Stars (Berlin: Springer)

Paxton, B., Bildsten, L. & Dotter, A., et al. 2011, ApJS, **192,** 3

Schwarzschild, M. 1958, in Structure and Evolution of the Stars (Princeton: Princeton Univ. Press), 96

AAS | IOP Astronomy

Understanding Stellar Evolution

Henny J.G.L.M. Lamers and Emily M. Levesque

Chapter 11

Polytropic Stars

In Section 4.8, we showed that under certain physical conditions the equation of state (EoS) can be written as $P \sim \rho^{\gamma}$. If a star has such a polytropic EoS, its density and pressure structure follows from the hydrostatic equilibrium (HE) equation. This solution of the HE equation does not provide information about the temperature structure. This is the case for degenerate stars. We also showed that a polytropic EoS applies to fully convective stars and to stars with a constant ratio $P_{\rm rad}/P_{\rm gas}$. In these stars, the gas also obeys the ideal gas law, $P \sim \rho T$; as a result, the P and ρ structure, derived from the HE equation for polytropic gas, also specifies the temperature structure. This temperature structure then specifies the energy flow through the star and the stellar luminosity. In this section, we describe the models of polytropic stars. Polytropic models were historically very important because they were the first stellar models before the age of computers. They provide insight into the structure of stars and give scaling laws such as the mass–luminosity relation and the mass–radius relation.

11.1 The Structure of Polytropic Stars: $P = K\rho^{\gamma}$

If the EoS can be expressed as a polytrope $P = K\rho^{\gamma}$, the pressure is independent of temperature and thus analytical solutions to the HE equation exist for certain values of γ.

In this case, the HE can be combined with the mass continuity equation

$$\left.\begin{array}{l} \dfrac{r^2}{\rho}\dfrac{dP}{dr} = -Gm \rightarrow \dfrac{d}{dr}\left(\dfrac{r^2}{\rho}\dfrac{dP}{dr}\right) = -G \cdot \dfrac{dm}{dr} \\ \qquad\qquad\qquad\qquad \dfrac{dm}{dr} = 4\pi r^2 \rho \end{array}\right\} \frac{1}{r^2}\frac{d}{dr}\left(\frac{r^2}{\rho}\frac{dP}{dr}\right) = -4\pi G\rho. \qquad (11.1)$$

Using the polytropic relation $P = K\rho^{\gamma} \rightarrow dP/dr = K\gamma\rho^{\gamma-1}\, d\rho/dr$, which yields

$$\frac{1}{r^2}\frac{d}{dr}\left(r^2 K\gamma\rho^{\gamma-2}\frac{d\rho}{dr}\right) = -4\pi G\rho \rightarrow \frac{K\gamma}{4\pi G}\frac{1}{r^2}\frac{d}{dr}\left(r^2\rho^{\gamma-2}\frac{d\rho}{dr}\right) = -\rho. \tag{11.2}$$

The equation becomes simpler with two substitutions:

1. $\gamma = 1 + 1/n \rightarrow n = 1/(\gamma - 1)$. (11.3*a*)

2. $\rho = \rho_c\theta^n$, where $\theta = (\rho/\rho_c)^{1/n}$ is a dimensionless variable. (11.3*b*)

These substitutions transform the HE equation into

$$\boxed{\left[\frac{(n+1)K}{4\pi G\rho_c^{(n-1)/n}}\right]\frac{1}{r^2}\frac{d}{dr}\left(r^2\frac{d\theta}{dr}\right) = -\theta^n}. \tag{11.4}$$

The term in square brackets is a constant α^2, with α in centimeters. So

$$\alpha = \left[(n+1)K/4\pi G\rho_c^{(n-1)/n}\right]^{1/2}. \tag{11.5}$$

Q (11.1) *Although θ is dimensionless, can you think of what it may describe in physical terms? Why was the symbol θ chosen?*
Hint: suppose the gas has a polytropic EoS and obeys the ideal gas law,
Q (11.2) *What is the range of θ?*
Q (11.3) *What does a model with $n = 0 \rightarrow \gamma = \infty$ describe in physical terms?*
What does a model with $n = \infty \rightarrow \gamma = 1$ describe in physical terms?

For mathematical reasons (and to make the equation look nicer), we may define

$r = \alpha\xi$ with ξ = dimensionless and α in centimeters. (11.6)

The HE equation can then be written as

$$\frac{1}{\xi^2}\frac{d}{d\xi}\left(\xi^2\frac{d\theta}{d\xi}\right) = -\theta^n \quad \rightarrow \quad \theta = f(n, \xi). \tag{11.7}$$

This is the **Lane–Emden equation,** named after the American physicist Jonathan Lane (1819–1880) and the Swiss physicist Robert Emden (1862–1940), who was married to astronomer Karl Schwarzschild's sister Klara. Lane introduced the concept in 1870 and Emden described it in the present form in 1907 in his famous book "Gas Spheres" (Emden 1907).

This equation describes the density structure $\theta = f(\xi)$ or $\rho = f(r)$ of a polytropic star in a dimensionless form, with only one parameter: $n = 1/(\gamma - 1)$. There are analytical solutions for $n = 0$ or $\gamma = \infty$ and for $n = 1$ or $\gamma = 3/2$. For any other value of n, the equation can easily be solved numerically, with $\xi = 0$ at the center of the star. The radius of the star is defined as the value of ξ, where $\rho = 0$, which means that $\theta = 0$.

Figure 11.1 shows the density structure in terms of $\theta^n = \rho/\rho_c$ versus r/R *for* $n = 3$ ($\gamma = 4/3$) and $n = 3/2$ ($\gamma = 5/3$).

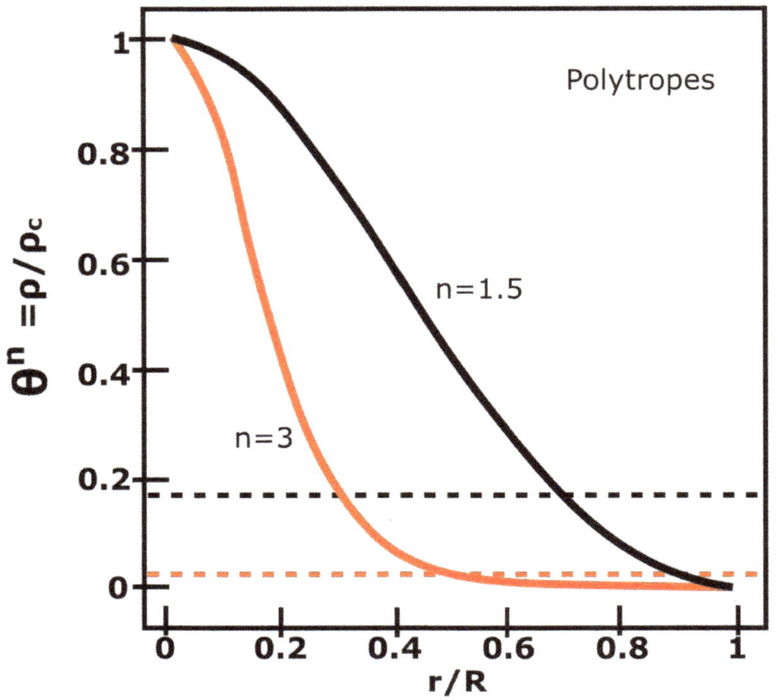

Figure 11.1. Density structure of polytropes for $n = 3$ ($\gamma = 4/3$) (red) and $n = 1.5$ ($\gamma = 5/3$) (black). The dashed lines show the mean density of the two models.

Polytropic models were historically important because they could be easily calculated before the age of computers. Eddington (1882–1944) calculated the first polytropic model for the Sun and stars, based on the assumption that $P_{\rm rad}/P_{\rm gas}$ is about constant ($\gamma = 4/3$), and obtained the mass–luminosity relation for stars. He showed that $L \sim \mu^4 M^3$ in 1926 (Eddington 1926).

11.2 Stellar Parameters of Polytropic Models

The solution of the Lane–Emden equation yields $\theta(\xi)$ with $\rho = \rho_c\theta^n$ and $r = \xi/\alpha$. The inner boundary condition at the center of the star is $\theta = 1$ at $\xi = 0$.

The radius of the star is defined by $\rho = 0$, so $\theta(\xi_1) = 0$, and ξ_1 is the value of the ordinate where $\theta = 0$. This means that

$$R = \alpha\xi_1 \equiv \alpha R_{\rm n}. \tag{11.8}$$

The mass of the star is given by

$$M = 4\pi \int_0^R \rho r^2\, dr = 4\pi\alpha^3\rho_c \int_0^{\xi 1} \xi^2\theta^n\, d\xi. \tag{11.9}$$

Using Equation (11.7) results in

Table 11.1. Physical Parameters of Polytrope Models.

n	γ	R_n	M_n	D_n	B_n
1.00	2.00	3.14	3.14	3.290	0.233
1.50	5/3	3.65	2.71	5.991	0.206
3.00	4/3	6.90	2.02	54.18	0.157

$$M = -4\pi\alpha^3\rho_c\xi_1^2\left(\frac{d\theta}{d\xi}\right)_{\xi 1} \equiv 4\pi\alpha^3\rho_c M_n, \tag{11.10a}$$

with

$$M_n = -\xi_1^2\left(\frac{d\theta}{d\xi}\right)_{\xi 1}. \tag{11.10b}$$

The mean density and the central density are related by

$$\bar{\rho} = \frac{3M}{4\pi R^3} \equiv \frac{\rho_c}{D_n} \quad \rightarrow \quad \rho_c = \bar{\rho} D_n. \tag{11.11}$$

The central pressure is given by $P_c = K\rho_c^{1+1/n}$, where K can be eliminated by means of its dependence on α^2 in Equation (11.5) and α can be expressed in terms of M and ρ_c via Equation (11.10). These substitutions result in

$$P_c = (4\pi)^{1/3} B_n G M^{2/3} \rho_c^{4/3} \sim GM^2/R^4. \tag{11.12}$$

This shows that the basic parameters M, R, P_c, and $\bar{\rho}$ can all be expressed in one physical parameter, ρ_c, plus the constants M_n, R_n, B_n, and D_n that depend only on n. Table 11.1 shows the dependences and the parameters for several values of n.

With these parameters, the models are fully specified with $P(r)$, $\rho(r)$, and $m(r)$ known. These models do not provide information on the temperature structure.

11.3 The Mass–Radius Relation of Polytropic Stars

Polytropic stars have a mass–radius relation that depends on n or γ. This follows from Equation (11.10a), which shows that $M \sim \alpha^3\rho_c$. Equation (11.8) shows that $R \sim \alpha$ and Equation (11.5) shows that $\rho_c^{(1-n)/n} \sim \alpha^2$. Combining these equations shows that the **mass–radius relation of polytropic stars** is

$$R \sim M^{(1-n)/(3-n)}. \tag{11.13}$$

Table 11.2 shows the properties of polytropic models for specific types of stars.

Remember that for an ideal gas that also obeys a polytropic EoS, such as stars with a constant ratio $P_{\rm gas}/P_{\rm rad}$ and fully convective stars, the temperature is $T \sim P/\rho \sim \rho^{\gamma-1}$. The mass–luminosity ratio of these stars will be discussed in Sections 11.3.5 and 14.3, respectively.

Table 11.2. Properties of Polytropic Stars

n	γ	Property	M–R Relation	Consequence
0	∞	Incompressible	$R \sim M^{1/3}$	Constant ρ
1	2	$P \sim \rho^2 \rightarrow T \sim \rho$	$R \sim M^0$	Constant R
1.5	5/3	Nonrelat. degeneracy	$R \sim M^{-1/3}$	Volume $\sim 1/M$
		Fully convective (ideal gas)		
3.0	4/3	Relativ. degeneracy	$R \sim M^{\infty}$	$M \sim R^0$
		Constant P_{gas}/P_{rad} (ideal gas)		
∞	1	Isothermal: $P \sim \rho$	$R \rightarrow \infty$	Infinite radius

11.3.1 Polytropes with $\gamma = 1$ and $n = \infty$: Isothermal Stars

The solution of the Lane–Emden equation for isothermal stars is not relevant because they have an infinite radius. Fully isothermal stars therefore cannot exist; however, these models can still be used to describe parts of a star. For instance, after the core H-fusion phase, when H in the core is exhausted and H-fusion occurs in a shell around it, the helium core is almost isothermal. This implies that we can use the polytropic model of $\gamma = 1$ for the helium core and attach a nonisothermal envelope to describe the properties of these stars.

11.3.2 Polytropes with $\gamma = 5/3$ and $n = 1.5$ and Fixed K: Nonrelativistic Degenerate Stars

Nonrelativistic completely degenerate stars behave as polytropes with $\gamma = 5/3$ because the EoS is $P = K_1(\rho/\mu_e)^{5/3}$ (Equation 4.20), with the value of K_1 fixed by atomic physics. This applies to white dwarfs and neutron stars. Substitution of the constant K_1 in the Lane–Emden equation results in the M-R relation for WDs

$$R = 0.012 \ M^{-1/3}(\mu_e/2)^{-5/3} \quad \text{if } M \text{ and } R \text{ are in solar units.} \qquad (11.14)$$

11.3.3 Polytropes with $\gamma = 5/3$ and $n = 1.5$ and Variable K: Fully Convective Stars

We have shown in Section 4.8.1 that fully convective stars behave as polytropes with $\gamma = 5/3$. One might then expect that they have an M-R relation similar to white dwarfs; however, this is not the case, because contrary to the case of white dwarfs the value of K in fully convective stars is not defined by the EoS of the gas but by the mass of the star. In other words, K is constant in a star but may vary from star to star depending on the stellar mass. We will discuss this in Section 12.7, which deals with fully convective pre-MS stars, and in Section 14.3, which deals with fully convective post-MS stars.

11.3.4 Polytropes with $\gamma = 4/3$ and $n = 3$ and Fixed K: Relativistic Degenerate Stars

A polytrope model with $n = 3$ (i.e., $P = K\rho^{4/3}$) and with a fixed constant K has the curious property that it can exist for only one particular mass. This follows from the substitution of Equation (11.5) for α into Equation (11.10) for M. This is the case for extreme relativistic degenerate stars, where the constant K_2 is given by Equation (4.22). The resulting mass is

$$M = 4\pi M_{n=3}(K_2/4\pi G)^{3/2} \text{ with } M_{n=3} = 2.02 \tag{11.15}$$

(see Table 11.1). We will show later that this results in upper limits for the mass of white dwarfs and neutron stars.

11.3.5 Polytropes with $\gamma = 4/3$ and $n = 3$ and Variable K: Eddington's Standard Model

We have shown in Section 4.8 that stars in which the ratio $1-\beta = P_{\rm rad}/P$ is constant behave as polytropes with $P \sim aT^4/3$ and $P \sim \rho\, T$, so $P \sim \rho^{4/3}$ (i.e., $\gamma = 4/3$ and $n = 3$). Sir Arthur Eddington assumed that $P_{\rm rad}/P = 1-\beta$ is constant in stars and derived the polytropic model for $n = 3$ (Eddington 1926). Let us follow his arguments.

$$P_{\rm rad} = \frac{a}{3}T^4 = (1-\beta)P = \frac{(1-\beta)}{\beta}P_{\rm gas} = \frac{(1-\beta)}{\beta}\frac{\mathfrak{R}}{\mu}\rho T. \tag{11.16}$$

so

$$T^3 = \left[\frac{1-\beta}{\beta}\frac{3\mathfrak{R}}{\alpha\mu}\right]\rho \rightarrow T = \left[\frac{1-\beta}{\beta}\frac{3\mathfrak{R}}{\alpha\mu}\right]^{1/3}\rho^{1/3} \tag{11.17}$$

and

$$P = \frac{\mathfrak{R}}{\mu}\frac{\rho T}{\beta} = \frac{\mathfrak{R}}{\beta\mu}\left[\frac{1-\beta}{\beta}\frac{3\mathfrak{R}}{\alpha\mu}\right]^{1/3}\rho^{4/3} \equiv K\rho^{4/3} \rightarrow K = \left[\frac{3\mathfrak{R}^4}{a\mu^4}\frac{1-\beta}{\beta^4}\right]^{1/3}. \tag{11.18}$$

Adopting Equation (11.10) for the mass and substituting α from Equation (11.5), we find

$$M = 4\pi M_{n=3}(K/\pi G)^{3/2}. \tag{11.19}$$

This results in

$$\frac{M}{M_\odot} = 18.2 \times \frac{\sqrt{1-\beta}}{\beta^2\mu^2}. \tag{11.20}$$

This shows that for stars with a constant ratio $P_{\rm rad}/P$ the value of β depends on μ and M.

The predicted ρ, P, and T structure of Eddington's standard model with a homogeneous composition of $\mu = 0.60$ and the Standard Solar Model (Guenther

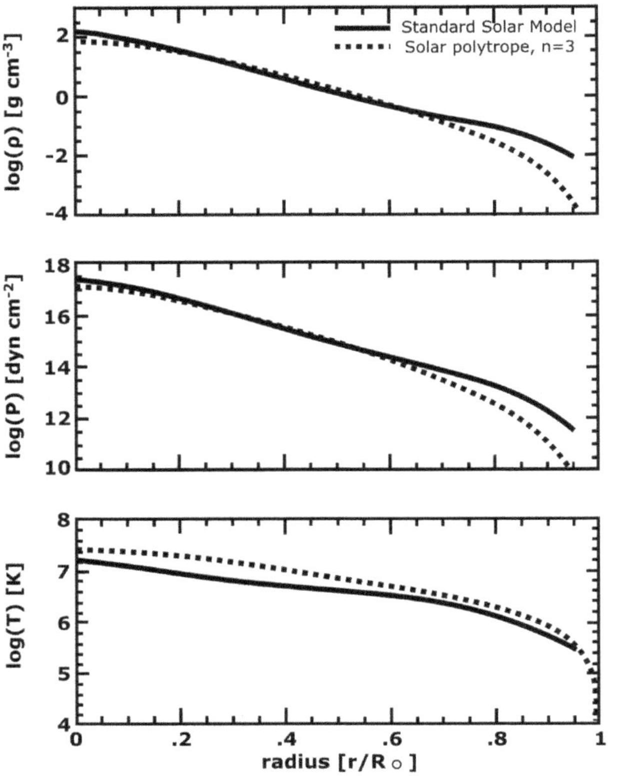

Figure 11.2. Comparison between the structure of the Sun predicted by the Standard Solar Model and the homogeneous Eddington's standard model (Courtesy of Chris Flynn).

et al. 1992) are compared in Figure 11.2. Note the good agreement over several orders of magnitude between the two models. The differences in the center are due to the increased He abundance in the Sun and the differences in the outer layers are due to the fact that the outer envelope of the Sun is convective.

We now consider the **luminosity of Eddington's standard model**.

The condition $P_{\text{rad}} = (1 - \beta)\, P$ implies that $dP_{\text{rad}}/dP = (1 - \beta) =$ constant. The gradient of the radiation pressure and the gradient of the gas pressure are described by Equations (6.10) and (3.3), respectively. Division of these two equations results in

$$\frac{dp_{\text{rad}}}{dP_{\text{gas}}} = \frac{\kappa}{4\pi c G}\frac{L_r}{M_r} \equiv 1 - \beta. \tag{11.21}$$

This is assumed to be valid at all depths as well as at the surface, which implies that

$$L = \frac{4\pi c G M}{\kappa}(1 - \beta) = L_E(1 - \beta), \tag{11.22}$$

Table 11.3. Parameters for Eddington's Standard Models as a Function of β for Homogeneous H-rich and He-rich Stars.

Properties		H-rich			He-rich		
β	$P_{\rm rad}/P_{\rm gas}$	K	$M/M_\odot$	$L/L_\odot$	K	$M/M_\odot$	$L/L_\odot$
0.999	0.001	5.2 (14)	1.6	6.1 (1)	1.8 (14)	0.3	1.2 (1)
0.99	0.010	1.2 (15)	5.2	2.0 (3)	4.0 (14)	1.1	6.9 (2)
0.95	0.053	2.1 (15)	12.6	2.5 (4)	7.2 (14)	2.6	8.3 (3)
0.85	0.176	3.5 (15)	27.3	1.6 (5)	1.2 (15)	5.5	5.4 (4)
0.50	1.000	1.1 (16)	143.9	2.8 (6)	3.6 (15)	29.1	9.5 (5)

Note. The values in parentheses are the powers of 10. The value of K is in cgs units.

with $(1-\beta) \sim M^2\mu^4\beta^4$ (Equation 11.20). Adopting $\kappa = \sigma_e = 0.20(1+X)$ (Equation 5.6) and inserting the constants results in

$$\frac{L}{L_\odot} = \frac{6.5 \times 10^4}{1+X}\beta^4\mu^4\left(\frac{M}{M_\odot}\right)^3. \tag{11.23}$$

This agrees with the dependence of $L \sim \mu^4 M^3$ that we already derived in Section 6.2 on simple scaling arguments. Table 11.3 shows ratio $P_{\rm rad}/P_{\rm gas} = (1-\beta)/\beta$, the value of K, the mass (Equation 11.20), and the luminosity (Equation 11.23) of H-rich MS stars with $X = 0.70$ and $\mu = 0.60$ and He-rich stars with $X = 0$ and $\mu = 4/3$ for various values of β. The table shows that in the most massive stars observed, $M \approx 150 M_\odot$, the radiation pressure is about equal to the gas pressure. It also shows that radiation pressure is much higher in a He-rich star than in a H-rich star of the same mass. This is due to the fact that the luminosity of stars in HE and RE scales approximately as $L \sim \mu^4$ (Equation (6.7)).

The importance of Eddington's approximation is that it allowed the calculation of the first stellar model and showed that parameters such as M, R, and L are coupled, with L depending on the composition via μ and β.

11.4 Summary

1. The density, pressure, and mass structure of stars with a polytropic EoS of the type $P = K\rho^\gamma$ for any value of γ or $n = 1/(\gamma - 1)$ is fixed. The mass–radius relation depends on n as listed in Table 11.2.
2. For degenerate stars, the value of K is fixed by atomic constants and so the M-R relation is fixed. Nonrelativistic degenerate stars have $R \sim M^{-1/3}$, with the constant given in Equation (11.14). Fully relativistic degenerate stars exist for only one value of mass, given in Equation (11.15).
3. Fully convective stars have a polytropic EoS with $\gamma = 5/3$ or $n = 1.5$. For these stars, the constant K depends on the stellar parameters: M and R. Stars with a constant ratio $P_{\rm rad}/P = 1 - \beta$ have a polytropic EoS with $\gamma = 4/3$ or $n = 3$. For these stars, the constant K is set by the value of β.

4. In nondegenerate stars, the gas also obeys the ideal gas law $P \sim \rho T$, so the pressure and density structure of a polytropic model also determine the temperature structure and the energy flow.
5. A model with a constant ratio $P_{\rm rad}/P$ is called Eddington's standard model. The masses and luminosities of stars with different values of β are listed in Table 11.3. It shows that $L \sim \beta^4\mu^4 M^3$. Low-mass He-rich stars are about 25 times more luminous than H-rich stars of the same mass.

Exercises

11.1 Explain in simple physical terms why a star with $\gamma = 4/3$ has a more concentrated density structure than a star with $\gamma = 5/3$.

11.2 Suppose that zero-age main-sequence stars are described by the Eddington standard model with electron scattering as the dominant opacity.
- (a) What would be the mean values of β and $P_{\rm rad}/P_{\rm gas}$ for stars of 1 and $60 M_\odot$?
- (b) What is the predicted luminosity of these stars?
- (c) Compare this with the data in Appendix D and comment on the comparison.

References

Eddington, A. S. 1926, The internal constitution of stars (Cambridge: Cambridge Univ. Press)

Emden, R. 1907, Gaskugeln: Anwendungen der mechanische Wärme theorie auf kosmologische und meteorologische Probleme (Berlin: Teubner)

Guenther, D. B., Demarque, P., Kim, Y.-C., & Pinsonneault, M. H., 1992, ApJ, **387,** 372

Henny J.G.L.M. Lamers and Emily M. Levesque

Chapter 12

Star Formation

In this chapter, we discuss the process of star formation. Stars form out of cold molecular interstellar clouds when the inward gravitational force is stronger than the outward force due to gas pressure and magnetic pressure. This leads to an initial collapse on a free-fall timescale and the subsequent fragmentation of the cloud into clumps. The released potential energy is partly used to dissociate the molecules and partly lost by infrared radiation. When the molecules are dissociated and the gas is subsequently ionized, the free-fall is halted and the contraction of the clumps proceeds on the much longer Kelvin–Helmholtz timescale. Initially, the energy is transported mainly by convection but as the internal temperature increases, radiation takes over and the star becomes a pre-main-sequence star that evolves toward the main sequence. After a short phase of Li- and D-fusion a star like the Sun settles on the main sequence, where H-fusion is the energy source. Stars with a mass below $0.08 M_{\odot}$ do not reach H-fusion and become brown dwarfs. During a significant fraction of this evolution, the star is surrounded by an accretion disk, which results in characteristic spectral features.

The fragmentation of a cloud results in the formation of stars over a large mass range from about 0.01 to $100 M_{\odot}$. We discuss the initial mass function, which is the number distribution of stars over this mass range. At the end of this section, we briefly discuss star formation in the early universe and the formation of Population II and III stars.

12.1 The Interstellar Medium

The interstellar medium consists of gas in different phases (i.e., with different temperatures and densities). The most important phases are listed in Table 12.1.

The mean value of the pressure factor ($nT = P/k$) in the galactic disk is $nT \approx 3 \times 10^3$ Kcm^{-3}. Note that most of the components are in equilibrium with the mean pressure in the galactic disk, with the exception of the molecular clouds. The high pressure in the molecular clouds might suggest that they will expand into the lower

doi:10.1088/978-0-7503-1278-3ch12

Table 12.1. Mean Properties of Phases of the Interstellar Medium in the Galactic Disk (Cox 1990)

Component	T (K)	n (cm^{-3})	nT (Kcm^{-3})
Molecular clouds	10–20	10^3–10^4	10^4–10^5
Cold neutral gas	50–100	20–50	10^3–10^4
Warm neutral gas	10^3–10^4	0.2–0.5	10^3
Warm ionized gas	10^4	0.1–1	10^3–10^4
Hot gas	10^6–10^7	10^{-2}–10^{-4}	10^3–10^4

pressure environment; however, we will show that this is not the case because it is prevented by their self-gravity.

12.2 The Jeans Mass for Gravitational Contraction

To understand star formation, we have to study the conditions necessary for interstellar clouds to contract. Consider for simplicity a spherical homogeneous cloud of mass M, radius R, temperature T, and density ρ (see Figure 12.1). The easiest way to test if such a cloud is stable or contracting is by testing the virial equilibrium. If the cloud is in hydrostatic equilibrium (i.e., neither expanding nor contracting), the Virial Theorem (Equation (3.14)) requires $E_{\text{kin}} = -½E_{\text{pot}}$ with

$$E_{\text{kin}} = (3/2)kT(M/\mu m_{\text{H}}) \tag{12.1}$$

and

$$E_{\text{pot}} = -\int_0^M \frac{Gm_r}{r}\,dm_r = -\frac{3}{5}\frac{GM^2}{R}, \tag{12.2}$$

where we used $r = (m_r/M)^{1/3}$ for a constant density medium. If $-½E_{\text{pot}} > E_{\text{kin}}$, then gravity wins and the cloud will contract. This is the case if

$$\frac{3kT}{2\mu m_{\text{H}}}M > \frac{3}{10}\frac{GM^2}{R} \rightarrow \boxed{M > M_J \equiv \frac{5kT}{\mu m_{\text{H}}}\frac{R}{G},} \tag{12.3}$$

where M_{J} is the **Jeans mass**, named after the British physicist Sir James Jeans (1877–1946) who described it in 1904. Since we adopted a constant density $\rho = \mu n m_{\text{H}}$, we can also express M_{J} in terms of T and n. For this purpose, we eliminate the radius R by the substitution of

$$R = \left(\frac{3}{4\pi\mu m_{\text{H}}}\right)^{1/3}\left(\frac{M}{n}\right)^{1/3}. \tag{12.4}$$

This yields

$$M_{\text{J}} = \left(\frac{3}{4\pi}\right)^{1/2}\left(\frac{5k}{Gm_{\text{H}}^{4/3}}\right)^{3/2}\left(\frac{T^3}{\mu^4 n}\right)^{1/2} \tag{12.5}$$

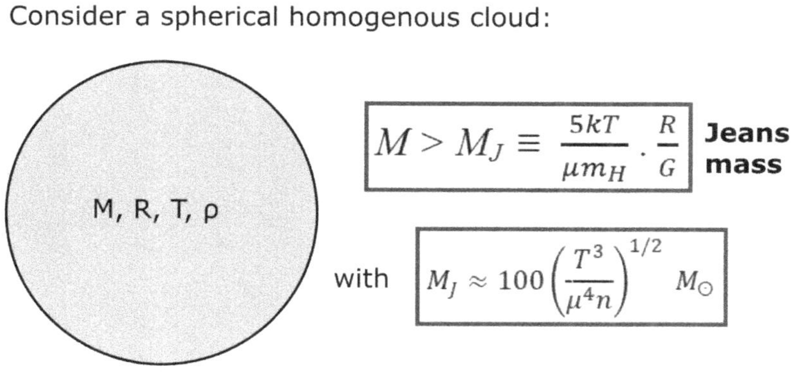

Figure 12.1 Jeans mass of a spherical homogeneous cloud. The cloud contracts if $M > M_J$.

or quantitatively

$$M_J \approx 100 \left(\frac{T^3}{\mu^4 n}\right)^{1/2} M_\odot, \tag{12.6}$$

with T in K. A higher temperature corresponds to a larger M_J, meaning that warmer clouds must be more massive in order to collapse. Stars that formed from contracting clouds in the early universe, when temperatures were much higher than at present, must have been more massive than stars forming today. This will briefly be discussed later in Section 12.13.

Q (12.1) *What is the typical Jeans mass of a cold cloud, consisting mainly of* H_2 *molecules, with a density of* $n = 10^3$ cm^{-3} *and* $T = 10$ K?

Giant molecular clouds in the Galactic disk, consisting largely of H_2 molecules, have a temperature of $T \sim 10$ K, a typical density of $n \sim 10^3$ cm^{-3}, a mass of the order of 10^6 to $10^7 M_\odot$, and a radius of about 30 pc. They are the major regions of star formation.

12.3 The Collapse of Molecular Clouds

A cloud of a given mass will contract and start forming stars if it exceeds the Jeans mass, which depends on density as $M_J \sim n^{-1/2}$ (Equation 12.6). The cloud contraction and star formation are triggered by compression due to shocks, which increases the density. The three major triggering mechanisms are as follows.

1. Clouds passing the density wave of a spiral arm in a spiral galaxy. This explains why star formation mainly occurs in the spiral arms.
2. Compression of clouds by a shockwave generated in a nearby supernova explosion.
3. Cloud–cloud collisions in colliding or merging galaxies. This explains the starbursts observed in merging galaxies.

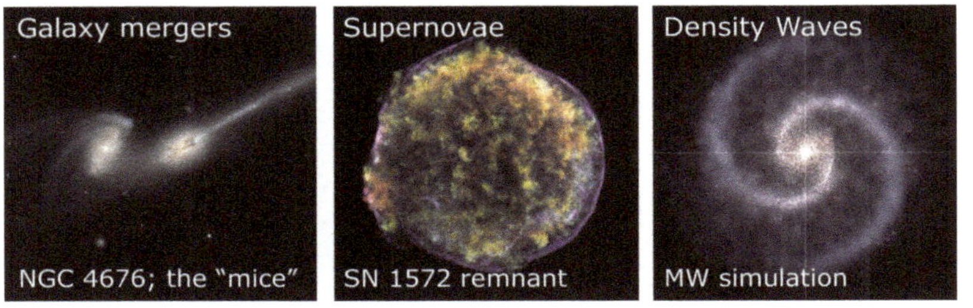

Figure 12.2 Three triggering mechanisms for star formation: merging galaxies, shocks from supernovae ejecta, and density waves in the disks of spiral galaxies. (Courtesy of NASA.[1])

Figure 12.2 shows the three mechanisms for triggering star formation.

When a cloud has a mass higher than the Jeans mass, it will collapse. This will occur approximately on a free-fall timescale (Equation (9.1)), which is

$$\tau_{\rm ff} \approx (G\rho)^{-\frac{1}{2}} \approx 1 \times 10^8 (\mu n)^{-\frac{1}{2}} \,{\rm yr} \tag{12.7}$$

or about 10^6 yr if $\mu n \sim 10^4$ cm^{-3}. If clouds had no cooling mechanism, they would contract *adiabatically* with temperature increasing as $T \sim \rho^{2/3}$ (Equation (4.33)). Because the Jeans mass is $M_J \sim (T^3/\rho)^{1/2}$, its value would *increase* as the density increases. At some point, the Jeans mass would reach the cloud mass: i.e., the cloud would be in hydrostatic equilibrium and the contraction would stop. Fortunately, however, clouds *do* have a cooling mechanism; if they did not, star formation would be extremely inefficient and we would not exist!

Clouds cool by radiative losses. For this to be effective, they have to emit IR photons; massive clouds are optically thick for UV and visual photons but optically thin for photons at longer wavelengths. The main cooling agents of molecular clouds are emission by molecules and emission by dust.

Cooling by molecules:

Collisions between molecules in high-density gas result in collisional excitation to higher rotation or vibration levels. An excited molecule can de-excite to a lower rotation or vibration level by emitting a photon (photo de-excitation). The dominant transitions are in the IR or submillimeter range. These photons can leave the cloud because the cloud is optically thin at IR wavelengths. The net effect is that the kinetic energy of molecules is transferred into excitation, which results via photo de-excitation in IR and submillimeter photons that escape the cloud, making this a cooling mechanism.

Cooling by dust:

If the density in a cloud is high and the kinetic temperature is low enough ($T <$ 1000 K), dust may form. Collisions of dust particles with molecules will heat dust

[1] http://www.spacetelescope.org/images/heic0206b/http://chandra.harvard.edu/photo/2005/tycho/
http://beltoforion.de/article.php?a=spiral_galaxy_renderer&hl=en

grains. The same occurs if photons (UV, optical, or near-IR) are absorbed by dust. Dust grains emit almost like blackbodies with a radiation temperature less than ~ 1000 K. This results in a large IR flux that leaves the cloud. Molecular clouds are therefore very strong IR emitters! The net effect, once again, is that the kinetic energy of molecules and optical photons is converted into IR radiation by dust. This radiation then escapes the cloud, making this another cooling mechanism.

12.4 Fragmentation of Molecular Clouds

The cooling mechanisms prevent the adiabatic heating of a cloud when it collapses. The potential energy gained by the contractions is immediately emitted, so the collapse proceeds approximately isothermally. The increasing density of isothermal clouds implies that the Jeans mass decreases so that substructures of the cloud can start to contract. This results in a *fragmentation of the original cloud*, which splits up into sub-clouds that may split up again. This fragmentation is helped by the fact that molecular clouds already have a clumpy structure to begin with. This structure is probably due to magnetic effects (McKee & Ostriker 2007).

The fragmentation continues on faster and faster timescales, because the free-fall timescale decreases as $\sim n^{-1/2}$ as the fragments get denser (Equation (9.1)). Eventually, the density in the fragments will become so high that they become optically thick for IR radiation. When that happens, the cooling mechanism is switched off and the collapse continues adiabatically. This results in a T-rise and a subsequent increase of the Jeans mass to the actual mass of the fragments. At this point the contraction stops and the fragments reach hydrostatic equilibrium. Observers call these fragments **clumps.**

Clumps will eventually evolve into stars. The mass distribution of the clumps sets the **initial mass function (IMF)** of the resulting stars. The IMF describes the number distribution of the stars as a function of their mass. Observations show that the IMF of the clumps has the same shape as the IMF of the stars (Elmegreen 2002; the stellar IMF will be discussed in Section 12.12).

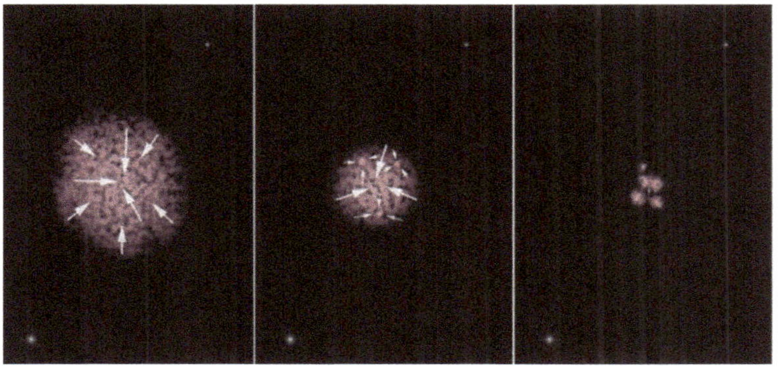

Figure 12.3 Schematic figure of the fragmentation of a collapsing molecular cloud. As the cloud contracts almost isothermally, the Jeans mass decreases and fragments of the cloud start to contract, giving rise to a clumpy structure in the cloud. The clumps evolve into stars.

Not all of the mass in a molecular cloud contributes to the formation of stars. In general, it is estimated that only about 10% of the mass of the initial cloud ends up in stars, although this fraction may be higher in starburst galaxies (Kruijssen 2011). The rest remains in the form of gas that is ejected later by the stellar winds of massive stars and by supernova explosions.

Figure 12.3 shows a sketch of the process of fragmentation and star formation in a collapsing cloud. It shows that large molecular clouds form groups of stars, rather than one individual star. The groups may be gravitationally bound and evolve into a star cluster or be unbound and evolve into an association. An animation of this process, based on a numerical simulation by Matthew Bates, can be found on the web.[2]

12.5 The Minimum Mass of Stars

The minimum mass of stars is set by the minimum mass, $M_{\rm min}$, of the fragments or clumps that can collapse and form stars. A clump collapses if its mass is smaller than the Jeans mass. Since we are looking for the minimum mass, we adopt $M_{\rm min} = M_{\rm J}$, with $M_{\rm J}$ inversely proportional to the density. The key question is, can a small clump cool by radiation? The smaller the $M_{\rm J}$, the higher the density and the higher the optical depth of the clump. A combination of Equations (12.3) and (12.6) shows that for a given temperature the optical depth, $\tau \sim n \times R$, increases with decreasing $M_{\rm J}$. When τ is so large that the clump is optically thick for IR radiation, photons can only escape from the boundary of the clump. This reduces the cooling efficiency drastically.

We have seen above that the collapse requires the presence of a cooling mechanism, viz. the ability to radiate the released potential energy. This condition can be expressed by comparing the amount of potential energy that is released per second, $-dE_{\rm pot}/dt$, with the maximum energy that can be radiated per second, $L_{\rm max}$. Efficient cooling requires that

$$L_{\rm max} > -dE_{\rm pot}/dt. \tag{12.8}$$

If a cloud is optically thick, then the radiation can only leave the clump via the surface, so $L_{\rm max} = 4\pi R^2 \sigma T^4$. The temperature of the dense clumps is about $T \sim 10$ K. The release of potential energy is

$$-dE_{\rm pot}/dt \sim \left(GM^2/R\right)/\tau_{\rm ff} \qquad \text{with} \qquad \tau_{\rm ff} \approx (G\rho)^{-1/2}. \tag{12.9}$$

Substituting this expression into condition (12.8) and using $\rho \approx M/R^3$ results in an expression of the type $M^{5/2} < CR^{9/2}T^4$, where C is a constant. The clump collapsed, so its mass must have been higher than the Jeans mass, given by Equation (12.3). For a given T, the Jeans mass $M_{\rm J}$ is proportional to R. Substituting the constants involved, we derive an estimate of the minimum mass of a clump that satisfies condition (12.8)

$$M_{\rm min} \sim 10^{-2} M_\odot. \tag{12.10}$$

[2] https://www.youtube.com/watch?v=3z9ZKAkbMhY

This is **the minimum star mass.** If the mass of the clump is smaller, it cannot cool efficiently. Objects with lower mass, e.g., Jupiter-like planets, cannot have formed from contracting clumps in star-forming clouds. They must have been formed by the accumulation and merging of rocky and icy clumps in the original circumstellar disks.

12.6 The End of the Free-fall Phase

The free-fall collapse of a clump that exceeds the minimum mass of $\sim 10^{-2}\, M_\odot$ ends when the cooling mechanism stops working and the contraction results in rapid heating that brings the clump into hydrostatic equilibrium. At the end of free-fall, the center of a clump reaches equilibrium first, while the surrounding gas keeps falling onto the core. From now on, the clump is considered a **protostar**. Up to this point, the temperature was so low and the density so high that hydrogen was molecular and the cooling mechanism was active. This ends when H_2 is dissociated and most of the H is subsequently ionized. We can estimate the size of a clump when all H_2 is dissociated and H is ionized by comparing the energy gained in the collapse with that needed for the dissociation and ionization.

The energy needed for dissociation of an H_2 molecule is $x_{H2} = 4.5\,\text{eV}$, and the ionization energy of an H atom is $x_H = 13.6\,\text{eV}$, with 1 eV = 1.602 10^{-12} ergs. The total energy needed for dissociation and ionization of a cloud consisting of H_2 molecules with composition $X = 0.7$ and $Y = 0.30$ is

$$E_{\text{dis}} = \frac{MX}{m_H}\{0.5\,x_{H2} + x_H\} = 1.3 \times 10^{58}\left(\frac{M}{M_\odot}\right)\text{eV} \approx 2 \times 10^{46}\frac{M}{M_\odot}\ \text{ergs}. \quad (12.11)$$

This energy is provided by the contraction in the form of potential energy

$$\Delta E_{\text{pot}} \approx AGM^2\left\{\frac{1}{R_{\text{end}}} - \frac{1}{R_{\text{begin}}}\right\} \approx A\frac{GM^2}{R_{\text{end}}} \approx 4 \times 10^{48} A \frac{(M/M_\odot)^2}{R_{\text{end}}/R_\odot}\ \text{ergs}. \quad (12.12)$$

The constant A is 3/5 for a constant density cloud (Equation (12.2)), but during the collapse the protostar will become centrally concentrated so the factor $A = 3/5$ is a lower limit. Let us assume for simplicity that $A \approx 2$.

In fact, the collapsing cloud loses most of the gained energy in the form of far-IR photons. Only a fraction $f < 1$ of ΔE_{pot} is used for dissociation and ionization. If the clump was contracting quasi-hydrostatically, then f would have been 0.5 according to the virial theorem. In reality it is smaller; detailed calculations show that $f \approx 1/3$ (McLaughlin & Pudritz 1997). If we equate the energies needed for dissociation and ionization, E_{dis}, with $f\,\Delta E_{\text{pot}}$ (Equation (12.12)), we find that the protostar is ionized when it reaches a radius of the order of

$$R/R_\odot \approx 100\, M/M_\odot. \quad (12.13)$$

We see that a protostar of $1M_\odot$ has a radius of $\sim 100 R_\odot$ at the end of the fast contraction and a star of $0.5M_\odot$ has a radius of $\sim 50R_\odot$. We will show later that the effective temperature of the stars when they reach equilibrium is $\sim$3000 K, so the luminosities are $\sim 10^3$ and $\sim 2 \times 10^2\, L_\odot$, respectively, for protostars of 1 and 0.5 $M_\odot$.

We can estimate the mean temperature inside the star by applying the virial theorem because, at the end of the collapse, the star has reached HE

$$\frac{3}{2}\frac{M}{\mu m_{\rm H}}k\bar{T} = \frac{A}{2}\frac{GM^2}{R} \quad \rightarrow \quad \bar{T} \approx \frac{A}{3}\frac{\mu m_{\rm H}}{k} \cdot \frac{GM}{R}, \tag{12.14}$$

with $M/R \approx 10^{-2}\, M_\odot/R_\odot$ and $\mu \approx 1/2$. This gives $\overline{T} \approx 7 \times 10^4$ K, so fusion cannot yet start at the end of the free-fall phase.

The mean density of a protostar of $1M_\odot$ and $100R_\odot$ is $\rho \approx 10^{-6}\,\mathrm{g\,cm^{-3}}$. At such low T and density, the absorption coefficient is very high: $\kappa \approx 10^2$ to $10^3\,\mathrm{cm^2g^{-1}}$ (see Figure 5.1). This means that the energy transport by radiation would be very inefficient and would require a high value of $|dT/dr|_{\rm rad}$ (Section 7.1), so the star is almost completely convective according to the Schwarzschild criterion.

At the end of the free-fall phase, when the protostar has become ionized, it will be in hydrostatic equilibrium and fully convective with $R/R_\odot \sim 100\, M/M_\odot$.

12.7 The Contraction of a Convective Protostar: The Descent along the Hayashi Track

When the dissociation and ionization of the star is complete, the star is in hydrostatic equilibrium. The star does not have nuclear fusion yet, but it has a temperature gradient so it radiates. The star must therefore contract to cover this energy loss.

We have argued that at this phase the star is fully convective. We will show in Section 14.3 that fully convective stars occupy a nearly vertical strip in the HRD at $T_{\rm eff} \approx 3000$ K. This is called the **Hayashi line** after the Japanese astronomer Chusiro Hayashi (1920–2010), who derived it in 1961 for red giants (Hayashi 1961) and in 1966 for gravitationally contracting stars (Hayashi 1966). Fully convective stars will evolve almost vertically upward along the Hayashi line if they expand (because $L \sim T_{\rm eff}^4 R^2$ with $T_{\rm eff} \approx$ constant), and downward if they contract. Protostars are contracting, so their luminosity decreases. This part of their evolutionary path in the HRD is called the **Hayashi track.** The tracks are shown in Figure 12.4. Note that the lines are approximately vertical at $T_{\rm eff} \approx 3000$ K, but deviate to lower $T_{\rm eff}$ at high luminosity. At the lowest points of the tracks the protostars go from convective equilibrium into radiative equilibrium, as we will see later.

During the descent along the Hayashi track the temperature and density in the star rises. When it reaches a mean T of a few times 10^6 K the opacity drops drastically (see the curve for log $\rho \approx 0$ in Figure 5.1). This implies that the Schwarzschild criterion for convection is no longer satisfied, so the energy can now be transported by radiation. Because the protostar is no longer convective, the downward evolution along the Hayashi track stops.

We can estimate the radius of the protostar at the end of the Hayashi track by assuming that the mean gas temperature should be of the order of a few times 10^6 K, say 3×10^6 K, to be in the range of $\kappa \approx \sigma_e \approx 0.3\,\mathrm{cm^2g^{-1}}$ (see Figure 5.1). During the convective contraction, when $\overline{T} \sim M/R$ (Equation (12.14)), the mean temperature increased from about 7×10^4 K to 3×10^6 K, which implies an increase by a factor of

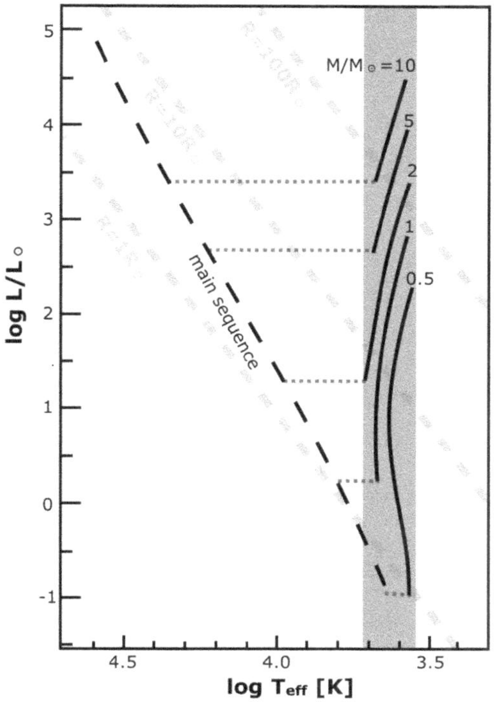

Figure 12.4 Hayashi tracks (full lines) for protostars of $0.5 < M/M_\odot < 10$ for $X = 0.70$, $Y = 0.28$, and $Z = 0.02$. The gray band indicates the region of the Hayashi line for fully convective stars. The light gray dashed lines are lines of constant radius. (Reproduced from Kippenhahn & Weigert 1990, with permission. © ESO.)

$\sim$50, so the radius of a $1M_\odot$ protostar has gone down from $\sim 100R_\odot$ to $\sim 2.0R_\odot$. At this radius, the luminosity of a convective star with $T_{\rm eff} \approx 3000$ K is $L = 4\pi R^2 \sigma T_{\rm eff}^4 \approx 0.5L_\odot$ for a $1M_\odot$ protostar. A more massive protostar starts and ends the contraction with a larger radius and higher luminosity than a protostar of $1M_\odot$.

We can use these values to estimate the duration of the Hayashi phase of a 1 $M_\odot$ protostar. Because the protostar is in HE, the contraction occurs on the Kelvin–Helmholtz timescale

$$\tau_{\rm Hayashi} \approx \tau_{\rm KH} \approx \frac{-(0.5)\Delta E_{\rm pot}}{L} \approx \frac{0.5AGM^2}{\bar{L}R_{\rm end}}, \tag{12.15}$$

with $A \approx 2$. Because the radius and luminosity decrease drastically during the descent along the Hayashi track, we adopt the logarithmic mean radius of $\bar{R} \approx (R_{\rm top}R_{\rm bottom})^{1/2} \approx 15R_\odot$ and the resulting mean luminosity $\bar{L} = 4\pi\sigma T_{\rm eff}{}^4\,\bar{R}^2 \approx 15L_\odot$ as characteristic values in Equation (12.15). This results in an estimate of $\tau_{\rm Hayashi} \approx 1 \times 10^6$ yr for a $1M_\odot$ protostar.

Equation (12.15) shows that $\tau_{\rm Hayashi} \sim M^2/LR$. This reduces to $\tau_{\rm Hayashi} \sim M^2/R^3$ because $T_{\rm eff} \approx$ constant on the Hayashi track, with $R \sim M$ at the end of the free-fall phase and the beginning of the descent along the Hayashi track, so $\tau_{\rm Hayashi} \sim 1/M$.

The descent along the Hayashi track of a protostar of $M_\odot$ takes about 1 million years. The more massive the protostar, the shorter the timescale: $\tau_{\rm Hayashi} \sim 1/M$.

12.8 The Contraction of a Radiative Pre-main-sequence Star: From the Hayashi Track to the Main Sequence

At the end of the Hayashi phase of a star, when $\bar{T} \sim 10^6$ K, the convection gradually stops and the protostar goes into radiative equilibrium. The protostar in radiative equilibrium has not yet started nuclear fusion, so it will keep contracting to cover the loss of energy by radiation. Because it is in radiative and hydrostatic equilibrium, it will roughly obey the mass–luminosity relation (Section 6.2). This means that its evolution track will now be approximately horizontal as the star moves toward the main sequence. This phase is called **the pre-main-sequence phase,** abbreviated as PMS phase.

Q (12.2) *Why does the convection stop when $\bar{T} \sim 10^6$ K?*

We can estimate the duration of this PMS contraction phase in the same way as before, by comparing the released potential energy with the luminosity using the virial theorem. During the PMS contraction, the star keeps approximately the same luminosity as it had at the end of the Hayashi contraction phase. At the end of the PMS phase, when H-fusion starts on the main sequence, the radius is approximately the main-sequence radius. We will show later that the radius on the MS is

$$R_{\rm MS}/R_\odot \approx (M/M_\odot)^{0.7}. \tag{12.16}$$

So

$$\tau_{\rm PMS} = \frac{-(0.5)\Delta E_{\rm pot}}{L} \approx 0.5A\frac{GM_\odot^2}{R_\odot L_\odot}\left(\frac{M}{M_\odot}\right)^2\left(\frac{R}{R_\odot}\right)^{-1}\left(\frac{L}{L_\odot}\right)^{-1} \tag{12.17}$$
$$\approx 6\times 10^7\,(M/M_\odot)^{-2.5}\,{\rm yr},$$

where we used $A \approx 2$, $R \sim M^{0.7}$, and $L = 0.5L_\odot$ for a protostar of $1M_\odot$ and $L \sim (M/M_\odot)^{3.8}$ as observed for MS stars (Section 2.5). The PMS phase of a star of 1 $M_\odot$ is about 6×10^7 yr, whereas the Hayashi phase lasts 1×10^6 yr. We see that the PMS phase of stars lasts considerably longer than the Hayashi phase.

The duration of the pre-main-sequence phase for stars of solar metallicity is listed in Table 12.2.

Q (12.3) *Why is the phase of the Hayashi track much shorter than the PMS phase?*

Figure 12.5 (left) shows the predicted evolution tracks of PMS stars in the L–$T_{\rm eff}$ diagram. Notice that the tracks for radiative contraction are not exactly horizontal in the HRD; instead, L increases by about a factor of 3. This is partly due to the decrease of κ as the star contracts and gets hotter (remember that $L \sim M^3/\kappa$) and

Table 12.2 The PMS Contraction Time (in yr) for Solar Metallicity Stars

$M_\odot$	τ_{PMS} (estimated)	τ_{PMS} (model)
1	6×10^7	2.9×10^7
3	4×10^6	7.2×10^6
9	3×10^5	2.9×10^5
25	2×10^4	7.1×10^4
60	2×10^3	2.8×10^4

Note. The last column shows the results of detailed models (Bernasconi & Maeder 1996).

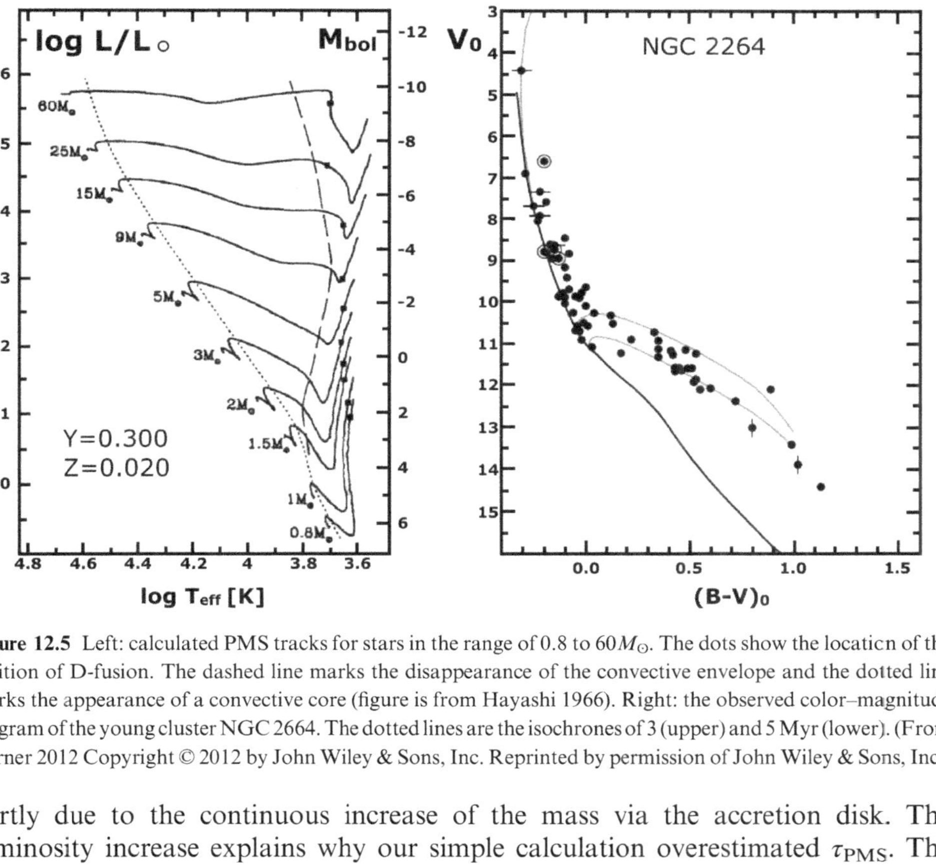

Figure 12.5 Left: calculated PMS tracks for stars in the range of 0.8 to $60 M_\odot$. The dots show the location of the ignition of D-fusion. The dashed line marks the disappearance of the convective envelope and the dotted line marks the appearance of a convective core (figure is from Hayashi 1966). Right: the observed color–magnitude diagram of the young cluster NGC 2664. The dotted lines are the isochrones of 3 (upper) and 5 Myr (lower). (From Turner 2012 Copyright © 2012 by John Wiley & Sons, Inc. Reprinted by permission of John Wiley & Sons, Inc.)

partly due to the continuous increase of the mass via the accretion disk. The luminosity increase explains why our simple calculation overestimated τ_{PMS}. The dashed line marks the disappearance of the convective envelope shortly after the star left the Hayashi track. The dotted line marks the appearance of a convective core just before the star reaches the MS. When a star reaches the MS and starts H-fusion it adjusts its radius, luminosity, and T_{eff} slightly to the new structure. This produces the little curl at the end of the PMS track.

Because the timescales for evolution on the Hayashi track and the PMS track both decrease with increasing mass, stars with the same age but with different masses will lie along an isochrone that crosses the tracks from upper left to lower right. The right panel of Figure 12.5 shows the observed color–magnitude diagram of the young cluster NGC 2664 with an age of about 4 Myr. The stars are located between the predicted isochrones of 3 and 5 Myr, shown as upper and lower dotted lines, respectively.

12.9 T Tauri Stars and Herbig Ae-Be Stars

Low-mass stars ($M \leq 2M_\odot$) on the radiative contraction track are called **T Tauri stars.** They are pre-main-sequence stars of types F, G, and K with strong emission lines formed in an accretion disk around these stars.

Higher-mass stars ($M \geqslant 2M_\odot$) on the radiative contraction track are called **Herbig Ae-Be stars.** They are pre-main-sequence stars of types B and A with strong emission lines formed in an accretion disk around these stars

T Tauri stars are divided into four classes that can be distinguished observationally by their spectral energy distribution. These classes correspond to sequential evolutionary phases. They are shown in Figure 12.6.

- **Class 0** corresponds to the youngest phase of $\sim 10^4$ yr when the star is still accreting almost spherically. The spectral energy distribution is that of a thick cold sphere that radiates mainly in the infrared.
- **Class I** objects, with an age of $\sim 10^5$ yr, are embedded in a thick accreting disk (with jets) that produces a strong infrared excess.
- **Class II** objects, with an age of $\sim 10^6$ yr, have a less optically thick accretion disk and the radiation from the central star becomes visible. Class II stars are on the radiative PMS track.
- **Class III** objects, showing weak emission lines and a small infrared excess, have an optically thin accretion disk and are close to the main sequence.

12.10 The Destruction of Lithium and Deuterium

At a temperature of about 10^6 K, the small fraction of initial deuterium, formed in the Big Bang, is destroyed by the reaction

$$^2\mathrm{H} + {}^1\mathrm{H} \rightarrow {}^3\mathrm{He} + \gamma + 5.5\ \mathrm{MeV}. \tag{12.18}$$

This occurs when the star is approximately on the 10^5 yr isochrones in the HRD. Sometime later, at a core temperature of about 2.5×10^6 K (i.e., when the star is near the 10^6 yr isochrones), Li is destroyed by the reaction chain

$$^6\mathrm{Li}\,(p, \gamma)\,{}^7\mathrm{Be}\,(e^-, \nu)\,{}^7\mathrm{Li}\,(p, \gamma)\,{}^8\mathrm{Be} \rightarrow 2\,{}^4\mathrm{He} + \text{energy}. \tag{12.19}$$

The stellar surface abundance of Li and D provides an important diagnostic tool for studying the formation of low-mass stars. This is because Li is only depleted at the stellar surface if the convective envelope was deep enough to include the Li-destruction zone at 2.5×10^6 K. Models show that this is only the case in PMS stars of $M < 1.4M_\odot$. At higher mass, the Li-destruction occurs when the star is no longer on the Hayashi track, so convection does not affect the surface composition of Li. This is shown in the calculated pre-main-sequence evolution tracks of Figure 12.7.

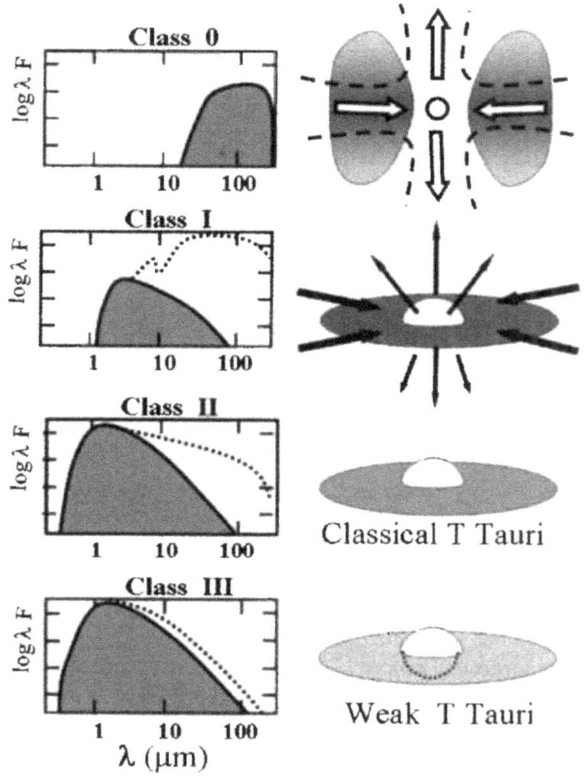

Figure 12.6 Four classes of T Tauri stars. Right: a sketch of the geometrical model. Left: the associated energy distributions. The gray area indicates the stellar flux, which is partly obscured in class 0 and class I objects, and the dotted line indicates the contribution by the disk. (Reproduced from Bouvier & Zahn 2002, with permission. © EAS, EDP Sciences, 2002.)

12.11 Stars That Do Not Reach H-fusion: Brown Dwarfs with $M < 0.08\ M_\odot$

Hydrogen fusion requires a minimum T_c of about 6×10^6 K. We have shown in Section 8.10 that this requires some minimum mass. This minimum mass is $0.08M_\odot$. If the mass is lower, the core becomes degenerate and the contraction stops before the required T_c is reached. Stars with $M < 0.08M_\odot$ therefore do not reach H-fusion. They are called **brown dwarfs.** These stars get their energy from Li burning (between 0.06–$0.08M_\odot$) and D-fusion (from $0.06M_\odot$ down to $\sim 0.01M_\odot$).

12.12 The Stellar Initial Mass Function

The **initial mass function (IMF)** describes the number distribution of stars of different masses between about 0.1 and $100M_\odot$. The IMF has been derived from observations of star clusters of different ages and from field stars. If clusters are used, the drifting of low-mass stars out of the clusters has to be taken into account. If field stars

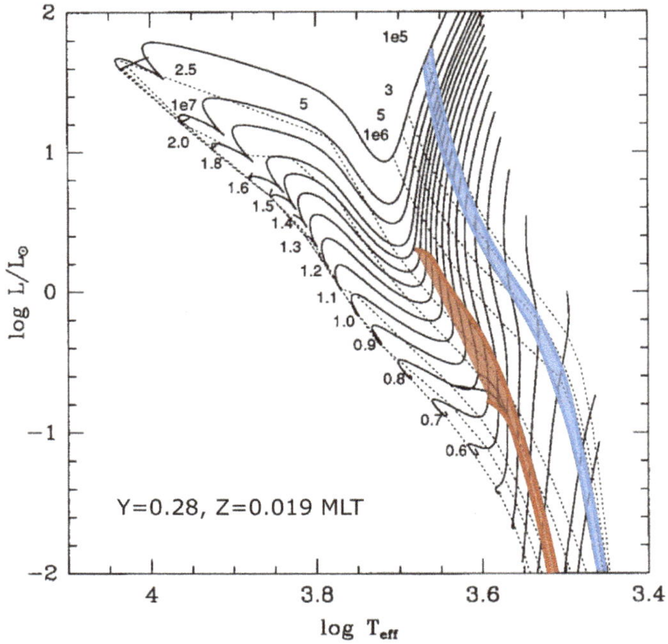

Figure 12.7 Pre-main-sequence evolution tracks of stars of $0.3 < M < 2.5M_{\odot}$. The dotted lines are isochrones with the ages indicated; from 10^5 yr (right isochrones) to 10^7 yr (left isochrones). The blue band is the location of the stars, where D is destroyed and the red band is the location where Li is destroyed. (Reproduced from D'Antona & Mazzitelli 1994.)

are used, the number of stars in any magnitude limited sample must first be corrected for four effects:

- the distance to the stars that are observed, which depends on their absolute magnitude and the magnitude limit of the sample,
- the conversion of absolute magnitude to stellar mass,
- the evolutionary effect for late mass-losing stages of evolution (to convert the present mass into the initial mass), and
- the lifetime effect, to correct the number of stars in a certain evolutionary phase for the duration of that phase.

In this way, the observed distribution of the stars as a function of apparent magnitude, spectral type, and evolutionary phase has been converted into an initial mass function.

The Austrian–American astronomer Ed Salpeter (1924–2008) derived the first stellar IMF in 1955 (Salpeter 1955) and found that

$$N(m)dm \sim m^{-2.35}. \tag{12.20}$$

This is called the **Salpeter IMF.** Later studies, based on better data, showed that the IMF cannot be represented by one single power law but has a more complex shape. The best known modern IMFs are those of Kroupa (2001) and Chabrier (2003).

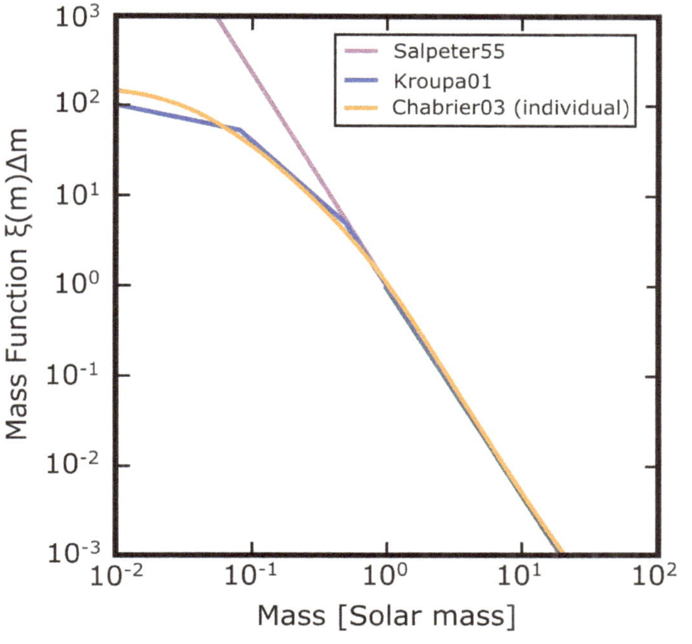

Figure 12.8 Initial mass functions of Salpeter (purple; Salpeter 1955), Kroupa (blue; Kroupa 2001), and Chabrier (orange; Chabrier 2003), normalized to $m = 1M_{\odot}$.

Kroupa IMF

$$\begin{aligned} N(m)dm &= C_1\, m^{-2.3} && \text{for} \quad m > 0.5M_{\odot}, \\ N(m)dm &= C_2\, m^{-1.3} && \text{for } 0.08 < m < 0.5M_{\odot}, \\ N(m)dm &= C_3\, m^{-0.3} && \text{for } m < 0.08M_{\odot}, \end{aligned} \tag{12.21}$$

with the ratios between C_1, C_2, and C_3 chosen in such a way that the three power laws fit at the connecting points of 0.08 and $0.5M_{\odot}$, so $C_2 = 2C_1$ and $C_3 = 25C_1$.

Chabrier IMF

$$\begin{aligned} N(m)dm &= D_1 \frac{1}{m} \exp\left[-\left\{\log\left(\frac{m}{0.08}\right)\right\}^2 / 0.952\right] && \text{for } m < 1M_{\odot} \\ N(m)dm &= D_2\, m^{-2.3} && \text{for } m > 1M_{\odot}, \end{aligned} \tag{12.22}$$

with the constants D_1 and D_2 adjusted to match the two mass ranges, so $D_2 = 3.539\ D_1$.

These three IMFs are plotted in Figure 12.8. It shows that all three IMFs agree at $m > 1M_{\odot}$, and that the Chabrier IMF and Kroupa IMF are very similar at $m < 1M_{\odot}$, but that the Salpeter IMF predicts many more low-mass stars than the more recent ones.

Table 12.3. The Relative Contributions to the Total Mass, to the Total Number of Stars, and the Mean Mass per Star in Three Mass Bins for a Kroupa IMF

Mass Range	0.01–0.1$M_\odot$	0.1–1.0$M_\odot$	1–100$M_\odot$	0.01–100$M_\odot$
N/N_{tot}	0.446	0.493	0.061	1.000
M/M_{tot}	0.059	0.414	0.527	1.000
$\overline{m}/M_\odot$	0.051	0.315	3.253	0.376

Notes. The table shows the following. (1) Stars with $M > 1M_\odot$ contribute more than 50% to the total mass, but only 6% to the total number of stars. (2) The lowest-mass stars with $M < 0.1M_\odot$ contribute only 6% to the total mass, but as much as 44% to the total number of stars. However, these stars are difficult to detect due to their very low luminosity. (3) The mean stellar mass in the mass range of 0.01 to 100$M_\odot$ is only 0.376$M_\odot$.

The shape of the IMF implies that the relative contributions to the number of stars and to their total mass is very different for different mass intervals. This is shown in Table 12.3.

12.13 Star Formation in the Early Universe

We have seen above that cooling plays a crucial role in star formation.

- The equilibrium between heating and cooling determines the temperatures of star-forming clouds.
- The temperature and density of these clouds determine the Jeans mass, which is the minimum mass of a cloud that can collapse and fragment into clumps.
- The minimum mass of stars that can be formed out of clumps depends on the ratio between the cooling time and the free-fall time: a clump must be able to radiate the potential energy gained during the collapse (Section 12.4).

The main cooling process in the Galactic disk is by molecules and by dust. In the very early universe, the metallicity, Z, was practically zero, because the interstellar medium was not yet enriched by the nuclear products from previous generations of stars. This implies that there was no dust and there were no molecules except H_2. So H_2 was the only molecule available for cooling.

Figure 12.9 shows the effect of cooling on star formation in the early universe. The left panel shows the equilibrium temperature of clouds at $Z = 0$ as a function of density n in a cold dark-matter-dominated universe at redshift $z = 31$ (Bromm 2002; Bromm & Larson 2004). The figure shows two distinct branches. At densities $n < 1$ cm^{-3}, the H_2 fraction is very small, viz. $< 10^{-5}$, and the cooling of contracting clouds is very inefficient. Therefore, T rises almost adiabatically as the cloud contracts so $T \sim n^{2/3}$ (Equation (4.33)) until it reaches a virial temperature of about 5000 K. At $n > 1$ cm^{-3}, the molecular fraction increases and cooling by H_2 becomes important. This results in an almost isobaric decrease of T at increasing n, i.e., $T \sim n^{-1}$. When the gas is cooler than about 500 K, collisions are no longer efficient to excite the H_2 molecules to their lowest excited rotational energy level of $E/k \approx 500$ K, so the cooling by radiative de-excitation stops. However, particles in the tail of

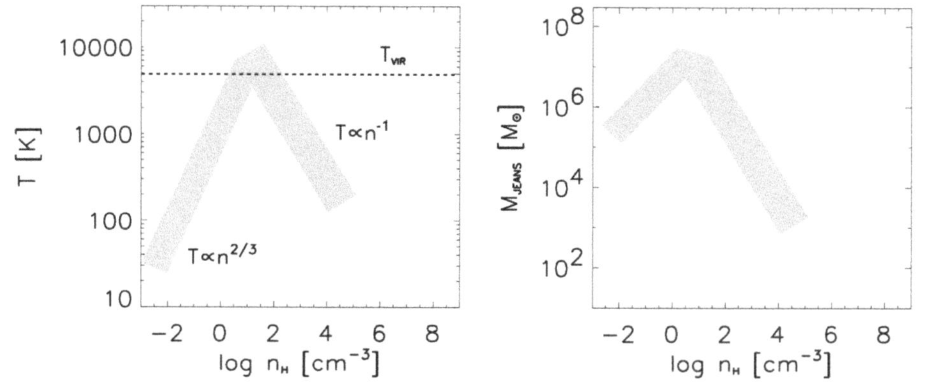

Figure 12.9 Conditions of zero-metallicity clouds in the early universe at redshift $z \sim 30$. Left: the equilibrium temperature as a function of density. Right: the resulting Jeans mass. (Figure is based on Bromm 2002 and Bromm & Larson 2004.)

the Maxwell distribution can still excite H_2 to its lowest excited level and cool the gas down to about 300 K.

For every value of T and n, the Jeans mass can be calculated (Equations (12.5) and (12.6)). The result is shown in the right panel of Figure 12.6. The Jeans mass of a quasi-hydrostatic clump of $T \sim 300$ to 500 K and $n \sim 10^4$ cm^{-3} and $\mu = 2$ is about 200 to $1000 M_\odot$, so star formation in the early universe, when $Z < 10^{-6}$, must have resulted in considerably more massive stars than present star formation in the Galactic disk. These predicted very massive stars with very low metallicity are called **population III stars.** The metal poor stars with $Z \sim 10^{-4}$ to 10^{-5} in old globular clusters are called **population II stars.** The metal-rich stars with $Z \sim 10^{-2}$ in the Galactic disk are called **population I stars.**

12.14 Summary

1. We summarize the important phases in the process of star formation, with the characteristic timescales and characteristic stellar parameters for a star of $1 M_\odot$.

Collapse of cloud = free fall phase $\tau_{ff} \sim 10^5$ yr:
Cooling by molecules and dust IR radiation.
$\bar{T}$ is low (about 10 K) and is about constant
Fragmentation of cloud into clumps.
Start: when cloud mass exceeds Jeans mass.
End: when H_2 is dissociated and H is ionized.
Characteristic parameters: $\bar{T} \sim 10^5$ K, $R/R_\odot \sim 120\ M/M_\odot$.

Protostar = pseudo-hydrostatic contraction $\tau_{KH} \sim 1 \times 10^6$ yr:
Fully convective because of low $\bar{T}$ and high κ.
$T_{eff} \approx 3000$ K $\approx$ constant.

Hayashi track: approximately vertical track downward in HRD.
Start: when the collapsing cloud is dissociated.
Characteristic parameters: $R \approx 120R_\odot$, $T_{\rm eff} \approx 3000$ K, $L \approx 10^3 L_\odot$.
End: when $\bar{T} \sim 10^6$ K $\rightarrow$ low $\kappa \rightarrow$ radiative equilibrium.
Characteristic Parameters: $R \approx 2.5R_\odot$, $T_{\rm eff} \approx 3000$ K, $L \approx 0.5L_\odot$.

Pre-main sequence = pseudo-hydrostatic contraction $\tau_{\rm KH} \sim 7 \times 10^6$ yr:
Radiative equilibrium because of high $\bar{T} \rightarrow$ low κ.
L about constant.
Approximately horizontal track leftward in HRD.
Short phases of D-fusion and Li-fusion.
Start: when the protostar reaches radiative equilibrium ($\bar{T} \sim 10^6$ K).
Characteristic parameters: $R \approx 2.5R_\odot$, $T_{\rm eff} \approx 3000$ K, $L \approx 0.5\ L_\odot$.
End: when H-fusion starts = on main sequence.
Characteristic parameters: $R \approx 1R_\odot$, $T_{\rm eff} \approx 6000$ K, $L \approx 1L_\odot$.

2. Fragmentation of collapsing clouds in the Galactic disk results in the formation of stars in the range of about 0.01 to $100M_\odot$. The lower boundary is set by the limit where the potential energy gained by contraction cannot be lost by radiation because the cloud is optically thick for IR radiation. The upper boundary is set by the Eddington limit for radiation pressure (Section 6.3).
3. Stars with $M < 0.08M_\odot$ do not reach H-fusion and remain brown dwarfs.
4. The initial mass function shows that more than half of the stars have $M < 0.1M_\odot$ and that the mean mass is about 0.4 $M_\odot$. Stars more massive than 1 $M_\odot$ contribute half of the total mass but only 6% of the total number of stars.
5. In the early universe, when cooling was inefficient due to the lack of dust and molecules (except H_2), only massive stars with $M \sim 10^2$ to $10^3 M_\odot$ could be formed.

Exercises.

12.1 Assume that the different components of the ISM are in pressure equilibrium.
   (a) What would be the Jeans mass of clouds forming out of each of the following?
      – cold neutral gas;
      – warm neutral gas;
      – warm ionized gas.
   (b) How is it possible that molecular clouds are not in pressure equilibrium with their surroundings?
   (c) Giant molecular clouds have masses of the order of $10^5 M_\odot$. Show that they are forming stars.

12.2
   (a) Estimate the radii at the beginning and end of the Hayashi contraction phase and at the beginning and end of the pre-main-sequence contraction for stars of 0.1, 0.3, 1.0, 3, 10, 30, and 100 $M_\odot$.

(b) Estimate the duration of the Hayashi contraction phase and of the pre-main-sequence contraction for stars of 0.1, 0.3, 1.0, 3, 10, 30, and $100 M_{\odot}$.

12.3 Explain the difference in the Li surface abundance of stars on the main sequence that had Li destruction during the Hayashi contraction phase or during the pre-main-sequence contraction phase.

References

Bernasconi, P. A. & Maeder, A. 1996, A&A, 307, 829

Bouvier, J., & Zahn, J.P. (ed) 2002, EAS Pub. Ser. 3, Star Formation and the Physics of Young Stars: Summer School on Stellar Physics 10, (Les Ulis: EDP Sciences)

Bromm, V., Coppi, P. S., & Larson, R. B. 2002, ApJ, **564, 23**

Bromm, V., & Larson, R. B. 2004, ARA&A, **42, 79**

Chabrier, G. 2003, ApJL, **586, L133**

Cox, A. N. 1990, in Allen's Astrophysical Quantities (4th ed.; Berlin: Springer), 523

D'Antona, F. & Mazzitelli, I. 1994, ApJS, **90, 467**

Elmegreen, B. G. 2002, ApJ, **564, 773**

Hayashi, C. 1961, PASJ, **13, 442**

Hayashi, C. 1966, ARA&A, **4, 171**

Kippenhahn, R., & Weigert, A. 1990, Stellar Structure and Evolution (Berlin: Springer)

Kroupa, P. 2001, MNRAS, **322, 231**

Kruijssen, D. 2011, MNRAS, **414, 1339**

Maeder, A. 2009, Physics, Formation and Evolution of Rotating Stars (Berlin: Springer)

McKee, C. F., & Ostriker, E. C. 2007, ARA&A, **45, 565**

McLaughlin, D.E. & Pudritz, R. E. 1997, ApJ, **476, 750**

Salpeter, E. E. 1955, ApJ, **121, 161**

Turner, D. G. 2012, AN, **333, 174**

Henny J.G.L.M. Lamers and Emily M. Levesque

Chapter 13

H-fusion in the Core: The Main Sequence Phase

When a star starts fusing H in its center, it is located on the zero-age main sequence in the HRD. The luminosity, temperature, and radius depend on the mass in a specific way, which is described by the properties of stars that are chemically homogeneous. We will show that these properties can be derived by means of homology relations for stars that have similar density structures. Homology relations also illustrate how the properties of stars depend on their initial chemical compositions. This is important for understanding the differences between old metal-poor stars and younger metal-rich stars. We discuss the evolution of the stars in the HRD during the core H-fusion phase, from the beginning at the **zero-age main sequence (ZAMS)** to the **terminal age of the main sequence (TAMS).** We will show that the stars increase in luminosity and radius.

13.1 The Zero-age Main Sequence (ZAMS): Homology Relations

When a PMS star arrives on the main sequence at the start of the core H-fusion, it is still chemically homogeneous, apart from minor changes due to D-fusion. The location of homogeneous stars in the HRD is called the **zero-age main sequence, ZAMS.** The properties of stars on the ZAMS can be understood on the basis of homology considerations because they all have the same chemical composition throughout the star. Homologous stars are those that have the same scaled density structure. So two stars, 1 and 2, are homologous if

$$\left[\frac{\rho}{\rho_c}\left(\frac{r}{R}\right)\right]_1 = \left[\frac{\rho}{\rho_c}\left(\frac{r}{R}\right)\right]_2. \tag{13.1}$$

The values of ρ_c and R can be different for both stars but the structure of ρ/ρ_c as function r/R is the same. Because of hydrostatic equilibrium, the same homology then also applies for P/P_c. If the equation of state is that of an ideal gas (i.e., radiation pressure can be ignored), the T/T_c - structure is also homologous, so

doi:10.1088/978-0-7503-1278-3ch13

$$\left[\frac{P}{P_c}\left(\frac{r}{R}\right)\right]_1 = \left[\frac{P}{P_c}\left(\frac{r}{R}\right)\right]_2 \quad \text{and} \quad \left[\frac{T}{T_c}\left(\frac{r}{R}\right)\right]_1 = \left[\frac{T}{T_c}\left(\frac{r}{R}\right)\right]_2. \tag{13.2}$$

Homologous stars differ only by their scaling factors ρ_c, T_c, P_c and M, R, L. Homology enables the predictions of trends in stellar properties. If we know R, L, and $T_{\rm eff}$ for a star of mass M (for instance, from a stellar model or from an observed double-lined spectroscopic binary), we can predict R, L, and $T_{\rm eff}$ of other stars with a similar structure but different M.

We already encountered homologous stars when we discussed polytropic stars in Chapter 11; however, that was for a specific EoS. Here, we discuss more general homology relations.

13.1.1 The Mass–Luminosity Relation for ZAMS Stars

We already derived, in Section 6.2, the scaling relations for the central temperature and luminosity of stars in radiative equilibrium

$$T_c \sim \frac{\mu M}{R} \quad \text{and} \quad L \sim \frac{\mu^4 M^3}{\kappa}. \tag{13.3}$$

We now consider the mass luminosity relation more closely for ZAMS stars.

For **massive stars, $M > 2M_\odot$**, electron scattering with $\sigma_e \sim (1 + X)$ is the dominant opacity, so

$$\boxed{L \sim \mu^4 M^3/(1+X)} \rightarrow L \sim M^3, \tag{13.4}$$

where the last equation is for ZAMS stars that have the same composition. The electron density is almost independent of the metallicity, Z, but strongly dependent on the He abundance. Ignoring the small fraction of Z, we can approximate $(1 + X) = (2-Y)$ and $\mu = (2 - 5Y/4)^{-1}$, so

$$L \sim M^3(2-Y)^{-1}(2 - 1.25\,Y)^{-4} \tag{13.5}$$

for massive ZAMS stars. This is important for understanding the brightening of massive stars during the H-fusion phase, when the He content gradually increases. An increase in He abundance from $Y = 0.25$ to 0.35 results in an increase in L by a factor of 1.5 and an increase from $Y = 0.25$ to $Y = 0.50$ increases L by a factor of 2.8. A homogeneous He star, $Y = 1$, is brighter by about a factor of 50 compared to a star with $Y = 0.25$.

For **lower-mass stars, $0.5 < M < 2M_\odot$**, the main opacity source is free–free and bound–free absorption, which is described by Kramers' law: $\kappa \sim (1 + X)Z\rho T^{-7/2}$ (Equation (5.8)).

Substitution of $\rho \sim M/R^3$ and $T \sim \mu M/R$ transforms Equation (13.4) into

$$\boxed{L \sim \mu^{7.5} M^{5.5}/ZR^{0.5}.} \tag{13.6}$$

We will see below that $R \sim M^x$ with $x \simeq 0.8$, so we predict for low-mass ZAMS stars

$$L \sim \mu^{7.5}M^{5.1}/Z \quad \rightarrow \quad L \sim M^{5.1}, \tag{13.7}$$

where the last equation is for ZAMS stars that have the same composition.

Figure 13.1 shows the M-L relation from detailed models for a composition of $X = 0.72$, $Y = 0.266$, and $Z = 0.014$. The figure shows that $L \sim M^3$ is a good approximation in the mass range of $2 < M < 30M_\odot$ and $L \sim M^{5.1}$ is a good approximation for $0.7 < M < 2M_\odot$. At $M > 30M_\odot$ the M-L relation becomes flatter because of the increasing contribution of radiation pressure to the EoS and the increasing mass fraction of the convective core. At $M < 0.7M_\odot$ the M-L relation becomes flatter because of an increasing mass fraction of the convective envelope (see Figure 7.6).

13.1.2 The Mass–Radius Relation for ZAMS Stars

The mass–radius (M-R) relation can be derived from the M-L relation if we can derive an expression for L. Since all ZAMS stars have H-fusion, we can derive a homology relation based on the nuclear energy production.

The energy production per gram per second is $\epsilon \sim \rho T^\nu$, with $\nu \simeq 4$ for the pp-chain and $\nu \simeq 18$ for the CNO cycle (Section 8.4). The nuclear luminosity is given by

$$L = \int_0^R \epsilon\rho 4\pi r^2 \, dr \sim \epsilon M \text{ with } \epsilon \sim \rho T^\nu. \tag{13.8}$$

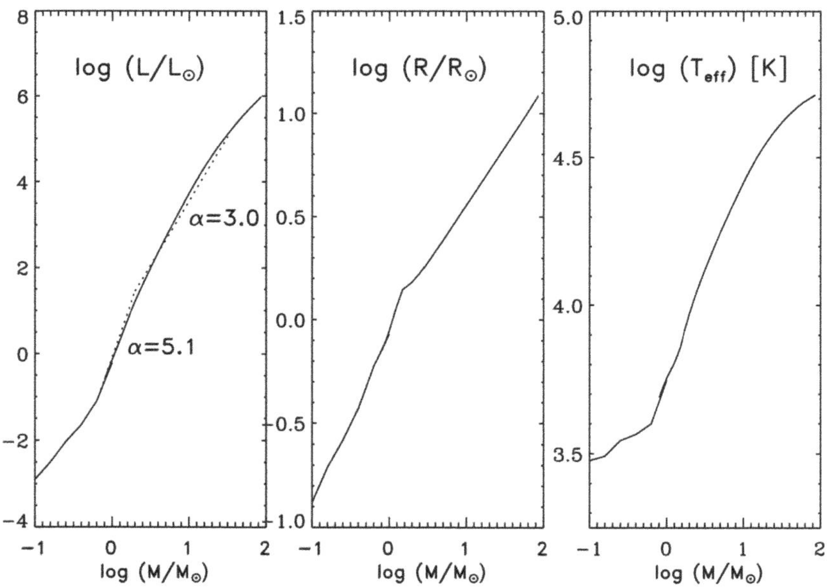

Figure 13.1. Predicted M-L, M-R, and M-$T_{\rm eff}$ relations for ZAMS stars of $X = 0.72$ and $Z = 0.014$. The slopes of the M-L relation in the range of $0.7 < M < 2M_\odot$ and $2 < M < 30M_\odot$, derived from homology relations, is shown by dotted lines. (Based on data from Ekström et al. 2012 and Pols 2011.)

Using Equation (13.3) for the scaling of T, we find

$$L \sim M\epsilon \sim M\rho T^{\nu} \sim M(M/R^3)\mu^{\nu}M^{\nu}/R^{\nu} \sim \mu^{\nu}M^{2+\nu}/R^{3+\nu}. \quad (13.9)$$

Comparing the above expression for L with $L \sim \mu^4 M^3/\kappa$ (Equation (13.3)) for stars in radiative equilibrium, we get

$$\frac{\mu^{\nu}M^{2+\nu}}{R^{3+\nu}} \sim \frac{\mu^4 M^3}{\kappa}. \quad (13.10)$$

This yields an expression for R

$$R \sim \mu^{\frac{\nu-4}{3+\nu}}\kappa^{\frac{1}{3+\nu}}M^{\frac{\nu-1}{3+\nu}}. \quad (13.11)$$

For **massive ZAMS stars**, where $\kappa \sim 1/(1+X)$ and the H-fusion happens via the CNO cycle, $\nu \simeq 18$ and $\kappa = \sigma_e$, so we find

$$\boxed{R \sim M^{0.81}\mu^{2/3}(1+X)^{0.05}} \rightarrow R \sim M^{0.81}, \quad (13.12)$$

where the last equation is for ZAMS stars with the same composition.

For **lower-mass ZAMS stars,** where $\kappa \sim 1/(1+X)$ and the H-fusion happens via the pp-chain, $\nu \simeq 4$ and $\kappa = \sigma_e$, so we find

$$\boxed{R \sim M^{0.43}\mu^{0}(1+X)^{0.13}} \rightarrow R \sim M^{0.43}, \quad (13.13)$$

where the last equation is for ZAMS stars with the same composition.

We see that the M-R relation of ZAMS stars can be approximated by $R \sim M^x$, with the predicted value of x decreasing from 0.8 to 0.4 toward lower-mass stars. The middle panel of Figure 13.1 shows the M-R relation derived from detailed models. Contrary to our simple prediction, the exponent x is larger at lower mass than at higher mass. This is due to the increasing role of the convective envelope toward lower-mass stars (see Figure 7.6) and the transition from $\kappa = \sigma_e$ to $\kappa = \kappa_{\rm bf}$ toward cooler stars.

13.1.3 The Mass–Temperature Relation for ZAMS Stars

The most interesting homology relation for ZAMS stars is the relation between M and $T_{\rm eff}$ because $T_{\rm eff}$ is one of the parameters in the HR-diagram. We can find the homology relation for $T_{\rm eff}$ by combining the predicted M-L and M-R relations using $T_{\rm eff} \sim [L/R^2]^{1/4}$. We have shown that $L \sim M^{\alpha}$ with $\alpha \geqslant 3$ and that $R \sim M^x$ with $x < 1$. This immediately tells us that $T_{\rm eff}$ increases with increasing M and L. This is the reason that, in the HRD, the ZAMS runs from upper left to lower right.

For **massive ZAMS stars** with $\kappa = \sigma_e$ and $\nu = 18$, we find

$$\boxed{T_{\rm eff} \sim M^{0.35}\mu^{0.83}(1+X)^{-0.5}} \rightarrow T_{\rm eff} \sim L^{0.12} \rightarrow L \sim T_{\rm eff}^{8.5}. \quad (13.14)$$

For **lower-mass ZAMS stars** with $\kappa = \sigma_{\rm e}$ and $\nu = 4$, we find

$$\boxed{T_{\rm eff} \sim M^{0.54}\mu^{1}(1+X)^{-0.32}} \rightarrow T_{\rm eff} \sim L^{0.18} \rightarrow L \sim T_{\rm eff}^{5.5}. \quad (13.15)$$

We see that $T_{\rm eff}$ decreases with decreasing M and L. We also see that the decrease is a stronger function of M for lower-mass stars than for higher-mass stars; this is why the ZAMS in the HR diagram is steeper for massive stars than for lower-mass stars. These relations cannot be applied for stars with $M < 1M_\odot$ because these stars are partly convective. We will show later (Section 14.3) that stars with very deep convective envelopes approach an almost constant value of $T_{\rm eff} \approx 3000$ K (see Figure 13.1). This results in a steepening of the ZAMS in the HRD at low temperatures. Figure 13.2 shows the location of the ZAMS in the HRD derived from detailed stellar models with $X = 0.72$, $Y = 0.266$, and $Z = 0.014$.

13.2 The Influence of Abundances on the ZAMS

The dependence on metallicity Z. The results above show that for massive ZAMS stars $T_{\rm eff}$ is rather insensitive to metallicity (Z), but for intermediate mass ZAMS stars a *lower Z implies a higher $T_{\rm eff}$ and a larger L.*

The dependence on helium abundance Y. The dependence of $T_{\rm eff}$ and L on X and μ implies that $T_{\rm eff}$ and L both increase for ZAMS stars with increasing He abundance. Figure 13.3 shows this effect schematically. The ZAMS of low-metallicity stars and the ZAMS of stars with higher initial He abundance are to the left and slightly

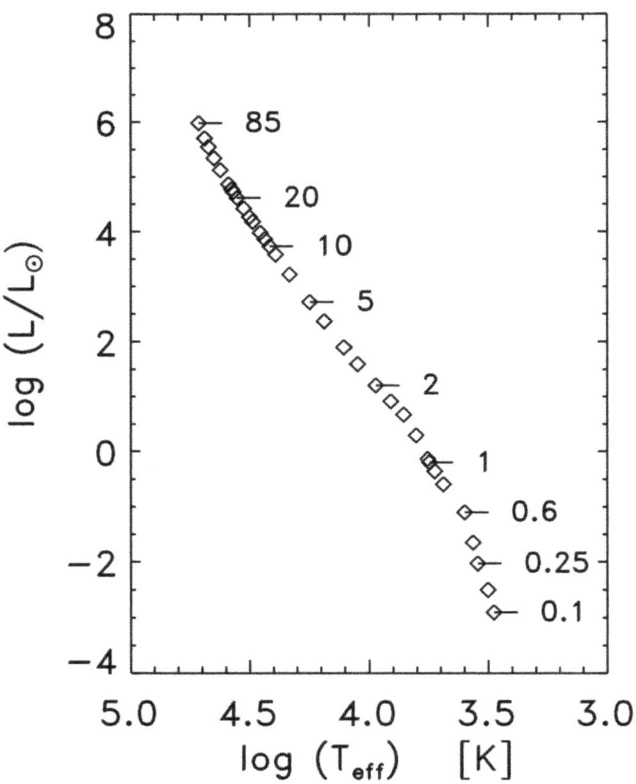

Figure 13.2. Location of the ZAMS in the HRD for stars of $X = 0.72$, $Y = 0.266$, and $Z = 0.014$. The masses of some stars are indicated. (Based on data from Ekström et al. 2012 and Pols 2011.)

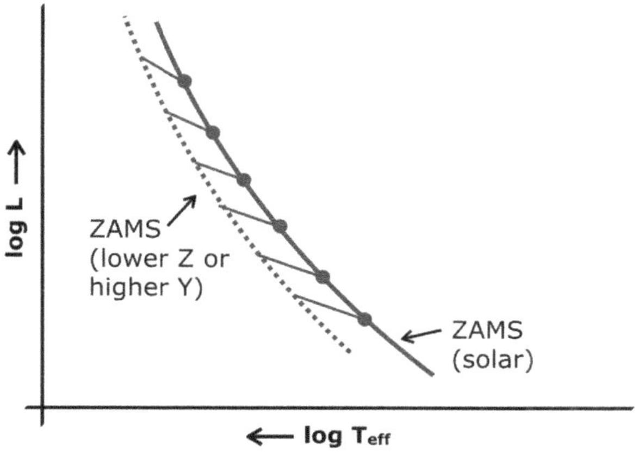

Figure 13.3. Schematic representation of the effect of the initial He and metal abundance on the location of the ZAMS in the HRD.

upward compared to the ZAMS for solar metallicity stars. This is important for understanding the observed HRDs or CMDs of Galactic globular clusters, which often show multiple main sequences.

Q (13.1) *Explain in simple physical terms why both T_{eff} and L increase with decreasing metallicity.*

Q (13.2) *Explain in simple physical terms why both T_{eff} and L increase with increasing He abundance.*

The left panel of Figure 13.4 shows the effect of a decreased metallicity on the location of the ZAMS, based on detailed models. The right panel shows the effect of an increased He abundance, based on the homology relations derived above. In all cases, a star moves to the left and upward in the HRD when the metallicity is decreased or the He abundance is increased.

13.2.1 The ZAMS for Helium Stars

Figure 13.5 shows the ZAMS of He-stars, compared with those of H-rich stars, as an extreme example of the effect of the He abundance on the location of the ZAMS. This is an interesting comparison because we will show later that Horizontal Branch stars are almost pure He-stars. Note that, when comparing stars of the same mass, the He-rich stars are much brighter (by about a factor of 30 to 100) and much hotter (by about a factor of 4 to 10) than H-rich stars of the same mass. The higher luminosity is due to the μ-effect ($L \sim \mu^4$ or $L \sim \mu^{7.5}$ in Equations (13.3) and (13.6)). The higher temperature is due to the higher L and smaller R by about a factor of 5. This radius decrease is not properly predicted by the homology relations because the internal structure of He-stars differs drastically from that of H-stars. This is due to several effects. The extreme temperature sensitivity of He-fusion ($\epsilon \sim T^{30}$) results in a larger convective core in He-stars than in H-stars. At the same time, the higher luminosity results in a higher ratio of P_{rad}/P. This also contributes to the failure of the homology comparison between He-stars and H-stars.

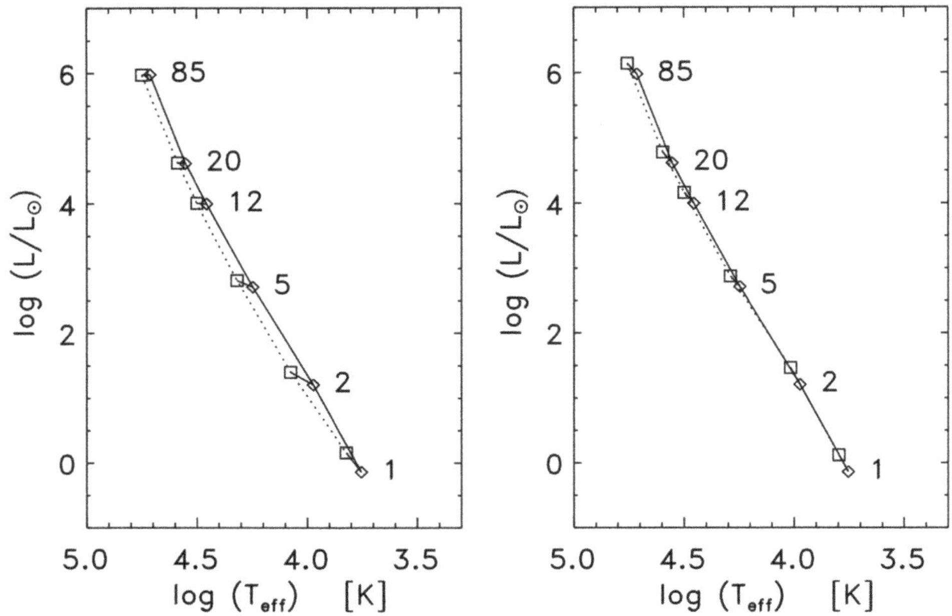

Figure 13.4. Left: the location of the ZAMS in the HRD for stars with solar metallicity, $Z = 0.014$ (full line), and $Z = 0.002$ (dotted line). Right: the effect of an increase in He abundance from $Y = 0.27$ (full line) to 0.37 (dotted line). (Figure based on data from Ekström et al. 2012.)

The results show that we can expect Horizontal Branch stars, which are low-mass ($M \approx 0.8 M_{\odot}$), almost pure He-stars at the beginning of this phase, to be much hotter and brighter than normal MS stars of the same mass (see Figure 2.7).

13.3 Evolution during the Main-sequence Phase

The main-sequence phase is the longest phase in the evolution of a star. During this phase the star changes slightly because its internal chemical composition is changing. The changes in the chemical composition of the star depend on the presence or absence of convection in the core. A star without a convective core creates a chemical gradient from the very center throughout the region where H-fusion occurs. A star with a convective core has a chemically homogeneous core that becomes more He-rich over time. This has an effect on the way the star reacts when H-fusion in the core comes to an end.

13.3.1 Nuclear Fusion as a Thermostat

During core H-fusion, the temperature in the core changes very little because the energy production by fusion is a strong function of T, with $\epsilon \sim T^4$ for the pp-chain in low-mass stars and $\epsilon \sim T^{18}$ for the CNO cycle in more massive MS stars. Even a small change in T would result in a large change in ϵ and L, which is not allowed by the HE requirement. **Nuclear fusion therefore acts like a thermostat** in the center of a star: the stronger the temperature dependence of ϵ, the more accurate the thermostat.

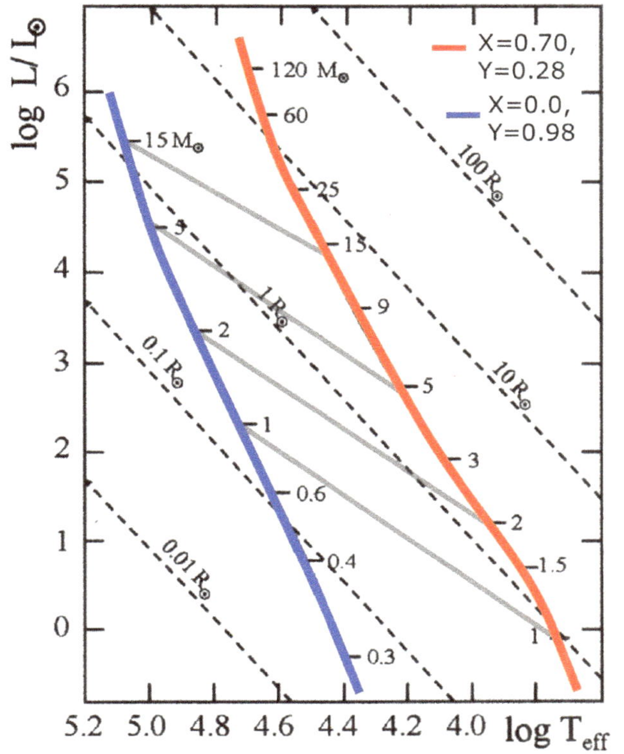

Figure 13.5. The ZAMS of H-rich stars of X = 0.70, Y = 0.28, Z = 0.02 (red) compared with that of He-rich MS stars of X = 0, Y = 0.98, Z = 0.02 (blue). Lines of equal radius are indicated by dashed lines. (Reproduced from Maeder 2009. © Springer-Verlag Berlin Heidelberg 2012.)

13.3.2 Changes in L and R during the MS phase

The nearly constant temperature in the star implies that the total thermal energy, $E_{\rm th}$, remains about constant. If T_c remains about constant during the MS phase but μ_c increases because of the increasing He content in the center, then $P_c/\rho_c \sim T_c/\mu_c$ must decrease. The ρ_c must therefore increase faster than P_c as more and more H is converted into He in the core. The density can only increase if the core shrinks.

If $E_{\rm th}$ remains about constant, the virial theorem requires that the total potential energy of the star also remains about constant. Since the $|E_{\rm pot}|$ of the core increases by core contraction, $|E_{\rm pot}|$ of the envelope must decrease so the envelope must expand. **Thus, as μ increases in the center, the radius of a star must increase.** Table 13.1 shows the changes in T_c, P_c, ρ_c, and μ_c at the beginning, the middle, and the end of the MS phase of a star of $1M_\odot$. It shows that P_c and ρ_c increase by factors of 5 and 8, respectively, between the beginning and end of core H-fusion, whereas T_c increases by only 14%.

The luminosity of a star increases during the MS phase due to the increase in the average value of μ inside the star ($L \sim \mu^4$). The change in $T_{\rm eff}$ during core H-fusion depends on the changes in L and R and cannot be derived from simple physical

Table 13.1. The Parameters of a 1 $M_\odot$ Model at Three Different MS Ages (Data are from Schaller et al. 1992)

Age Phase	$L/L_\odot$	$R/R_\odot$	$T_{\rm eff}$	T_c	P_c	ρ_c	X_c	Y_c	Z_c	μ_c
0 Gyr ZAMS	0.69	0.66	6540	13.6	1.24	78	0.680	0.301	0.02	0.71
4.5 Gyr Now	1.00	1.00	5820	15.9	2.36	157	0.302	0.677	0.02	0.88
9.4 Gyr TAMS	1.55	1.25	5820	18.7	6.73	572	0.000	0.980	0.02	1.32

Note. Units of $T_{\rm eff}$ are in K, T_c are in MK, P_c are in 10^{17} dyn cm^{-2}, and ρ_c are in g cm^{-3}.

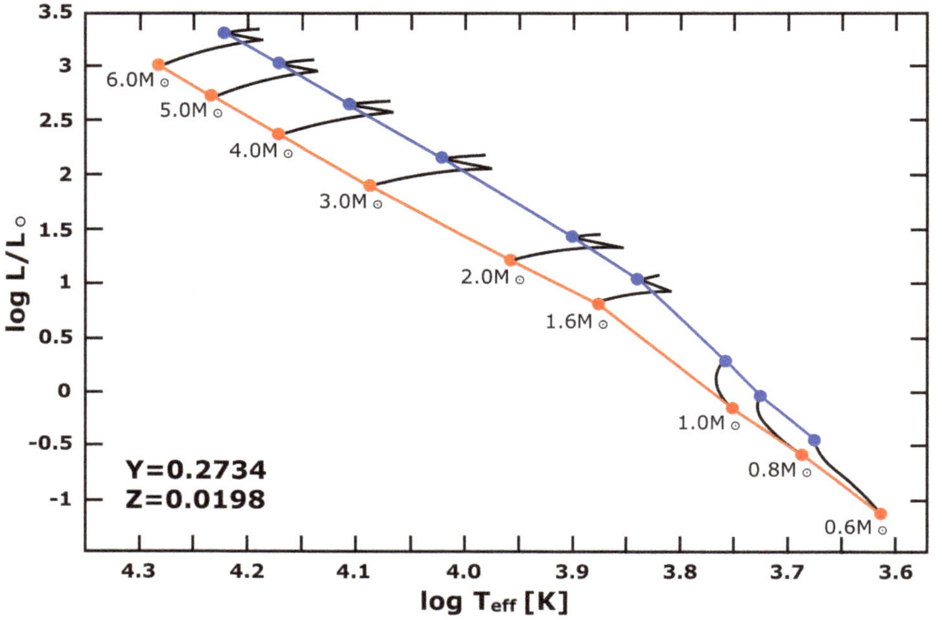

Figure 13.6. Evolutionary tracks of stars between 0.5 and 6$M_\odot$ during the core H-fusion phase. The ZAMS is indicated by a red line, while the TAMS is indicated by a blue line. The dots indicate the end of core H-fusion. (From Salaris & Cassisi 2005. Copyright © 2005 by John Wiley & Sons, Inc. Reprinted by permission of John Wiley & Sons, Inc.)

arguments. The homology relations that we used to derive the properties of ZAMS stars do not apply to stars with a changing chemical profile. In general, the HE condition, together with the radiative equilibrium condition, determine the structure of the star with a given chemical profile, therefore they determine L and R. The effective temperature has to adjust to L and R because $T_{\rm eff} \sim (L/R^2)^{1/4}$, **so there is no direct correlation between changes in T_c and $T_{\rm eff}$.**

Figure 13.6 shows evolutionary tracks of stars between the ZAMS and TAMS, which marks the end of the core H-fusion. Note that all stars increase in brightness during the MS phase due to the μ^4-effect. Also note that the radius of all stars increases during the MS phase as explained above. The radius increases strongest in massive stars because the severe sensitivity of $\varepsilon \sim T^{18}$ results in a very strong thermostat effect.

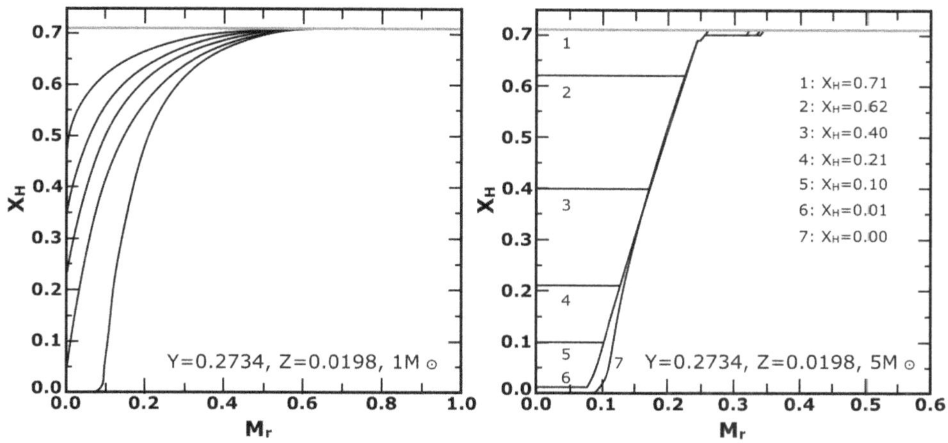

Figure 13.7. Evolution of the H-profile during the MS evolution of a low-mass star of $1M_\odot$ without a convective core (left) and a star of $5M_\odot$ with a convective core (right) during several phases. In the right figure, the central composition is indicated during these phases. The gray horizontal line is the H-profile of the ZAMS. The deepest line is the profile at the TAMS. (From Salaris & Cassisi 2005. Copyright © 2005 by John Wiley & Sons, Inc. Reprinted by permission of John Wiley & Sons, Inc.)

The combination of the changes in L and R results in a decrease in $T_{\rm eff}$ for stars with $M > 1.2M_\odot$ and an increase in $T_{\rm eff}$ for lower-mass stars. (The exact values of $T_{\rm eff}$ and R depend on the details of the convection theory used in the computations.) Note the short blue loop at the end of the MS tracks for stars more massive than about 1.2 $M_\odot$ and the lack of this loop for lower-mass stars. This is due to differences in the chemical profiles at the end of the MS phase, as will be explained in Section 13.4.

13.3.3 The Role of Convective Cores in Stars with $M > 1.2M_\odot$

In stars with $M > 1.2M_\odot$, H is fused via the CNO cycle, which is very sensitive to T. The nuclear energy is therefore generated in a very small central volume. This means that the energy flux is very high in and around the core of these stars. If that energy had been transported by radiation, it would have required a steep temperature gradient with $|dT/dr|_{\rm rad} > |dT/dr|_{\rm ad}$, so the Schwarzschild criterion implies that the core is convective. In Section 7.9, we showed that a high value of $L/4\pi r^2$ was one of the reasons for convection.

MS stars with $M > 1.2M_\odot$ have a convective core.

MS stars with $M < 1.2M_\odot$ have a radiative core.

N.B. The convective region in the center of stars with $M \gtrsim 2M_\odot$ is larger than the region where most of the energy is generated by nuclear fusion (see Figure 7.6.). In these stars, the fusion occurs in the center of the convective core because of the extreme sensitivity of the reaction rate to the temperature, $\epsilon \sim T^{18}$. The mass fraction of the convective core of ZAMS stars with $M \gtrsim 1.2M_\odot$ increases with increasing mass because the condition of a high value of $L/4\pi r^2$ is reached in a larger fraction of the stellar mass.

During the MS evolution the mass fraction of the convective cores of massive stars shrinks. This is the result of two effects. At the boundary, $r_{\rm conv}$, of the convective core, Y increases inside $r_{\rm conv}$ but remains unchanged outside, so a μ-gradient develops. According to the Ledoux criterion, this makes these layers more stable against convection (see Section 7.2). At the same time, the absorption coefficient $\kappa = \sigma_e \sim (1 + X)$ decreases inside the convective core, so $|dT/dr|_{\rm rad}$ decreases because $-dT/dr \sim \kappa$ (Equation (6.1)). This also makes these layers more stable against convection. We will see later, in Section 24.3, that the decrease of the mass of the convective core has a strong effect on the changing surface abundances in massive stars during the MS phase.

Convection in the core of a star has three effects:

1. Convection brings fresh H from all over the convective region into the center, so more H can be fused. This extends the MS lifetime of stars with convective cores.
2. As an MS low-mass star of $M < 1.2M_{\odot}$ ages, its chemical profile becomes more peaked (higher Y) toward the center, whereas in more massive stars it remains flat in the convective core due to mixing, with Y increasing with time.
3. The decrease in the mass of the convective core during the MS phase in massive stars implies that the chemical profile is flat in the center with increasing Y in a decreasing fraction of the mass.

These effects result in an evolution of the chemical profile that is very different for stars with M smaller or larger than $1.2M_{\odot}$. Figure 13.7 shows the evolution of the H-profile of a star of $1M_{\odot}$ and a star of $5M_{\odot}$ during MS evolution. In low-mass stars, the H-profile gets gradually steeper. In stars with a convective core, the profile remains flat, but gets deeper with time. The extent of the horizontal parts indicates the extension of the convective core. Note that this decreases drastically with time, as explained above.

13.3.4 The MS Evolution of the Sun

During the MS phase, while H-fusion continues, the structure of a star gradually changes. It is interesting to study these changes in the Sun.

Table 13.1 shows the changes in the stellar parameters for a solar model of $Z = 0.02$ between $t = 0$ (ZAMS), 4.5 Gyr (now), and $t = 9.4$ Gyr (TAMS) when core H-fusion ends. We see that

- L and R increase but $T_{\rm eff}$ decreases, and
- T_c, P_c, and ρ_c increase, with ρ_c increasing the most and T_c increasing the least.

Figures 13.8a and 13.8b show differences in structure in terms of $R(m)$, $T(m)$, $P(m)$, $L(m)$, and $X(m)$ for the solar model between $t = 4.5$ Gyr (now) and 9.2 Gyr ($\approx$TAMS).

Q (13.3) *Study the relation between X and L. Indicate the region where 70% of the luminosity is generated and explain this.*

Q (13.4) *Compare the run of r(m) in both models and explain the difference. Indicate the size of the region that contains the inner 10% of the mass and the outer 2% of the mass. Explain the difference.*

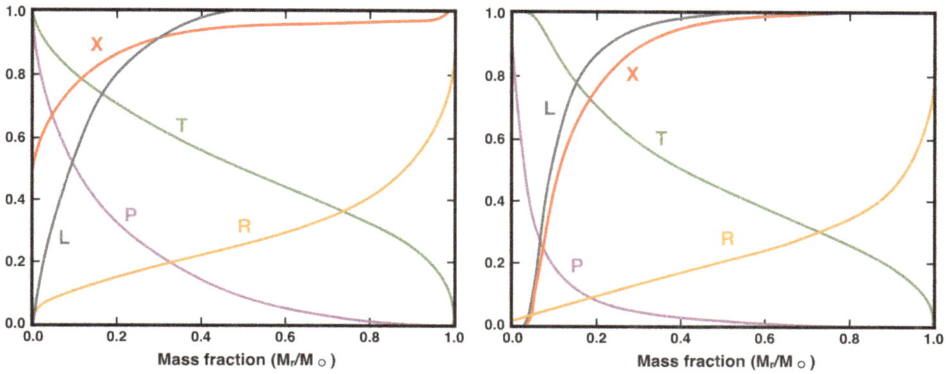

Figure 13.8. Internal structure of a $1M_\odot$ star at $t = 4.5$ Gyr (left) and $t = 9.2$ Gyr (right). The lines show the run of X (red), $L(m)/L$ (gray), $T(m)/T_c$ (green), $P(m)/P_c$ (purple), and $r(m)/R$ (orange). (Reproduced from Iben 1967.)

13.4 The End of the MS Phase: The TAMS

When H-fusion stops in the core, the core must contract to compensate for the radiative losses. As noted previously, this phase is called the **terminal age of the main sequence (TAMS).**

In **stars of $M \gtrsim 1.2M_\odot$**, the whole core contracts because H is exhausted in the whole core due to convection (see Figure 13.7). The core gets hotter so $E_{\rm th}$ increases. Virial equilibrium requires that the star contracts, which produces a small leftward loop in the HRD at the end of the H-fusion phase (see Figure 13.6). This contraction ends when the region around the He-core has reached a sufficiently high T and ρ to start H-fusion in a shell around the He core.

In **stars of $M \lesssim 1.2M_\odot$**, the contraction is more gradual because the chemical profile allows H-fusion to move gradually outward. Meanwhile, the core contracts (just like in stars with $M > 1.2M_\odot$), but this slower contraction does not result in a decrease of the stellar radius, so these stars do not make a leftward loop in the HRD.

13.5 The MS Lifetime

The duration of the core H-fusion phase, $\tau(H)$, depends on the mass of the star and its luminosity during the MS phase. If the same mass fraction of all stars took part in H-fusion, stellar lifetimes would be proportional to M/L with $L \sim M^{3.8}$ so we might expect $\tau(H) \sim M^{-2.8}$. Instead, the mass fraction that is used in core H-fusion increases with mass because of two effects.

(a) The central temperature increases with the stellar mass (Equation (3.9)), so H-fusion is possible in a larger mass fraction of the star.

(b) In stars with $M > 1.2M_\odot$, core convection brings material into the region where fusion occurs.

Figure 13.9 shows the duration of the core H-fusion phase as a function of mass for stars with solar abundance, $Z = 0.014$, and with low metallicity, $Z = 0.002$. Note

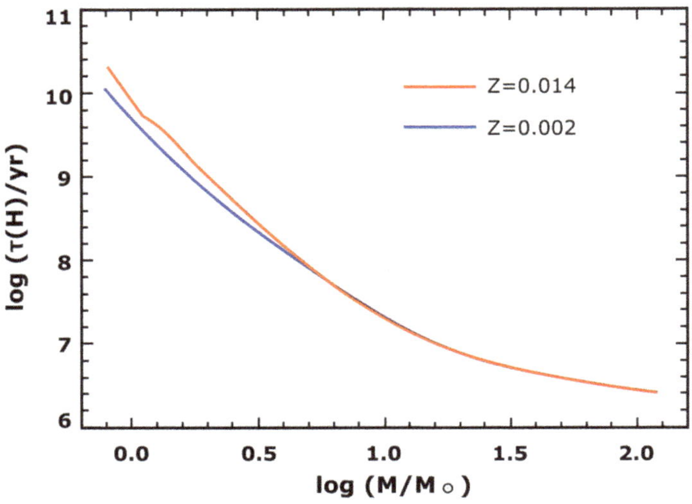

Figure 13.9. Main-sequence lifetime as a function of mass for stars with $Z = 0.014$ (solid line) and $Z = 0.002$ (dashed line). (Based on data from Ekström et al. 2012 and Georgy et al. 2013.)

that the lower metallicity stars have a shorter main-sequence lifetime. This is due to their higher luminosity, as explained in Section 13.2.

The relation between $\tau(H)$ and mass for the models of $Z = 0.014$ can be approximated to better than 10% accuracy by

$$\log \tau(H)/\text{yr} = 9.9551 - 3.3674x + 0.4794x^2 + 0.3676x^3 - 0.1013x^4, \quad (13.16)$$

where $x = \log(M/M_\odot)$.

13.6 Summary

1. At the start of the core H-fusion phase, stars occupy a narrow strip in the HRD called the ZAMS. The location of the ZAMS depends on the initial chemical composition and can be explained by homology relations.
2. The ZAMS of stars with lower metal abundance or higher He abundance is shifted to the left and upward in the HRD.
3. He-rich and H-poor ZAMS stars are about 30 to 100 times brighter and T_{eff} is about 4 to 10 times higher than H-rich ZAMS stars of the same mass.
4. As the mean particle weight μ in the core increases during the MS phase, the core must contract and the envelope expands. This results in an increase in radius. This increase is higher for massive stars that fuse H via the CNO cycle than for low-mass stars that fuse H via the pp-chain.
5. During core H-fusion stars with $M \gtrsim 1.2M_\odot$ move upward and to the right in the HRD and stars with $M \lesssim 1.2M_\odot$ move upward and to the left.
6. Core convection in massive stars results in a short leftward loop in the HRD at the TAMS.

Exercises

13.1 Explain in physical terms the difference in main-sequence lifetime between stars of $Z = 0.014$ and $Z = 0.002$ plotted in Figure 13.9.

13.2 We will show later (Section 25) that differential rotation in a fast rotating star may result in severe mixing during the main-sequence phase.
Calculate the increase or decrease of L, R, and $T_{\rm eff}$ of stars more massive than $2M_{\odot}$ when the composition has changed from $X = 0.70$, $Y = 0.28$, $Z = 0.02$ to $X = 0.49$, $Y = 0.49$, $Z = 0.02$.

13.3 The first generation of stars had a very low metallicity and a slightly lower He abundance than the Sun.
   a. Calculate the increase or decrease of L, R, and $T_{\rm eff}$ of massive stars of $X = 0.75$, $Y = 0.25$, and $Z = 0$ compared to $X = 0.70$, $Y = 0.28$, and $Z = 0.02$.
   b. What is the effect of this abundance on the mass of the convective core?
   c. How does this affect the lifetime of massive MS stars?

13.4 **Computer exercise**
Use the MESA code to calculate the mass fraction of the convective cores in ZAMS stars of 0.3, 1, 3, 10, 30, and $100M_{\odot}$, with solar abundances and with globular abundances (0.01 times solar) and comment on the differences.

13.5 **Computer exercise**
Use the MESA code to calculate the location in the HRD of ZAMS stars of 0.3, 1, 3, 10, 30, and $100M_{\odot}$, with solar abundances (i.e., $Z = 0.014$), SMC abundances (0.1 times solar), and globular abundances (0.01 times solar), and comment on the differences.

References

Ekström, S., Georgy, C., & Eggenberger, P. et al. 2012, A&A **537** 146

Georgy, C., Ekstrom, S., & Eggenberger, P., et al. 2013, A&A **558** 103

Iben, I. 1967, ApJ **147** 624

Maeder, A. 2009, Physics, Formation and Evolution of Rotating Stars (Berlin: Springer)

Pols, O. 2011, Stellar Structure and Evolution, Utrecht University Lecture Notes, 111, https://www.astro.ru.nl/~onnop/education/stev_utrecht_notes/

Salaris, M. & Cassisi, S. 2005, Evolution of stars and stellar populations (Chichester: John Wiley and Sons)

Schaller, G., Schaerer, D., Meynet, G. & Maeder, A. 1992, A&A **96** 269

AAS | IOP Astronomy

Understanding Stellar Evolution

Henny J.G.L.M Lamers and Emily M. Levesque

Chapter 14

Principles of Post-main-sequence Evolution

Before explaining the evolution of low-mass stars and massive stars, we must first discuss some physical principles of post-main-sequence evolution. These principles are important for understanding the evolution of the cores of stars, the properties of isothermal cores, the influence of fusion in a shell, and the structure of fully convective stars. We will show that (a) there is an upper limit to the mass of isothermal cores of stars in hydrostatic equilibrium, (b) shell fusion acts like a mirror for the expansion of stellar cores and envelopes, and (c) fully convective stars occupy a nearly vertical strip in the HR diagram called the Hayashi line.

14.1 Isothermal Cores: The Schönberg–Chandrasekhar Limit

At the end of the MS phase, a star has a core with mass M_c and radius R_c that no longer produces nuclear energy. This implies $L(r) \approx 0$ for $r < R_c$. If $L(r < R_c) \approx 0$, then $dT/dr \approx 0$ in the core; otherwise, there would be transport of radiation. This shows that a core without fusion is approximately isothermal. The question is now whether an isothermal core can be in hydrostatic equilibrium.

An isothermal core in HE must have a steep density gradient to produce the pressure gradient, because at constant temperature $dP/dr \sim d\rho/dr$. In other words, the density gradient must provide the pressure gradient without the help of the temperature gradient. It turns out that an isothermal core can only exist if its mass is smaller than a certain fraction of the total mass of the star. We will derive this in a way that resembles our derivation of the Virial Theorem in Section 3.3.

Hydrostatic equilibrium requires

$$\frac{dP}{dr} = -\frac{Gm}{r^2}\rho. \tag{14.1}$$

Multiplication by $4\pi r^3$ and integration over the core, i.e., $0 < r < R_c$, gives

doi:10.1088/978-0-7503-1278-3ch14

$$\int_0^{R_c} 4\pi r^3\, dP = -\int_0^{R_c} \frac{Gm}{r}\rho 4\pi r^2\, dr = E_{\text{pot}}^c = -\alpha\frac{GM_c^2}{R_c}, \tag{14.2}$$

with $\alpha \gg 1$. Partial integration of the left term, as in the derivation of the virial theorem (Section 3.3), yields

$$\begin{aligned}\int_{R_c}^0 4\pi r^3\, dP &= 4\pi r^3 P\Big|_0^{R_c} - 3\int_{R_c}^0 P4\pi r^2\, dr \\ &= 4\pi R_c^3\, P(R_c) - \frac{3\mathcal{R}T_c}{\mu}\int_{R_c}^0 \rho 4\pi r^2\, dr = 4\pi R_c^3\, P(R_c) - \frac{3\mathcal{R}T_c}{\mu_c}M_c,\end{aligned} \tag{14.3}$$

where we have used the isothermal condition $P = \mathcal{R}\rho T_c/\mu_c$ with T_c = constant. Note that, contrary to the derivation of the VT, the first right term is not zero because the integration is not done to R (where $P(R) = 0$) but to R_c, so $P(R_c)$ is the pressure at the outer boundary of the isothermal core. Equations (14.2) and (14.3) show that

$$4\pi R_c^3 P(R_c) - \frac{3\mathcal{R}T_c}{\mu_c}M_c = -\alpha\frac{GM_c^2}{R_c}. \tag{14.4}$$

This provides a relation between the pressure at the boundary of the core and its radius, for a given core mass M_c and temperature T_c:

$$P(R_c) = \frac{3}{4\pi}\frac{\mathcal{R}T_c}{\mu_c}\frac{M_c}{R_c^3} - \frac{\alpha}{4\pi}\frac{GM_c^2}{R_c^4}. \tag{14.5}$$

This expression is of the type $y = Ax^3 - Bx^4$ if $x = 1/R_c$. It has a maximum at $x = 3A/4B$, where $y_{\max} = 3^3A^4/4^4B^3$. Substituting A and B gives the maximum pressure that an isothermal core can support.

$$P^{\max}(R_c) = C\frac{T_c^4}{\mu_c^4 M_c^2}, \tag{14.6}$$

where C is a constant. The pressure due to the weight of the envelope on top of the core should be less than this maximum value, otherwise the core cannot support it. The higher the core mass, the smaller this maximum pressure. We can also see that a higher core temperature allows for a larger maximum pressure.

Q (14.1) *Explain in physical terms why there is a maximum for $P(R_c)$.*

Let us assume that the isothermal core contains only a small fraction of the total stellar mass M, and that the radius R of the star is much larger than R_c, so the contribution of the isothermal core to the total mass of the star is small. The pressure at the bottom of the envelope can then be estimated in the same way as the estimate of the central pressure of any star in HE (Section 3.2)

$$P_{\text{env}}(R_c) \cong GM^2/R^4. \tag{14.7}$$

The temperature at the bottom of the envelope is the same as the isothermal core temperature T_c. We have derived in Section 3.2 that for a star in HE consisting of an ideal gas the central temperature is of the order of

$$T_c \sim \frac{\mu_{\rm env}}{\mathcal{R}} \frac{GM}{R} \quad \text{or} \quad \frac{M}{R} \sim \frac{\mathcal{R} T_c}{G\mu_{\rm env}}, \tag{14.8}$$

where $\mu_{\rm env}$ is μ in the envelope. Substitution of this into the expression for $P_{\rm env}(R_c)$ yields

$$P_{\rm env}(R_c) \sim \frac{T_c^4}{\mu_{\rm env}^4 M^2}. \tag{14.9}$$

The isothermal core can only be stable if $P_{\rm env}(R_c) < P^{\rm max}\,(R_c)$. This gives

$$\boxed{\frac{M_c}{M} \lesssim C\left(\frac{\mu_{\rm env}}{\mu_c}\right)^2} \quad \text{with} \quad C = 0.37. \tag{14.10}$$

This is called the **Schönberg-Chandrasekhar limit for isothermal cores** (Schönberg, M., & Chandrasekhar 1942).

At the end of core H-fusion when $\mu_c = 4/3$ and $\mu_{\rm env} \approx 0.6$ the remaining isothermal He core can only be stable if its mass is smaller than the maximum value of

$$\frac{M_c}{M} < 0.37\left(\frac{0.60}{1.33}\right)^2 \approx 0.08. \tag{14.11}$$

Here, we assumed that the envelope is not chemically enriched at all. In reality, the lower part of the envelope has an increased He abundance with $\mu_{\rm env} > 0.6$. If that is taken into account, the Schönberg–Chandrasekhar limit for isothermal cores is $M_c/M \approx 0.10$.

If the core is less massive than this, the subsequent H-shell fusion occurs around a stable He core. If the core is more massive than this, the isothermal He core cannot support the weight of the shell plus the envelope in HE, so it will contract during H-fusion in a shell around it. Contraction of the He core will create a temperature gradient that produces extra gas pressure and helps to stabilize the core. However, if the core is no longer isothermal, the temperature gradient results in an energy flux from the core, so once the core starts to contract it loses energy and has to keep shrinking until He-fusion starts in the core. This contraction occurs on the Kelvin–Helmholtz timescale of the core:

$$\tau_{\rm KH} \simeq \frac{GM_c^2}{R_c} / (L - L_{\rm shell}), \tag{14.12}$$

where $L_{\rm shell}$ is the energy generated by fusion in the shell around the core.

Q (14.2) *Why does* Equation (14.12) *contain the factor* $(L - L_{\rm shell})$ *instead of* L?

14.2 The Mirror Principle of Stars with Shell Fusion

Whenever a nondegenerate star in radiative equilibrium has fusion in a shell around the core, the shell appears to act like a mirror (see Figure 14.1).

- When the core contracts, the envelope expands.
- When the core expands, the envelope contracts.

There are two ways to explain this: using the virial theorem argument or using the pressure argument. In both arguments, the thermostatic behavior of a fusion shell plays a key role (see Section 13.3.1).

The virial theorem argument

For a nondegenerate star in radiative equilibrium to remain in thermal equilibrium, the energy generation by fusion must remain about constant. The fusion efficiency is very sensitive to T. Therefore, as the core contracts, the fusion shell can hardly contract with it, because doing so would increase its T and its energy production. The shell must remain at about the same T, so the mean temperature $\bar{T}$ of the star will not change very much. If $\bar{T}$ does not change, the total kinetic (thermal) energy will not change. The virial theorem then implies that the total potential energy should also remain constant, so if the core contracts (decreasing $E_{\rm pot}^{c}$) the envelope must expand (increasing $E_{\rm pot}^{\rm env}$), and vice versa.

The pressure argument

If the core contracts and the shell on top of the core follows, the temperature in the fusion shell rises. This would imply an increase of the energy generation rate unless the density in the shell decreases, but if the density decreases at a roughly constant temperature, the pressure on the shell, exerted by the weight of the envelope, must decrease. Therefore, the envelope must expand. The same argument can be used to show that the envelope must shrink if the core expands.

We conclude the following:

- Fusion shells act like mirrors to the expansion or contraction of the core and the envelope. This is due to the thermostatic action of fusion shells.

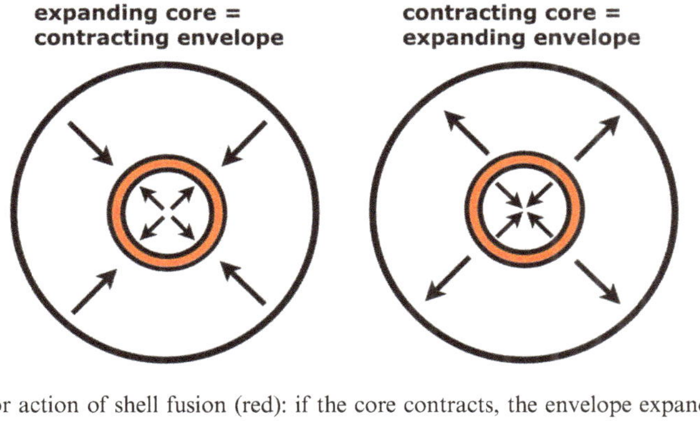

Figure 14.1. Mirror action of shell fusion (red): if the core contracts, the envelope expands and vice versa.

- The contraction of a nondegenerate He core surrounded by a fusion shell at the end of the MS phase will result in an expansion of the star and a decrease of $T_{\rm eff}$.
- We will show later, in Section 16.4, that the contraction of a degenerate core (due to the increasing mass) surrounded by a fusion shell will result in an increase in luminosity.

14.3 The Hayashi Line of Fully Convective Stars

Fully convective stars occupy a narrow, almost vertical, band at low $T_{\rm eff} \approx 3000$ to 4000 K in the HR diagram. We have seen that this applies to fully convective pre-MS stars (Figure 12.4). We will see later that it also applies to fully convective red giants, red supergiants, and AGB stars; all lie in roughly the same vertical band in the HR diagram. This band is called the **Hayashi line** (Hayashi 1961, 1966). The part of the evolutionary track of a star that follows the Hayashi line is called the **Hayashi track.**

The Hayashi line is the result of two effects:
- Convective stars have extended envelopes and large radii, so their $T_{\rm eff}$ is low.
- At low temperatures of $T_{\rm eff} \lesssim 3000$ K the opacity in the photosphere drops steeply toward lower $T_{\rm eff}$.

We will derive the physical principle of the Hayashi line by showing what happens if a radiative photosphere is attached to the top of a fully convective star. The star is supposed to be fully convective from the center to the bottom of the photosphere, where $\tau \simeq 1$. We will call this photospheric radius R_1, with R_1 very close to the stellar radius R.

A. **The pressure at $\tau = 1$, derived from the interior structure.**
A fully convective star obeys the polytropic relation $P = K\rho^{5/3}$, where K is a scaling constant that depends on the star and is to be determined. All convective stars are homologous (i.e., they all have the same P/P_c, T/T_c, and ρ/ρ_c structures if these are expressed in a function of r/R). We can derive how K depends on the mass and radius of the star by realizing that the central pressure is proportional to

$$P_c \sim \frac{GM^2}{R^4} \quad \text{and } P_c = K\rho_c^{5/3} \text{ with } \rho_c \sim M/R^3. \tag{14.13}$$

The first equation results from the condition of HE, the second from the polytropic condition, and the third from the homology condition, so the scaling constant K depends on the stellar parameters as

$$K = \frac{P_c}{\rho_c^{5/3}} \sim \frac{M^2}{R^4}\bigg/\rho^{\frac{5}{3}} \sim \frac{M^2}{R^4}\bigg/\frac{R^5}{M^{5/3}} \rightarrow K \sim M^{1/3}R. \tag{14.14}$$

The polytropic expression for P is valid at all depths, as well as at the top of the convection zone at R_1, where $P(R_1) \equiv P_1$, so the pressure at R_1 derived from the interior condition is

$$\boxed{P_1^{\text{int}} = K\rho_1^{5/3} \sim (M^{1/3}R)\,\rho_1^{5/3}}, \qquad (14.15)$$

where $\rho_1 \equiv \rho(R_1)$.

B. **The pressure at $\tau = 1$, derived from the photosphere.**
The temperature structure of a simple (gray) photosphere is given by

$$T^4 = \frac{3}{4}T_{\text{eff}}^4\left(\tau + \frac{2}{3}\right). \qquad (14.16)$$

This implies that the photosphere is almost isothermal because $0.84 < T/T_{\text{eff}} < 1.06$ at $0 < \tau < 1$. We therefore adopt, for simplicity, that the photosphere is isothermal with $T = T_1$. The pressure in the photosphere decreases exponentially with height as

$$\frac{dP}{dr} = \frac{\mathcal{R}T_1}{\mu}\frac{d\rho}{dr} = -\frac{GM}{R^2}\rho \;\rightarrow\; \frac{d\ln\rho}{dr} = -\frac{GM}{R^2}\frac{\mu}{\mathcal{R}T_1} = \frac{\mu g}{\mathcal{R}T_1} \equiv \frac{1}{\mathcal{H}}, \qquad (14.17)$$

where $\mathcal{H} \equiv \mathcal{R}T_1/\mu g \sim T_1R_1^2/GM$ is the density scale height of the photosphere. The density distribution in the photosphere is

$$\rho = \rho_1\, e^{-\frac{r-R_1}{\mathcal{H}}}. \qquad (14.18)$$

Adopting this density distribution $\rho(r)$, we can now find the pressure $P_1 \equiv P(R_1)$ at $\tau = 1$ derived from the photosphere. The optical depth $\tau = 1$ is defined by

$$\int_{R_1}^{\infty} \kappa\rho\, dr = 1 \;\rightarrow\; \bar{\kappa}\int_{R_1}^{\infty} \rho_1 e^{-\frac{(r-R_1)}{\mathcal{H}}} dr = \bar{\kappa}\rho_1\mathcal{H} = 1, \qquad (14.19)$$

where we used a mean value for $\bar{\kappa} \approx \kappa(\tau = 1)$. So, the density ρ_1 at $\tau = 1$ is $\rho_1 = 1/\mathcal{H}\bar{\kappa} \sim GM/TR^2\bar{\kappa}$ and the gas pressure at $R_1 \equiv R\ (\tau = 1)$ due to the photosphere is

$$\boxed{P_1^{\text{phot}} \sim \rho_1 T_1 \sim \frac{GM}{R^2}\frac{1}{\bar{\kappa}}} \quad \text{with} \quad \rho_1 \sim GM/TR^2\bar{\kappa}. \qquad (14.20)$$

C. **Matching the pressure of the interior to photosphere at $\tau = 1$.**
We now have two expressions for the pressure $P(R_1)$, one derived for the top of the polytropic interior and one derived for the bottom of the photosphere at $\tau = 1$. These two should be equal.

$$P_1^{\text{phot}} = P_1^{\text{int}} \rightarrow MR^{-2}\kappa^{-1} \sim M^2R^{-7/3}T^{-5/3}\kappa^{-5/3} \rightarrow R \sim M^3T^{-5}\kappa^{-2} \qquad (14.21)$$

(for clarity we have replaced the symbol $\bar{\kappa}$ with κ). Up to now, we have specified neither the energy source nor the luminosity. The luminosity does not have to obey the M-L relation that was derived in Section 6.2 because that was for stars in radiative equilibrium, whereas here we are dealing with convective stars.

The luminosity is set by R_1 and $T_1 \approx T_{\rm eff}$ because $L \sim R^2T^4$. This gives the luminosity of fully convective stars:

$$\boxed{L \sim M^6 T^{-6} \kappa^{-4}} \tag{14.22}$$

D. **The absorption coefficient κ in the photosphere of cool stars.**
For very cool stars with $T_{\rm eff} \lesssim 6000$ K, H^- is the dominant atmospheric opacity. This opacity is very small: at $T = 3000$ K and $\rho = 10^{-10}$ g/cm^3, which are the approximate values in an atmosphere of an AGB star, the opacity is only 5×10^{-5} cm^2/g^{-1} (compare this with $\kappa \approx 0.3$ cm^2/g in the interior of ionized stars). The H^- opacity is a very steep function of T

$$\kappa \simeq \kappa_0 \rho^{1/2} T^9 (Z/0.02) \quad \text{with} \quad \kappa_0 = 2.5 \times 10^{-31}\,\text{cm}^2/\text{g}. \tag{14.23}$$

We can write this as $\kappa = \kappa_0 \rho^a T^b$ with $a = 1/2$ and $b = 9$.

E. **The effective temperature of fully convective stars.**
Substituting the expression for κ into Equation (14.22) for L, we find

$$\log L = A \log T_{\rm eff} + B \log M + \text{constant}, \tag{14.24}$$

with $A = \dfrac{18a + 4b + 6}{3a - 1} \simeq 104$ and $B = \dfrac{2a + 6}{1 - 3a} \simeq -14$. We see that for fully convective stars

$$\boxed{L \sim T_{\rm eff}^{104} M^{-14}} \quad \text{and} \quad \boxed{T_{\rm eff} \sim L^{0.01} M^{-0.14}}. \tag{14.25}$$

This means that $T_{\rm eff}$ is almost independent of L and M! The Hayashi line is therefore almost vertical in the HR diagram and the predicted lines for various masses are very close together.

N.B.

- The temperature of the Hayashi line depends on metallicity because the photospheric opacity is proportional to Z (Equation (14.23)). The Hayashi line of low-metallicity stars is at a higher $T_{\rm eff}$ than the Hayashi line of solar metallicity stars.
- Detailed evolutionary models show that the Hayashi line bends a little bit to the right (toward cooler temperatures) at $L > 10^2 L_\odot$ (see figure 12.4).
- In the color–magnitude diagram with B-V on the x-axis the Hayashi line of fully convective stars bends much more strongly to the right because this color index depends very strongly on $T_{\rm eff}$ for cool stars (see Figures 2.5 and 2.7).
- The inverse dependence of L on M may appear counterintuitive; however, realize that this was derived for stars that are fully convective

without a core. We will see later (in Sections 16.2 and 16.4) that low-mass stars become *almost* fully convective but have a high density core when they reach the Hayashi line. Their luminosity is set by the mass of the core. As this core mass increases, the radius increases and the star *climbs* the Hayashi track in the HR diagram.

– Because stars on the Hayashi line have nearly the same $T_{\rm eff}$, their luminosity is set by the radius because $L \sim T_{\rm eff}{}^4 R^2$. We have seen in Section 12.5 that convective protostars are contracting (i.e., R is decreasing), so they evolve *down* along the Hayashi track.

Figure 14.2 shows predicted evolutionary tracks of stars with a large range of masses, from 0.8 to $25 M_\odot$. The stars with a minimum effective temperature occupy a nearly vertical band in the HRD: this is the Hayashi line.

14.3.1 Intuitive Explanation of the Hayashi Line

We have derived the location of the Hayashi line from mathematical expressions. Can we also understand the location of the Hayashi line based on simple physical insight?

Consider the two main ingredients:

– Fully convective stars are geometrically extended. This is because they have a small T-gradient, with $|dT/dr| = |dT/dr|_{\rm ad} < |dT/dr|_{\rm rad}$. With this small T-gradient, going from the hot inside of $T_{\rm c} \gtrsim 10^7$ K to the much cooler edge of the star, where $T \sim 10^4$ K $<<< T_c$, requires a long distance, so their radii are large and their effective temperatures are low. This is a general property of stars with convective envelopes: as the convective envelope of a star gets deeper, the stellar radius expands (see Figure 10.1).

– The opacity at low temperatures is due to H^-. This opacity is very small (e.g., $\kappa \sim 10^{-4}$ cm^2/g^{-1} at $T \simeq 3000$ K and $\rho \simeq 10^{-10}$ g/cm^{-2}) and decreases very steeply with decreasing temperature as $\kappa \sim T^9$.

Suppose a very cool star of constant L increases its size. The $T_{\rm eff}$ would then decrease and the outer layers would become (almost) transparent. Thus, *even* if the star's size increased, we could look deeper and deeper into the star up to the depth where $\tau = 1$ is reached at *almost constant T*. This is the main reason why the Hayashi line is approximately vertical in the HR diagram!

N.B. This argument does not apply to stars that contain dust in their envelope. For such stars (e.g., stars at the tip of the AGB with dense cold winds and class 0 T Tauri stars; see Figure 12.6), the dust is the main opacity source, and κ is large and has a weak dependence on T. These stars therefore radiate as black bodies with the temperature of the dust and the radius of the dusty wind. The radiation from the photosphere is shielded.

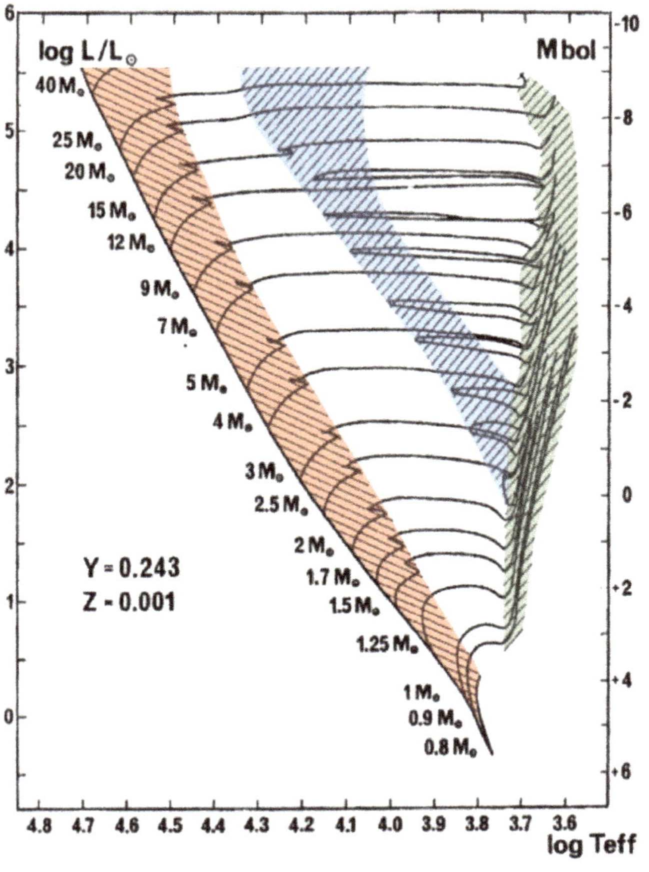

Figure 14.2. Predicted evolutionary tracks of low-metallicity stars. The core H- and core He-fusion phases are shown in red and blue, respectively. The green region indicates the Hayashi line of stars at $T_{\rm eff} \approx 4000$ to 5000 K with extended convective envelopes. (Reproduced from Schaller et al. 1992, with permission. © ESO.)

14.4 Summary

1. After the MS phase, when H-fusion stops in the core but continues in a shell, the core is nearly isothermal. Isothermal cores can only remain in hydrostatic equilibrium if the ratio $M_c/M < 0.10$, where M_c is the core mass. This is the Schönberg–Chandrasekhar limit. If $M_c > 0.10\ M$, the core must contract.
2. In stars with shell fusion, the shell acts like a mirror: if the core contracts, the envelope expands and vice versa. This is a consequence of the fact that fusion acts like a thermostat.
3. Fully convective stars with a radiative photosphere have an almost constant $T_{\rm eff}$ of about 3000 to 4000 K. This is due to the very strong T-sensitivity of the H^- absorption in cool photospheres. The location of these stars in the HRD is called the Hayashi line and the evolutionary tracks of (almost) fully convective stars are called Hayashi tracks. Fully convective stars evolve

upward along their Hayashi track if the radius expands (e.g., red giants, AGB stars, and red supergiants) and downward if the radius decreases (protostars).

Exercises.

14.1 Which one of these stars is more likely to have a contracting core during H-shell fusion: a star of $0.5 M_\odot$ or a star of $10 M_\odot$? Explain.

14.2 Estimate the timescale between the TAMS and the start of the core He-fusion for a post-MS star of $4 M_\odot$. Assume that half of its luminosity is obtained by core contraction, and the other half by shell H-fusion.
Hints: consider convection in the core and the Schönberg-Chandrasekhar limit; estimate the core radius at the TAMS in analogy to the Sun.

References

Hayashi, C. 1961, PASJ, **13, 442**
Hayashi, C. 1966, ARA&A, **4, 171**
Schaller, G., Schaerer, D., Meynet, G. and Maeder, A. 1992, A&AS, **96, 269**
Schönberg, M., and Chandrasekhar, S., 1942, ApJ, **96, 161**

Henny J.G.L.M. Lamers and Emily M. Levesque

Chapter 15

Stellar Winds and Mass Loss

The evolution of stars is strongly affected by mass loss, so we must first discuss the physics of stellar winds and the resulting mass-loss rates before continuing the description of post-MS stellar evolution. Stars may lose mass due to a stellar wind. The mass-loss rates are particularly high in luminous hot stars, where the wind is driven by radiation pressure in the spectral lines of ions, and in cool low gravity stars such as red giants and AGB stars. In these stars, the mass loss is driven by radiation pressure on dust grains, helped by the pulsation of the stars. The mass-loss rate of hot stars and cool supergiants is so large that it changes their evolution. Mass loss strips away the outer layers of stars. After severe mass loss, the original deeper layers that have been chemically modified appear at the surface. This allows observational tests of the predicted nuclear fusion and mixing processes. In this section, we discuss the two major mass-loss mechanisms. We also summarize expressions for the observed mass-loss rates in various types of stars. Mass-loss rates are indicated by the symbol $\dot{M} \equiv -dM/dt$ and are expressed in units of $1 M_\odot \mathrm{yr}^{-1} = 6.3 \times 10^{25}$ g s^{-1}. A mass-loss rate of 3×10^{-6} $M_\odot \mathrm{yr}^{-1}$, which is typical for a massive O-star, corresponds to the loss of one Earth mass per year.

15.1 Types of Winds

The theory and observations of stellar winds are described in the book "Introduction to Stellar Winds" (Lamers & Cassinelli, 1999). There are four basic types of stellar winds.

a. **Line-driven winds** of hot stars are driven by radiation pressure on highly ionized atoms of C, N, O, and Fe-group elements. This mechanism is responsible for the fast ($\sim 10^3$ km s^{-1}) winds of hot luminous stars: O- and B-type MS stars, hot giants and supergiants, Wolf–Rayet stars, and the central stars of planetary nebulae. The mass-loss rates can be as high as 10^{-5} $M_\odot \mathrm{yr}^{-1}$.
b. **Dust-driven winds** are driven by radiation pressure on dust grains. This mechanism also needs stellar pulsations to produce a high mass-loss rate. It is

doi:10.1088/978-0-7503-1278-3ch15

responsible for the low velocity ($\sim$10 km s^{-1}) winds of pulsating red giants and AGB stars. The mass-loss rate can be as high as 10^{-5} to 10^{-4} $M_\odot \mathrm{yr}^{-1}$.

c. **Coronal winds** are driven by gas pressure due to the high temperatures of stellar coronae. This mechanism is responsible for the solar wind and winds of cool MS stars and red giants. The mass-loss rates are too small to affect stellar evolution; for instance, the solar wind produces a mass-loss rate of $\sim 10^{-14}$ $M_\odot \mathrm{yr}^{-1}$, which implies a loss of only 10^{-4} $M_\odot$ during the MS lifetime of $\sim 10^{10}$ yrs.

d. **Alfven wave driven winds** are driven by magnetic waves. The magnetic field lines are rooted in the photosphere. Horizontal motions of the footpoints, due to convective cells, generate magnetic waves, which produce an outward pressure gradient that can accelerate ionized gas. This mechanism is (probably) responsible for the mass loss from red giants of $\sim 10^{-8}$ $M_\odot \mathrm{yr}^{-1}$ with speeds of $\sim 10^2$ km s^{-1}.

In this chapter, we discuss the two mechanisms, a and b, that produce the highest mass-loss rates and play a crucial role in stellar evolution.

15.2 Line-driven Winds of Hot Stars

Line-driven winds are due to the transfer of the momentum from the stellar radiation to the gas above the photosphere. Hot stars with $T \gtrsim 30{,}000$ K emit most of their light in the UV. This is also the wavelength regime where the ions of the most abundant elements (doubly or triply ionized C, N, O, Si, S, Fe, etc.) have their strongest spectral lines (i.e., where photons are most easily absorbed or re-emitted by electron transitions in these ions). These specific ions are accelerated by the repeated absorption or scattering of photons, so only a small fraction of all ions (typically $\sim 10^{-4}$) are accelerated; however, because of the frequent interactions of these specific ions with protons and electrons, the gas is dragged along.

15.2.1 A Simple Estimate of a Radiation-driven Mass-loss Rate

Suppose a particular abundant ion in the wind of an O-star (e.g., a C IV ion) has one very strong absorption line at a wavelength λ_0 that corresponds to the peak of the Planck function for that star. Assume that its absorption cross-section is so large that the line is optically thick and absorbs or scatters all photons at its wavelength. How much mass loss can one absorption line produce?

Suppose that the velocity increases from v = 0 at the photosphere to v_∞ at large distances. Due to the Doppler shift, all photons in the frequency range ν_0 to $\nu_0(1 + \mathrm{v}_\infty/c)$ are absorbed or scattered (Figure 15.1). The momentum of a photon is $E/c = h\nu/c$ and the total momentum of all photons leaving the star per second is L/c, so the photons that are absorbed or scattered in the wind transfer a total momentum per second to the gas:

$$L_{\mathrm{abs}}/c = \int_{\nu_0}^{\nu_0\left(1+\frac{\mathrm{v}_\infty}{c}\right)} (4\pi R_*^2 F_\nu/c)d\nu = L_\nu \Delta\nu/c, \tag{15.1}$$

where L_ν is the luminosity at ν_0 and $\Delta\nu = \nu_0 \mathrm{v}_\infty/c$. The total momentum loss of the

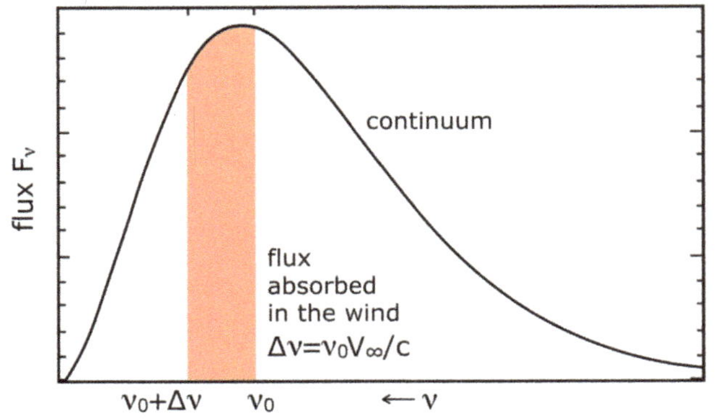

Figure 15.1. Radiative flux absorbed or scattered in a wind with a terminal velocity v_∞ by one strong spectral line of rest frequency ν_0 near the peak of the stellar energy distribution.

wind per second is $\dot{M}v_\infty$. The photons that are absorbed or scattered (with a luminosity L_{abs}), transfer a momentum L_{abs}/c per second to the gas. If the wind is driven by radiation pressure, the momentum of wind is equal to that of the absorbed radiation. This yields

$$\dot{M}v_\infty = L_{abs}/c = L_\nu \Delta\nu/c = L_\nu \nu_0 v_\infty/c^2. \tag{15.2}$$

At the peak of the Planck function $L_\nu \nu_0 \simeq 0.6L$ (which is a property of the Planck function), so we find that one strong spectral line can drive a mass-loss rate of

$$\boxed{\dot{M} \simeq 0.6\frac{L}{c^2} \simeq \frac{L}{c^2}} \text{ per optically thick spectral line.} \tag{15.3}$$

For a hot star of $L = 10^6\,L_\odot$, this corresponds to a mass-loss rate of about $4 \times 10^{-8}\,M_\odot\,\mathrm{yr}^{-1}$ per strong spectral line. If the spectrum contains N_{eff} optically thick spectral lines, with $N_{eff} \simeq 10^2$, then the mass-loss rate of that star is

$$\dot{M} \simeq N_{eff}\,L/c^2 \simeq 10^2\,L/c^2 \simeq 7 \times 10^{-6}\,M_\odot\,\mathrm{yr}^{-1}, \tag{15.4}$$

which is about the mass-loss rate of a typical luminous hot star. In reality, the winds of hot luminous stars are driven by a mixture of a large number of optically thick and thin lines (Castor et al. 1975).

An upper limit to the mass-loss rate of a radiation-driven wind can be derived by assuming that all photons leaving the photosphere are absorbed or scattered in the wind, resulting in

$$\dot{M}_{max}v_\infty < L/c \rightarrow \boxed{\dot{M}_{max} \simeq L/v_\infty c.} \tag{15.5}$$

The terminal wind velocities of hot stars are typically two or three times the escape velocity at the photosphere: $v_\infty \simeq 3\sqrt{2GM/R}$. For a typical O main-sequence star, this is about $v_\infty \simeq 2000\ \mathrm{km\,s^{-1}}$, so the maximum radiation-driven mass-loss rate of an

O star of $L \simeq 10^5 L_\odot$ is $\sim 10^{-6}\ M_\odot\ \mathrm{yr}^{-1}$; for a star of $10^6 L_\odot$, it is $10^{-5}\ M_\odot\ \mathrm{yr}^{-1}$. These values are close to the observed mass-loss rates of massive O and B supergiants.

N.B. In the estimates above, we have assumed that the outward-directed momentum of a stellar photon can be used only once. In reality, if there are many spectral lines and the photons are scattered multiple times, the total momentum transfer can be increased by at most a factor of $\simeq 3$. As a result, hot luminous stars may have mass-loss rates slightly higher than the value of $\dot{M}_{\mathrm{max}}$ derived above.

Q (15.1) Sketch the path of a scattered photon that transfers more momentum to the wind than $h\nu/c$.

15.2.2 Observed Wind Velocities and Mass-loss Rates of Hot Stars

The observed terminal velocities of winds from hot massive stars are in the range of about 2×10^2 to $3 \times 10^3\ \mathrm{km\,s^{-1}}$. Figure 15.2 shows that velocity scales with the escape velocity at the photosphere. The ratios are (Lamers et al. 1995)

$v_\infty \approx 2.6\ v_{\mathrm{esc}}$ if $T_{\mathrm{eff}} \gtrsim 21000$ K,
$v_\infty \approx 1.3\ v_{\mathrm{esc}}$ if $10,000 \lesssim T_{\mathrm{eff}} \lesssim 20,000$ K, and
$v_\infty \approx 0.7\ v_{\mathrm{esc}}$ if $8000 \lesssim T_{\mathrm{eff}} \lesssim 10,000$ K.

The jumps in the ratio $v_\infty/v_{\mathrm{esc}}$ are called **bi-stability jumps.** The difference between the three regimes is due to the fact that the ionization stages of the elements that drive the wind are dependent on T_{eff}. In the hottest stars, the driving is mainly done by triply ionized C, N, and O ions in stars of moderate temperature by doubly

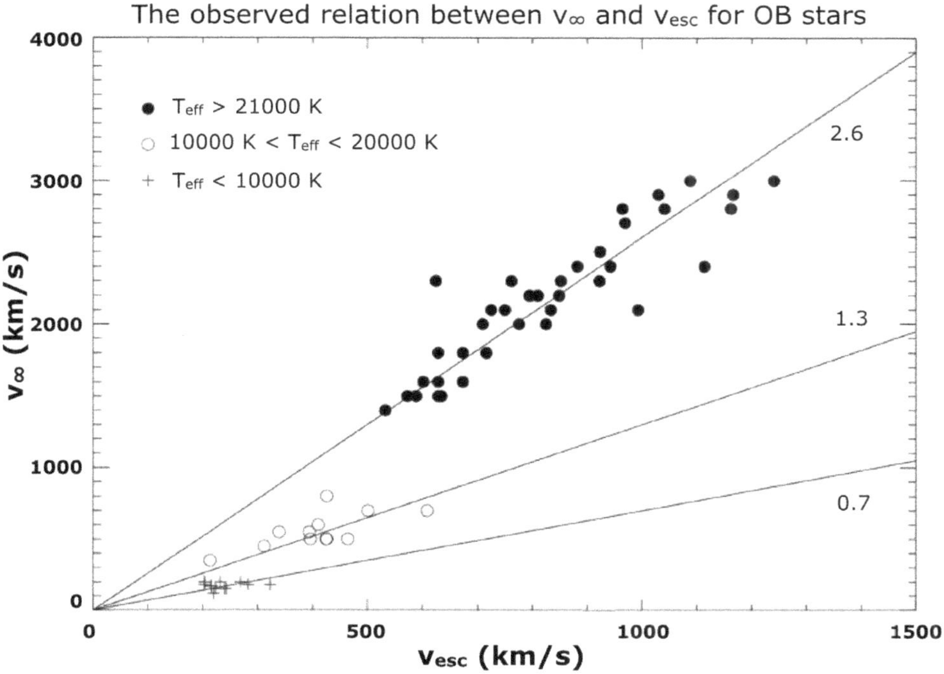

Figure 15.2. Observed relation between v_∞ and v_{esc} for line-driven winds of hot luminous stars. (Reproduced from Lamers & Cassinelli 1999. © Cambridge University Press. Reprinted with permission.)

ionized Fe-group elements, and in the coolest stars by singly ionized Fe-group elements (Vink et al. 2001).

Mass-loss rates of radiation driven winds are often expressed in terms of the **modified wind momentum**, defined by Kudritzki (1988) as

$$D \equiv \dot{M} v_{\infty} R^{1/2}, \tag{15.6}$$

which is expected to scale approximately as $L^{1/\alpha}$ with $\alpha \approx 2/3$ by the theory of radiation driven winds (Lamers & Cassinelli 1999). Figure 15.3 shows these relations for OB stars in the Galaxy, the LMC, and the SMC. The predictions roughly agree with observations.

Characteristic mass-loss rates for a number of Galactic massive main-sequence stars are listed in Table 15.1. Based on these rates, it is expected that stars with $M \gtrsim 30 M_{\odot}$ lose about 5% to 25% of their initial mass during the MS phase.

15.3 Dust-driven Winds of Cool Stars

Cool stars can have dust in their outer layers. The dust-to-gas ratio is typically $\sim 10^{-2}$ by mass, because dust cannot be formed from H and He but only from metals. Dust is an efficient absorber of radiation, so the absorption of stellar photons by dust grains produces a transfer of momentum (i.e., radiation pressure) from the photons to the dust grains. The outwardly accelerated dust grains collide with the gas atoms and molecules and drag these along, producing a stellar wind.

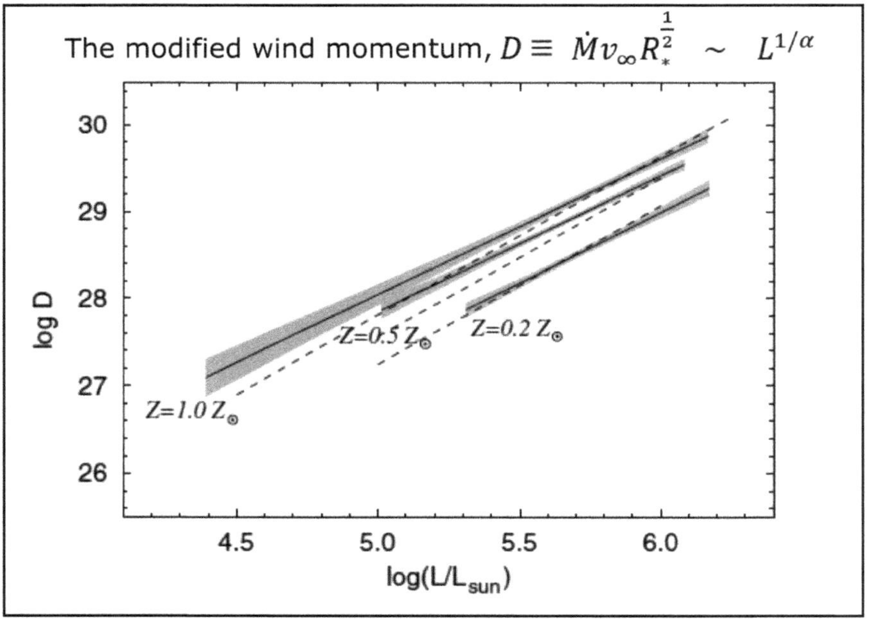

Figure 15.3. Observed modified wind momentum D in units of g cm s^{-2} ($R_{\odot}^{0.5}$) versus $L/L_{\odot}$. Gray bands indicate observed values (Mokiem et al. 2007) of Galactic, LMC, and SMC stars (from top to bottom, with the full line as the empirical mean relation). Dashed lines show the predictions by Vink et al. (2001) for the corresponding metallicities. (Reproduced from Mokiem et al. 2007, with permission. © ESO.)

Table 15.1. Characteristic Mass-loss Rates of Massive Galactic Main-sequence Stars.

$M/M_\odot$	$\log L/L_\odot$	$\log \dot{M}$ ($M_\odot \mathrm{yr}^{-1}$)	τ_{MS} (Myr)	$\Delta M/M$
30	5.15	−6.51	5.5	0.05
50	5.58	−5.86	3.8	0.10
80	5.97	−5.27	3.9	0.20
100	6.16	−4.99	2.7	0.27

15.3.1 The Minimum Luminosity for Dust-driven Winds

Suppose that the dust grains are spherical with a mean radius a and a mean density ρ_d. For icy particles $\rho_d \approx 1\ \mathrm{g\,cm^{-3}}$ and for silicates $\rho_d \approx 4\ \mathrm{g\,cm^{-3}}$. For simplicity, we adopt a mean value of $\rho_d \approx 2\ \mathrm{g\,cm^{-3}}$. So 1 g of gas contains $n_d = 10^{-2}/(4/3)\pi a^3 \rho_d$ dust grains, with a total cross-section of $n_d \pi a^2$. The total absorption coefficient of 1 g of dusty gas is

$$\kappa_d = n_d \pi a^2 Q \simeq \frac{3 \times 10^{-2}\, Q}{4a\rho_d} \simeq 10\ \mathrm{cm^2\,gram^{-1}\,of\ gas}, \tag{15.7}$$

where $Q \approx 10^{-2}$ is the efficiency factor for absorption and scattering, and the mean particle radius is $a \approx 0.05\,\mu$ (Lamers & Cassinelli 1999; Mathis et al. 1977). The wind can be driven by dust if the radiation pressure force exceeds gravity:

$$\frac{L\kappa_d}{4\pi r^2 c} > \frac{GM}{r^2} \rightarrow L > \frac{4\,\pi G M c}{\kappa_d} \rightarrow \frac{L}{L_\odot} \gtrsim 10^3 \frac{M}{M_\odot}. \tag{15.8}$$

This is the case for red giants, AGB stars, and red supergiants.

15.3.2 The Temperature of Dust

Circumstellar dust can only exist around cool stars; dust is destroyed at temperatures above ~1200 to 1500 K, depending on the composition of the dust. The temperature of dust depends on the equilibrium between the energy of the radiation that the dust grains absorb and their cooling by radiation. The energy absorbed per second by a spherical grain of radius a at distance r from the star of luminosity L is

$$e_{\mathrm{in}} = \pi a^2\, Q_{\mathrm{abs}} \frac{L}{\pi R^2} W(r/R) \approx \pi a^2 Q_{\mathrm{abs}} \frac{L}{4\pi r^2} = \pi a^2\, Q_{\mathrm{abs}}\, \sigma T_{\mathrm{eff}}^4 \frac{R^2}{r^2} \quad \text{if } r > 2R, \tag{15.9}$$

where $W(r/R) = \frac{1}{2}\{1-\sqrt{1-(R/r)^2}\}$ is the geometrical dilution factor that allows for the fact that the radiation intensity does not decrease as fast as $(R/r)^2$ at small distances when the star is not a point source ($W \approx (R/2r)^2$ at $r > 2R$). $Q_{\mathrm{abs}} \approx 10^{-2}$ is the efficiency factor for absorption by particles with a radius of $a \approx 0.05\ \mu\mathrm{m}$.

The energy radiated by a grain of temperature T_d per second is

$$e_{\rm out} = 4\pi a^2 Q_{\rm em} \sigma T_{\rm d}^4, \tag{15.10}$$

where $Q_{\rm em} \approx Q_{\rm abs}$ is the emission efficiency. Equating (15.9) and (15.10) gives the dust temperature at distance r from the star of

$$T_d^4 = T_{\rm eff}^4 \times \frac{1}{2}\left\{1-\sqrt{1-(R/r)^2}\right\} \tag{15.11}$$

for gray dust (i.e., Q is independent of wavelength). Taking into account the wavelength dependence of $Q \sim \lambda^{-p}$ with $p \approx 1$ yields

$$T_d(r/R) = T_{\rm eff}\,[W(r/R)]^{\frac{1}{4+p}} \approx T_{\rm eff}\,(2r/R)^{\frac{-2}{4+p}}. \tag{15.12}$$

Dust sublimates if it gets hotter than the condensation temperature, $T_{\rm cond}$, which is about 1000 to 1500 K depending on the type of dust. This means that dust can form only at

$$W(r/R) < (T_{\rm cond}/T_{\rm eff})^{4+p} \quad \text{or} \quad \frac{r}{R} > \frac{1}{2}\left(\frac{T_{\rm eff}}{T_{\rm cond}}\right)^{\frac{4+p}{2}} \quad \text{if} \quad r > 2R. \tag{15.13}$$

Table 15.2 shows the condensation radius, r_c (i.e., the minimum distance where dust can form), for various types of dust and effective temperatures. Note that for $T_{\rm eff} >$ 2500 K dust can only form at distances larger than about $2R$.

Q (15.2) *Equation (15.12) shows that the dust temperature is independent of the size of the dust grains. Explain this.*

15.3.3 The Role of Pulsation in Dust-driven Winds

We have shown that all cool stars with $L/L_\odot > 10^3\ M/M_\odot$ could, in principle, drive a wind by radiation pressure on dust grains (Equation (15.8)); however, we have also seen that dust can only form above the photosphere at a distance that depends on $T_{\rm eff}$, e.g., at $r > 3R$ if $T_{\rm eff} = 3000$ K. This creates a serious problem, since if there is no wind, the density at a distance of $2R$ is so low that dust cannot form. If there is no

Table 15.2 The Condensation Radius of Dust in Cool Star Winds.

Dust type $T_{\rm cond}$	Silicate 1100 K	Amorphous Carbon 1500 K	Graphite 1500 K
$T_{\rm eff}$	r_c/R	r_c/R	r_c/R
3000 K	2.42	3.42	4.03
2500 K	1.86	2.12	2.34
2000 K	1.37	1.24	1.29

Note. Data for carbon-rich dust are from Lamers & Cassinelli (1999) and data for silicates are from Bladh & Hoefner (2012).

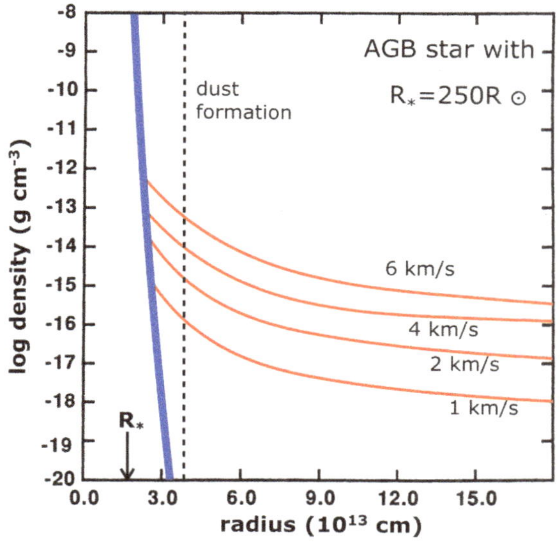

Figure 15.4. Time-averaged density structure $\rho(r)$ in the atmosphere of a pulsating AGB star of $R = 250R_\odot$ for different pulsation amplitudes of 1, 2, 4, and 6 km s^{-1}. The blue line is the density structure if the star does not pulsate. (Figure is based on Bowen 1988.)

wind dust cannot form, and if dust cannot form there is no wind! This does not apply to stars cooler than ~2000 K, where dust forms close to the star.

The problem can be solved if the star is pulsating (Lamers & Cassinelli 1999; Bowen 1988). Many red giants and AGB stars pulsate. Pulsation tosses the outer layers of the star radially outward, giving rise to a density distribution that decreases with distance much more slowly than in a hydrostatic atmosphere. This is shown in Figure 15.4, where the smoothed density structure of a pulsating AGB star with different pulsation amplitudes is compared to that of a nonpulsating star.

Figure 15.5 shows empirical evidence that the mass-loss rates of Mira variables are related to their pulsation. The mass-loss rate increases drastically with the period at $300 < P < 600$ days and saturates at a level of $\dot{M} \approx 10^{-4}\ M_\odot \mathrm{yr}^{-1}$ for longer periods.

Q (15.3) *Can you think of a mechanism that sets the upper limit for the dust-driven mass-loss rates of stars?*
(Hint: consider the optical depth of the wind).

15.4 Mass-loss Formulae for Stellar Evolution

15.4.1 Massive O, B, and A Stars

Vink et al. (2001) predicted mass-loss rates for luminous hot stars of $L > 3 \times 10^4\ L_\odot$, based on Monte Carlo simulations of radiation driven winds. These predictions fit observations very well (see Figure 15.3). They derived a formula of the type

$$\begin{aligned}\log \dot{M} \;=\; & A + B \log (L/10^5\, L_\odot) + C \log (M/30 M_\odot) + D \log (0.5 \mathrm{v}_\infty/\mathrm{v}_{\mathrm{esc}}) \\ & + E \log (T_{\mathrm{eff}}/T_{\mathrm{ref}}) + F\, (\log(T_{\mathrm{eff}}/T_{\mathrm{ref}}))^2 + G \log (Z/Z_\odot),\end{aligned} \tag{15.14}$$

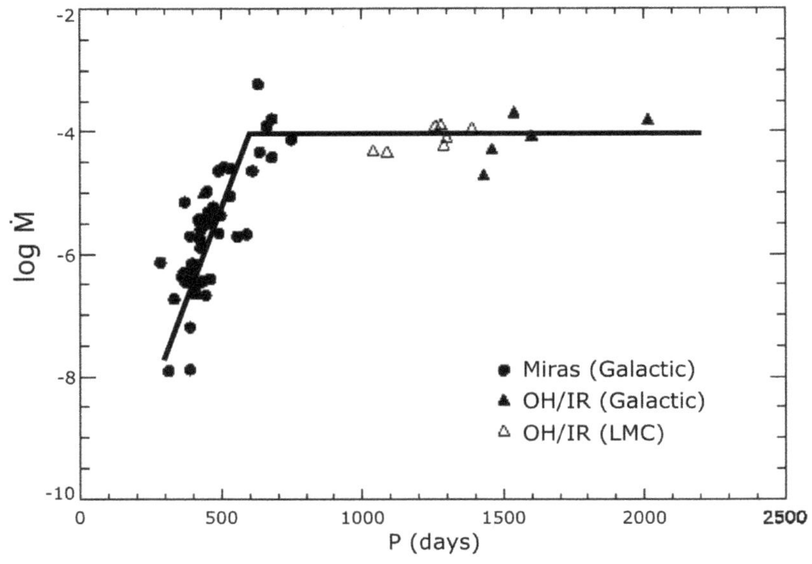

Figure 15.5. Mass-loss rates of Mira and OH/IR variables increase with longer pulsation periods, saturating at a mass-loss rate of about 10^{-4} M_{sun} yr^{-1}. (Reproduced from Vassiliades & Wood 1993.)

with $\dot{M}$ in $M_{\odot}yr^{-1}$, $Z_{\odot} = 0.02$, and $v_{\infty}/v_{esc} = 2.6$ if $T_{eff} > 21,000$ K and 1.3 if $10,000 < T_{eff} < 20,000$ K. The parameters in two different temperature ranges are listed in Table 15.3.

For the region $22,500 \lesssim T_{eff} \lesssim 27,500$ K an interpolation between the values of the two ranges is recommended. Figure 15.3 shows a comparison between these predictions and observations for luminous hot stars in the Milky Way, the LMC, and the SMC. The metallicity dependence arises from the fact that the winds are driven by radiation pressure in metal lines.

15.4.2 Wolf–Rayet stars

Wolf–Rayet stars are a late evolutionary stage of massive stars that have almost completely lost their H-rich envelope (Section 2.7). Nugis and Lamers (2000) derived the following empirical expression for the mass loss of Wolf–Rayet stars from the observations of 64 C-rich and N-rich Wolf–Rayet stars with $Y \gtrsim 0.3$

$$\dot{M} \simeq 2.8 \times 10^{-5} (L/10^5 L_{\odot})^{1.29} Y^{1.7} Z^{0.5}, \tag{15.15}$$

where Y is the helium mass fraction and $Z = 1 - X - Y$. The terminal velocities of these winds are between 2000 and 4000 km s^{-1}.

15.4.3 Nonpulsating Red Supergiants: The Reimers Relation

Reimers (1975) determined the mass-loss rates of six red giants and supergiants with $4 < M < 18 M_{\odot}$, $10^3 < L < 10^5$ $L_{\odot}$, $40 < R < 600 R_{\odot}$, and $3500 < T_{eff} < 5000$ K and derived an empirical relation that is known as the **Reimers relation**

Table 15.3. Mass-loss Rate Formula for Luminous Hot Stars (Vink et al. 2001)

	$27,500 \lesssim T_{\rm eff} < 50,000$ K	$12,500 < T_{\rm eff} \lesssim 22,500$ K
A	−6.697	−6.688
B	+2.194	+2.210
C	−1.313	−1.339
D	−1.226	−1.601
$T_{\rm ref}$ (K)	40, 000	20, 000
E	+0.933	+1.07
F	−10.92	0
G	+0.85	+0.85

$$\boxed{\dot{M} = 4 \times 10^{-13}\eta_R \frac{L}{L_\odot}\frac{R}{R_\odot}\frac{M_\odot}{M}\, M_\odot {\rm yr}^{-1},} \qquad (15.16)$$

where η_R is a correction factor that was added to adjust it to the mass-loss rates derived from observations of a larger number of cool stars of different spectral types. This relation is often used in stellar evolution codes to describe the mass-loss rates of cool stars.

This empirical relation implies that $\dot{M}M/R \sim \dot{M}(GM/R) \sim \dot{M}{\rm v}_{\rm esc}^2 {\sim} L$. Winds of cool supergiants and AGB stars have ${\rm v}_\infty \approx 10\,{\rm km\,s^{-1}} \ll v_{\rm esc}$. This shows that the kinetic energy of the winds is much smaller than the potential energy. The Reimers relation thus implies that for these stars a *fixed fraction of the stellar luminosity* is used to provide the *potential energy* of their wind that allows the gas to escape the gravity of the star. This is different from the case of the hot stars, where the *momentum* of the wind scaled with the *momentum* of the radiation.

Q (15.4) *What would $\eta_{\rm R}$ be if the Reimers relation was used to predict $\dot{M}$ of the Sun?*

An alternative empirical expression was proposed by Nieuwenhujzen & de Jager (1990).

$$\dot{M} = 1.2 \times 10^{-8}(L/L_\odot)^{1.64}(M/M_\odot)^{0.16}\,T_{\rm eff}^{-1.61}\,M_\odot {\rm yr}^{-1}. \qquad (15.17)$$

Van Loon et al. (2005) determined mass-loss rates from the spectral energy distribution of dusty stars in the LMC and derived the following empirical mass-loss formula for O-rich dust-enshrouded AGB stars and red supergiants:

$$\dot{M} = 2.2 \times 10^{-6}(L/10^4\,L_\odot)^{1.05}(T_{\rm eff}/3500\,{\rm K})^{-6.3}\,M_\odot {\rm yr}^{-1}. \qquad (15.18)$$

15.4.4 Pulsating Miras and AGB Stars

Vassilidadis & Wood (1993) derived an empirical mass-loss formula from the observed infrared flux of dusty winds (see Figure 15.5):

$$\begin{aligned}
\log \dot{M} &= -11.4 + 0.0123 \times P(\text{days}) && \text{if } P < 600\,\text{days and } M < 2.5M_\odot, \\
\log \dot{M} &= -11.4 + 0.0123\left\{P(\text{days}) - 100\left(\frac{M}{M_\odot} - 2.5\right)\right\}, \\
& && \text{if } P < 600\,\text{days and } M > 2.5M_\odot, \\
\log \dot{M} &= -4.0 && \text{if } P > 600\,\text{days},
\end{aligned} \tag{15.19}$$

all with $\log P_{\text{days}} = -2.07 + 1.94 \log(R_*/R_\odot) - 0.9 \log(M/M_\odot)$. The terminal velocity of these winds is about 5 to 20 km s^{-1}.

15.5 Summary

1. Luminous early-type stars and luminous cool stars have mass-loss rates high enough to affect their evolution. This is important for the late-stage evolution of low-mass stars and throughout the evolution of massive stars.
2. The winds of hot luminous stars are driven by radiation pressure in UV spectral lines. The mass-loss rates are of the order of 10^{-7} to 10^{-5} $M_\odot \text{yr}^{-1}$. These winds reach a terminal velocity of a few times v_{esc}, which corresponds to $v_\infty \approx 1000$ to 4000 km s^{-1}. Massive stars lose about 5% to 25% of their mass during the main-sequence phase.
3. Cool stars with $L/L_\odot > 10^3\, M/M_\odot$ have dust-driven winds. This is the case for luminous red giants and AGB stars. The mass-loss rates can be as high as 10^{-4} $M_\odot \text{yr}^{-1}$. Except for stars with $T_{\text{eff}} < 2500$ K, a star can only have a strong dust driven wind if the photosphere is extended by pulsation. The wind velocity of cool stars is of the order of 5 to 20 km s^{-1}.
4. There are several theoretical and empirical formulae that describe mass-loss rates as a function of stellar parameters. The theoretical formulae for the mass-loss rates of massive O, B, and A stars agree well with the observed mass-loss rates. The various empirical relations for the mass-loss rates of cool stars apply only to specific types of stars.

Exercises

15.1 The momentum of an efficient radiation driven wind is about equal to the wind momentum.
   - What is the ratio between the momentum of the wind and the momentum of the radiation of the stars listed in Table 15.1?
   - What is the ratio between the kinetic energy of the wind and the radiative energy of the stars listed in Table 15.1?
   - What is the ratio between the thermal energy of the wind and the radiative energy of the stars listed in Table 15.1?

15.2 Equation (15.11), which describes the temperature of dust that is irradiated by a star, can also be used to estimate the mean surface temperature of planets (with a small correction for reflection by clouds or ice). Check this for Mercury, Earth, and Neptune.

15.3 Compare the mass-loss rates predicted by Equations (15.16) and (15.17) for a set of MS stars (class V), giants (class III), and supergiants (class I) of $T_{\rm eff} \approx 6000$ and 4000 K (use the data of Appendix B). What do you conclude?

15.4 (a) Calculate the radiation driven mass-loss rate at the ZAMS and TAMS of stars of 20, 40, 60, and $120 M_\odot$ with solar metallicity ($Z = 0.014$) using stellar data from Appendix D.
(b) Take the mean value and estimate the fraction of mass that is lost from these stars during the main sequence.
(c) Do the same for metal-poor stars of $Z = 0.002$ and compare the results. What is the physical reason for the difference?

15.5 (a) Calculate the mass-loss rates for a variety of O, B, and A stars with $L > 3 \times 10^4 L_\odot$, for which the Vink et al. (2001) predictions apply (Section 15.4.1). Use stellar data from Appendix B.
(b) Calculate the value of the modified wind momentum D and verify if these predictions support the empirical statement that $D \sim L^{1/\alpha}$.
(c) Derive the value of α from the mass-loss predictions and compare it with the empirical value from Figure 15.3.

References

Bladh, S. & Hoefner, S. 2012, A&A, **546, 76**

Bowen, G. H. 1988, ApJ, **329, 299**

Castor, J. I., Abbott, D. C. & Klein, R. I. 1975, ApJ, **195, 157**

Kudritzki, R. P. 1988, The atmospheres of hot stars: modern theory and observations Radiation in Moving Gaseous Media, 18th Advanced Course of the Swiss Society of Astrophysics and Astronomy, ed. Chmielewsky (Sauverny-Versoix: Geneva Observatory)

Lamers, H. J. G. L. M. & Cassinelli, J. P. 1999, Introduction to Stellar Winds, (Cambridge: Cambridge Univ. Press)

Lamers, H. J. G. L. M., Snow, T. P. & Lindholm, D. M. 1995, ApJ, **455, 269**

Mathis, J. I., Rumpl, W. & Nordsieck, K. H. 1977, ApJ, **217, 425**

Mokiem, M. R., de Koter, A. & Vink, J. S., et al. 2007, A&A, **473, 603**

Nieuwenhuizen, H. & de Jager, C. 1990, A&A, **231, 134**

Nugis, T. & Lamers, H. J. G. L. M. 2000, A&A, **360, 227**

Reimers, D. 1975, MSRSL, **8, 369**

van Loon, J. Th., Cioni, M.-R. L., Zijlstra, A. A. & Loup, C. 2005, A&A, **438, 273**

Vassiliades, E. & Wood, P. R. 1993, ApJ, **414, 641**

Vink, J. S., de Koter, A. & Lamers, H. J. G. L. M. 2001, A&A, **369, 574**

Henny J.G.L.M. Lamers and Emily M. Levesque

Chapter 16

Shell H-fusion in Low- and Intermediate-mass Stars: Red Giants

Color–magnitude diagrams and Hertzsprung–Russell diagrams of old star clusters invariably show a significant fraction of stars on the red giant branch. These stars are brighter and cooler than the main-sequence stars in the same cluster. This indicates that stars increase their radii after the phase of core H-fusion. Red giants fuse H in a shell around the He core, so the expansion of the star is due to the contraction of the core and the mirror action of shell fusion. In the description of their evolution, stars with $M_i < 2M_\odot$ are referred to as **low-mass stars** and stars in the range of $2 < M_i < 8M_\odot$ are called **intermediate-mass stars**, because their post-MS evolutions are different. We discuss the evolution of stars of $M_i = 1$ and $5M_\odot$ in detail as examples of the two mass ranges to demonstrate the various effects. This makes it possible to connect computed phases of the internal evolution of stars to observed locations in the HRD. We will show that stars experience a convective dredge-up of nuclear products in the red giant phase. This changes the chemical composition of their photospheres.

16.1 The Start of the H-shell Fusion

When H is exhausted in the core, the core contracts. The layers around it also contract and their temperature increases to above $T > 10^7$ K, high enough for H-shell fusion. We have shown in Section 13.4 that H-shell fusion starts differently in stars with $M \gtrsim 1.2M_\odot$ and stars with $M \lesssim 1.2M_\odot$ (see Figure 13.6).

- In stars with $M \gtrsim 1.2M_\odot$, which had a convective core during the MS phase, the whole star contracts before H-shell fusion starts. This causes a small leftward loop in the evolutionary track. The shrinking of the radius ends when H-fusion is ignited in the shell.
- In stars with $M \lesssim 1.2M_\odot$, H-fusion in the shell starts gradually because the chemical profile is gradual. The transition from core H-fusion to H-shell

doi:10.1088/978-0-7503-1278-3ch16

fusion is smooth, so the envelopes of these stars do not contract and the evolutionary tracks do not show the small leftward loop.

16.2 The H-shell Fusion Phase of Low-mass Stars of 0.8–$2M_\odot$

We first describe the evolution of a $1M_\odot$ star as an example of the evolution of a star with $0.8 \gtrsim M \gtrsim 2M_\odot$.

In presenting the evolution of stars, we will discuss their tracks in the HRD combined with their **Kippenhahn diagram (KD).** This diagram was introduced by the German astrophysicist Rudolph Kippenhahn in 1965 (Kippenhahn 1965). It shows the changes in the interior stellar structure as a function of time (horizontal axis) and mass fraction (vertical axis), with various colors and shadings used to indicate regions of different fusion phases, convection, and composition. The combination of the KD with the HRD provides us with very good insights into stellar evolution.

Figure 16.1 shows the HRD and KD of a star of 1 $M_\odot$. The letters in the KD correspond to the locations of the same letters in the HRD. The following phases can be recognized.

A-B: The main-sequence phase of H-fusion in the core.

B-C: H becomes exhausted in the center ($X < 10^{-3}$) at point B. The core contracts and its temperature increases. Before it is high enough for core He-fusion, H-fusion is ignited in the shell around the core. The mass of the He core exceeds the Schönberg–Chandrasekhar limit (SC-limit) of $0.1M$ so the core keeps contracting. Because of the mirror action of the shell, the outer layers expand and the star moves to the right on the HRD. The expansion implies that an increasingly large fraction of the stellar mass becomes convective (see Figure 10.1). Phase B-C lasts about 2 Gyr. The star is now on the **sub-giant branch**.

C: At this point, about half of the stellar mass is convective: the star reaches the *Hayashi line.* From now on it is a **red giant.** Around this time the He core has contracted so much that it becomes degenerate!

Q (16.1) *Check in the systematic diagram of* Figure 8.10 *that a star of* $1M_\odot$ *with a core mass of about* $0.2M_\odot$ *can become degenerate before He is ignited.*

C-D: The H-shell fusion keeps adding mass to the degenerate core, which contracts because degenerate cores shrink when mass is added (see Equation (11.14)). Because of the mirror action of the H-fusing shell, the envelope expands. The star is on the Hayashi line so $T_{\rm eff}$ hardly changes, which implies that the expansion results in an increasing luminosity. So the star climbs the Hayashi line in the HRD during the red giant branch phase (RGB). As the core contracts, the density in the shell increases because it is directly on top of the degenerate core, so the shell-fusion becomes much more efficient. This produces the increasing luminosity required by HE and TE for stars on the Hayashi track. Note that the mass in the fusion shell gets smaller (narrower in the KD) because less mass is needed to produce the luminosity at higher fusion efficiency. The star climbs the RGB for about 0.5 Gyr.

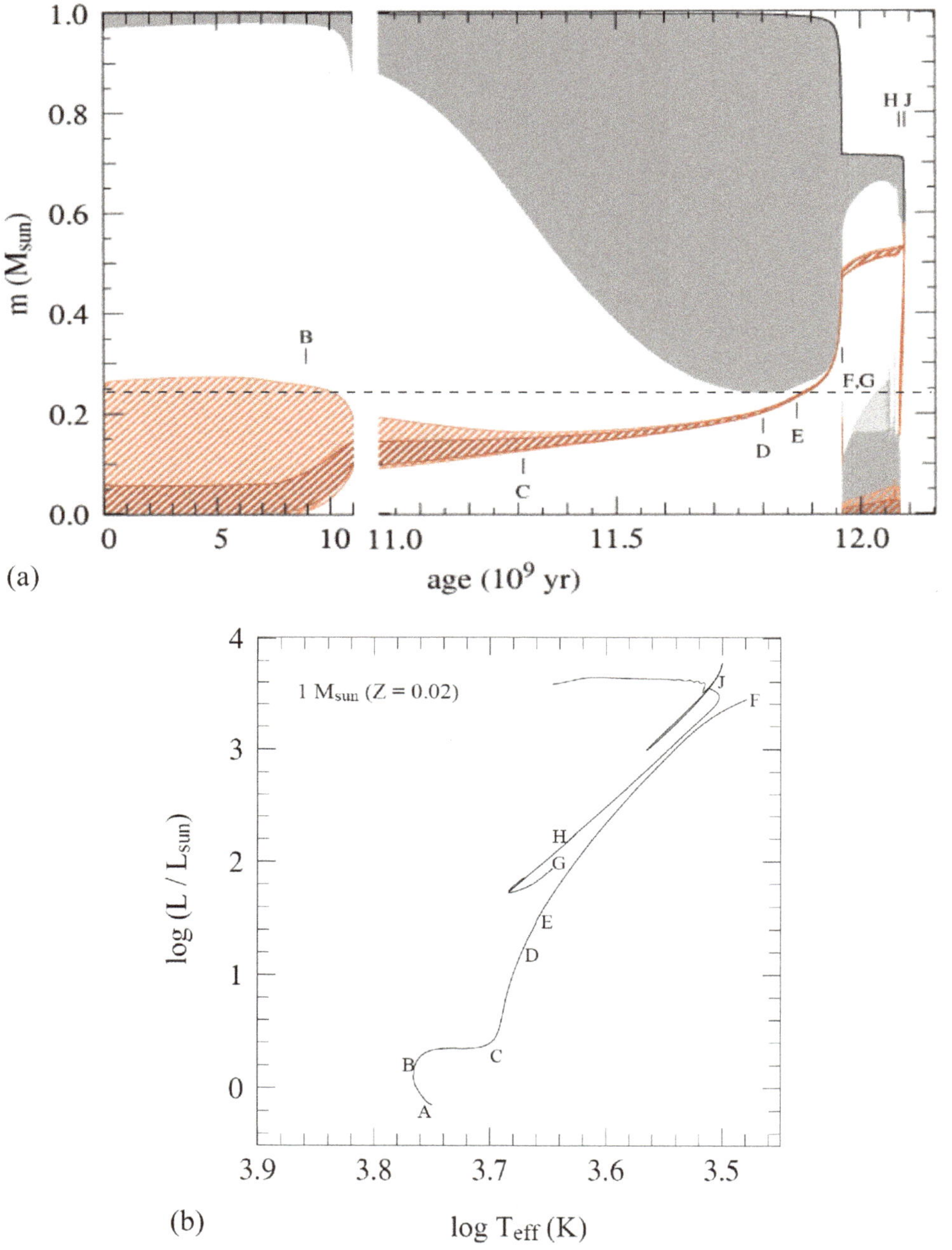

Figure 16.1. Kippenhahn diagram (top) and the HR-diagram (bottom), showing the evolution of a star of $1 M_\odot$. Hatched areas indicate fusion regions: dark hatched areas are for efficient fusion ($\varepsilon > 5\ L/M$) and light hatched areas are for inefficient fusion ($L/M < \varepsilon < 5\ L/M$). Gray regions indicate convection, while the light gray area indicates semi-convection. The horizontal dashed line indicates the deepest reach of the convective envelope, which results in the first dredge-up. (© Pols 2011.)

D: Near point D the outer convection reaches so deep into the star (to $m/M \approx 0.25$), that the products of H-fusion from the main-sequence phase (light red shaded area in the KD) are mixed to the surface. The surface abundance may now start to show evidence of a slight He enrichment (difficult to detect in spectra of cool stars), a slight increase in the N abundance, and a slight decrease of C and O. This is called **the first dredge-up.**

D-E-F: As the degenerate core gets more massive, it keeps shrinking, so the star keeps expanding and the luminosity keeps increasing. The H-shell fusion reaches hotter layers and becomes more efficient, producing the required luminosity. This results in faster and faster growth of the core, which in turn results in an even more rapid increase in luminosity, etc. This acceleration of the evolution can clearly be seen in the KD by the fast growth of the core mass. The increase in efficiency of the shell fusion can also be seen in this diagram because the mass of the shell decreases, whereas its energy output (luminosity) increases.

E: At this point, the shell has reached a mass zone, $m(r)/M = 0.25$, which was earlier reached by the deepest extent of the convective envelope. The convection has mixed fresh H from the outer layers down to this depth. Although the main H-fusion occurred in deeper layers, there had been some depletion of H due to inefficient fusion during the MS phase (see Figure 13.7). Thus, at phase E, the H-fusion shell finds itself in a layer with increased H content and a lower mean atomic weight than before. As a consequence, it starts fusing at a slightly lower rate. In fact, its luminosity temporarily decreases (because L depends on μ) and the star moves slightly down the Hayashi track as it shrinks a little. This produces a little loop in the track that is shown in Figure 16.2 for a star of 0.8 $M_\odot$. The star spends about 20% of its RGB time near this loop. This produces a small peak in the observed luminosity distribution of RGB stars at $M_V \approx 1.0$.

E-F: During this phase, the shell is burning in a region of higher H abundance, so the fusion can be slower and still produce the required luminosity. Therefore, phase E-F lasts longer than phase D-E, although the luminosity is higher.

F: At this point the degenerate core has reached a mass of $M \approx 0.45 M_\odot$. The contraction of the core has resulted in a temperature high enough for the ignition of He fusion in the degenerate core. The star leaves the Red Giant Branch.

Phases F-G-H and H-J are the core He-fusion and the double shell fusion phases, which will be discussed in Chapters 17 and 18, respectively. The gap in the track between points F and G indicates that the star is not in equilibrium during that transition (see Section 17.2).

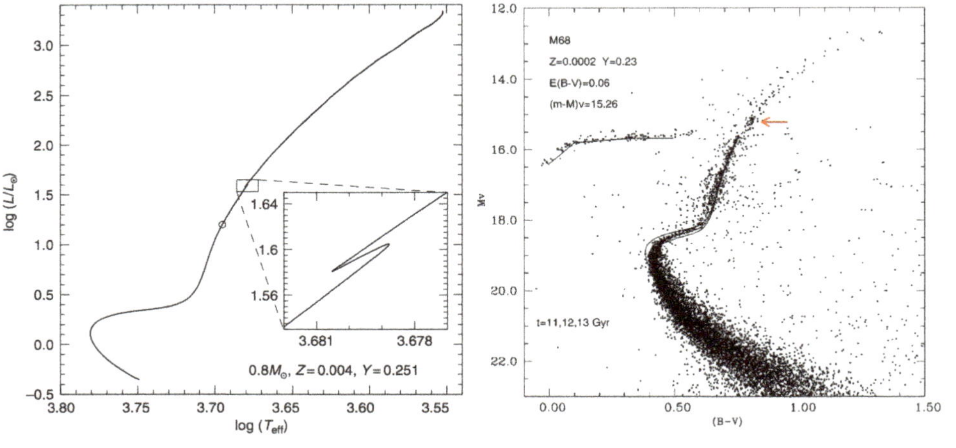

Figure 16.2. Left: the little loop in the RGB evolution of a star of 0.8 $M_\odot$ occurs when the H-fusion shell finds itself in a layer where the H abundance is larger because of envelope convection. This loop causes a small bump in the luminosity distributions of old globular clusters (Reproduced from Salaris & Cassisi 2005. Copyright © 2005 by John Wiley & Sons, Inc. Reprinted by permission of John Wiley & Sons, Inc.). Right: the CMD of globular cluster M68 with the location of the small density increase produced by this loop indicated by the red arrow (Reproduced from Salaris 1997).

Q (16.2) *The location of the red clump in the HRD of old clusters is very sensitive to convective overshooting. Can you explain this?*

16.3 The H-shell Fusion Phase of Intermediate-mass Stars of 2–8 $M_\odot$

The post-MS evolution of stars with an initial mass of $2 \lesssim M_i \lesssim 8 M_\odot$ is very similar to that of lower-mass stars. The tracks have the same general structure as in Figure 16.1. The stars go through the subgiant phase and the red giant phase when the star is largely convective and evolves upward on the Hayashi track; however, there are some differences.

a. In lower-mass stars, the He core is more massive than the Schönberg–Chandrasekhar (SC) limit, so the core keeps contracting when H-shell fusion starts. In more massive stars, the mass of the He core is initially smaller than the SC-limit, so the isothermal He core does not contract; however, due to the addition of He by the H-shell fusion, the core mass does later exceed the SC-limit and contract. The mirror action of the shell then results in an expansion of the star by increasing the depth of the convective envelope, just like in lower-mass stars.
b. Contrary to the lower-mass stars, the He core of stars with $2 \lesssim M_i \lesssim 8 M_\odot$ does *not* become degenerate during the H-shell fusion. This is due to the higher T and lower ρ at the start of H-shell fusion. We will show in chapter 17 that this has important consequences for the He ignition in the core.

d. The H-shell fusion ends when the mass of the He core is only slightly higher than the value of $M_c \approx 0.5M_\odot$ for the lower-mass stars.
e. The evolution timescales of the more massive stars are shorter than those of the lower-mass stars. The MS lifetimes of stars of 1 and $5M_\odot$ are, respectively, 9.0 Gyr and 82 Myr; the times to cross the HRD from the TAMS to the Hayashi line are 2.3 Gyr and 2.4 Myr; and the stars climb the Hayashi track in 160 Myr and 0.3 Myr, respectively.

Q (16.3) *Verify with* Figure 8.10 *that the core of a* $5M_\odot$ *star does not become degenerate during the H-shell fusion phase.*

16.4 The $M_{\rm core}$–L Relation for Red Giants

Stars with $M_i \lesssim 2M_\odot$ have a degenerate He core and a very extended envelope when they are on the RGB. This implies that the pressure on top of the H-fusion shell is small. The pressure at the bottom of the H shell is mainly due to the shell itself and the gravity from the core. This implies that the efficiency of the shell fusion is mainly determined by the core mass and not by the envelope mass. Detailed evolutionary models have shown that there is a strong and steep **$M_{\rm core}$–L relation for stars with a degenerate He core** (Pols 2011; Refsdal & Weigert 1970):

$$\boxed{L/L_\odot \simeq 2.3 \times 10^5 (M_c/M_\odot)^6}. \quad (16.1)$$

The luminosity is independent of the *total mass* of the star. Therefore, all evolutionary tracks of stars of different mass converge onto the Hayashi line of the RGB. In other words, one can derive the core mass from the luminosity of a star on the RGB, but not the total mass.

16.5 Metallicity Dependence of the Red Giant Branch

We have seen that fully convective stars are on the Hayashi line. Red giants are *almost* fully convective stars because their core contains only a small fraction of the stellar mass. We have shown in Section 14.3 that the location of the Hayashi line in the HRD depends strongly on the opacity in the photosphere. The H^- opacity in cool photospheres depends on metallicity because the metals provide the electrons for H^-. A higher metallicity provides more free electrons and a higher opacity. A higher opacity means that $\tau \simeq 1$ is reached at a lower density (i.e., further outward). As a result, the **Hayashi line for metal-rich stars is at a lower $T_{\rm eff}$** than that of metal-poor stars. This is why the metallicity of globular clusters can easily be derived from the location of the RGB in the HRD.

Figure 16.3 shows the CMD of the cluster M54, which is at the center of the Sagittarius Dwarf Elliptical Galaxy. The cluster had multiple star-formation periods with different metallicities due to infalling gas. The CMD clearly shows the presence of two main sequences: a heavily populated short one corresponding to an age of 12 Gyr and an extension to higher luminosity produced by a younger population with an age of <2 Gyr. It also shows two RGBs: a blue one with a color of (F606W-

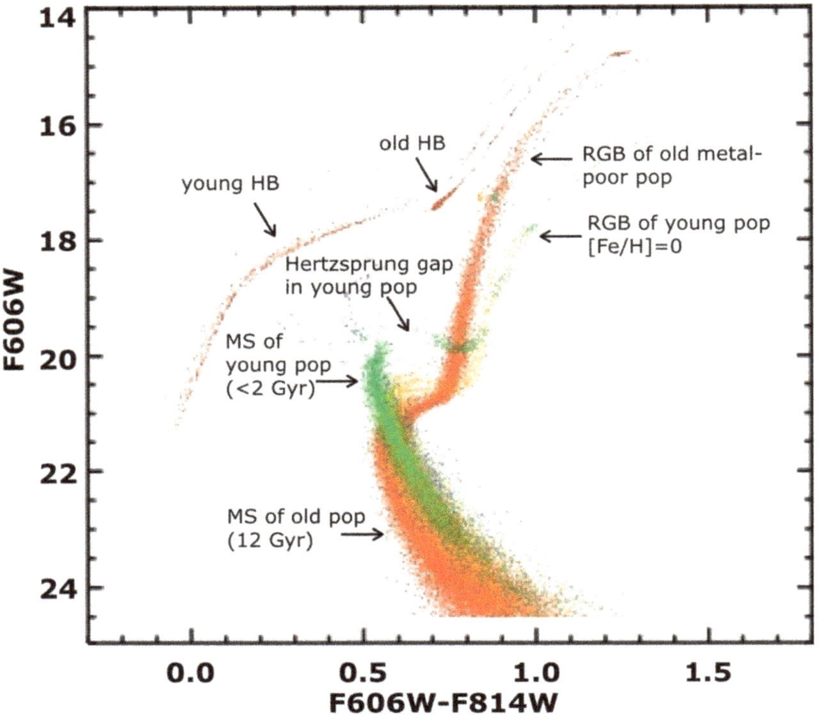

Figure 16.3. Color–magnitude diagram (CMD) of the cluster M54, which is the center of the Sagittarius Dwarf Elliptical Galaxy. The horizontal axis shows the color (F606W-F814W), (i.e., the difference in magnitude between these two HST filters), and the vertical axis shows the apparent magnitude in the F606W filter. Note the multiple main sequences (MS), the multiple red giant branches (RGB), and the multiple horizontal branches (HB) corresponding to populations of stars with different metallicities. Also note the Hertzsprung gap of the younger population and the lack of it for the older population. (Reproduced from Siegel et al. 2007.)

F814W) ≈ 0.8, which is due to the old metal-poor population with $Z \approx 0.016\ Z_\odot \approx 2 \times 10^{-4}$ and a redder more metal-rich RGB due to the younger population with $Z \approx Z_\odot$. The curvature of the RGB to the right in this diagram is due to the strong sensitivity of the (F606W-F814W) color index to $T_{\rm eff}$ (see Section 2.6). The CMD also shows multiple horizontal branches.

The distribution of the young population of stars in the HRD of the cluster M54 shows a gap between the end of the main sequence and the lower RGB. This is called the **Hertzsprung gap.** It is due to the short crossing time between the MS and the RGB. For low-mass stars this crossing occurs on the KH timescale of $\tau_{\rm KH} < 0.1\ \tau_{\rm MS}$. For stars more massive than about $3M_\odot$ the crossing time in the HRD is even shorter, so the probability of observing stars during that crossing is small.

16.6 Mass Loss during the Red Giant Phase

Red giants lose mass by means of a stellar wind. These stars are located in the HRD on the red giant branch (RGB). The winds of subgiants and low-luminosity RGB stars are possibly driven by gas pressure in a chromosphere combined with dissipation of magnetic Alfvén waves that originate in the convection zone (Lamers & Cassinelli 1999; Suzuki 2007). For stars higher up on the red giant branch, the winds are driven by a combination of stellar pulsation and radiation pressure on dust (see Section 15.3.3).

In most stellar evolution calculations, the Reimers relation (Equation (15.16)) is adopted with a free efficiency factor $\eta_R \simeq 0.25$ to 0.5, which appears to give reasonably good evolution predictions. Adopting this description, a star of $1M_{\odot}$ loses about $0.3M_{\odot}$ on the RGB (see the upper limit at phase F in the KD of Figure 16.1) and a star of $5M_{\odot}$ loses about $0.8M_{\odot}$ during the RGB phase.

16.7 Summary

1. At the end of the core H-fusion phase the core contracts and H-fusion starts in a shell around the He core.
2. In stars of $M \lesssim 2M_{\odot}$, the core mass is higher than the Schönberg–Chandrasekhar limit, so the core contracts. Due to the mirror action of the shell, the envelope expands by deepening the convective layers. Initially, the star moves horizontally in the HRD as a **subgiant,** but as it hits the Hayashi line it becomes a **red giant**. From then on, the expansion results in an increase in luminosity.
3. The H shell keeps moving outward in mass and adds helium to the contracting core. The He core keeps shrinking and becomes **degenerate.**
4. When the lower boundary of the convection zone reaches layers where fusion has occurred during the MS phase, the nuclear fusion products are transported to the surface. This **first dredge-up** increases the N abundance and decreases the C and O abundance in the atmosphere.
5. As the fusion shell moves outward in mass (but inward in radius) it reaches the deepest layers of the convection zone, which still have the initial H abundance. This results in a decrease of the fusion rate and a small drop in luminosity. This produces the **red clump** in the red giant branch of the HRD.
6. The post-MS evolution of stars with $2 \lesssim M \lesssim 8M_{\odot}$ is very similar to that of lower-mass stars, but they do not develop a degenerate core.
7. When the core has reached a mass of about $0.5M_{\odot}$, only slightly dependent on the stellar mass, He fusion is ignited.
8. The transition from the MS to the RGB is short and decreases with increasing stellar mass. This is the reason that the HRD of young clusters and field stars shows a gap between the MS and the RGB. This Hertzsprung gap is clearly present in the young population of cluster M54 (Figure 16.3).

Exercises

16.1 What is the radius of a star of $1M_\odot$ at the start and at the end of the RGB phase?

16.2 The Kippenhahn diagram in Figure 16.1 shows that the mass fraction of the convective envelope decreases between phases D and F, when the radius of the star increases. However, we have argued in Figure 10.1 that the radius of a star with a convective envelope increases when the convection reaches deeper into the star. Explain the apparent discrepancy between these two statements about the depth of the convection.

16.3 Assume that the mass-loss rate of a red giant branch star of $M_i = 1M_\odot$ is described by the Reimers relation (Equation 15.16) with $\eta_R = 0.5$. Calculate the mass-loss rates at points C, E, and F (in Figure 16.1) of the RGB phase and compare it with the Kippenhahn diagram.

16.4 **Computer exercise**

Write a simple program to calculate the luminosity, radius, mass-loss rate, core mass, and envelope mass as a function of time, when a star of 1Msun climbs the RGB track from point C to F in Figure 16.1.

Hints: (a) use Equation (16.1) for the relation between L and the core mass.

(b) calculate the increase in the core mass by the efficiency of the H-shell fusion.

(c) use the Reimers mass-loss rate with $\eta_R = 0.5$.

References

Kippenhahn, R. 1965, MitAG, **9, 53**

Lamers, H. J. G. L. M., and Cassinelli, J. P. 1999, Introduction to Stellar Winds (Cambridge: Cambridge Univ. Press)

Pols, O. 2011, in Stellar Structure and Evolution, Utrecht University Lecture Notes, 123

Refsdal, S., and Weigert, A. 1970, A&A, **6, 426**

Salaris, M. and Cassisi, S. 2005, Evolution of Stars and Stellar Populations (Chichester: John Wiley and Sons)

Salaris, M., Degl'Innocenti, S., and Weiss, A. 1997, ApJ, **479, 665**

Siegel, M. H.,, Dotter, A., and Majewski, S. R., et al 2007, ApJL, **667, L57**

Suzuki, T. K. 2007, ApJ, **659, 1592**

IOP Astronomy

Understanding Stellar Evolution

Henny J.G.L.M. Lamers and Emily M. Levesque

Chapter 17

Helium Fusion in Low-mass Stars: Horizontal Branch Stars

The Hertzsprung–Russell diagrams of old clusters show a concentration of stars on the horizontal branch. Stars on the horizontal branch are about a factor of 30 to 100 more luminous than stars of about the same mass at the tip of the main sequence. We will show that stars on the horizontal branch have He-fusion in their cores surrounded by a H-fusion shell. The cool end of the horizontal branch reaches the red giant branch. In some clusters the hot end extends to $T_{\rm eff} > 30{,}000$ K whereas, in other clusters, it reaches only to $T_{\rm eff} \approx 6000$ K. The extent to the blue side of the H-R diagram depends on the mass of the envelope that is left after the start of core He-fusion.

17.1 The Ignition of Helium Fusion in Low-mass Stars

When the mass of the He core reaches a value of about $0.5M_\odot$, slightly dependent on the total stellar mass, the contracting core reaches the ignition temperature of He-fusion at $T_c \sim 10^8$ K. Because of the core mass–luminosity relation of red giants, this happens at the tip of the RGB at $L \sim 10^3 L_\odot$ (see point F in Figures 16.1 and 17.1). We have seen that stars with initial masses of $M_i \lesssim 2M_\odot$ have a degenerate He core at the end of their RGB phase, whereas stars with $M_i \gtrsim 2M_\odot$ have a nondegenerate He core at the tip of the RGB. This results in different ignition processes for stars in these two mass ranges.

When He is ignited, the produced energy leads to a T increase.

- In a *nondegenerate core,* such a rise in T results in an increase in P, so the core expands. As the core expands, T and ρ decrease, which reduces the fusion efficiency, so the star remains in hydrostatic and thermal equilibrium. In this case, gravity acts like a regulator: the fusion does not get out of hand. This is similar to the regulated start of the core H-fusion in main-sequence stars.

doi:10.1088/978-0-7503-1278-3ch17

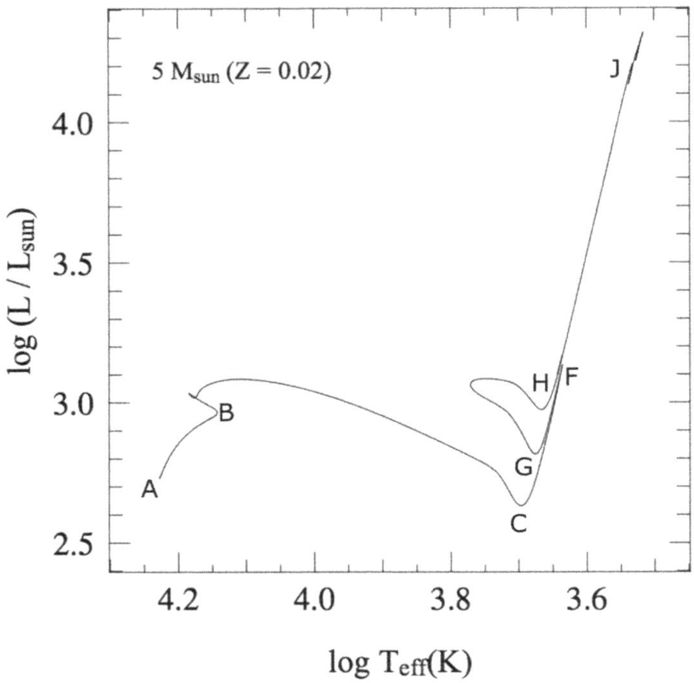

Figure 17.1. Evolution track of a star of $M_i = 5M_\odot$ with solar metallicity $Z = 0.02$. The letters refer to the same phases as those of figure 16.1 for a star of $M_i = 1M_\odot$. Phase F-G-H is the core He-fusion phase. Phase H-J is the double shell fusion phase, which will be discussed in chapter 18. (© Pols 2011.)

- In a *degenerate core*, the ignition of He-fusion also produces a rise in T; however, this does *not* result in an increased P, because P is independent of T in a degenerate core. Because T rises and the core does not expand, the energy production ϵ increases drastically; remember that $\epsilon \sim T^{30}$ for He-fusion. This leads to more efficient fusion, which yields still higher T, which leads to even more efficient fusion, etc. A degenerate core that is ignited acts like a bomb! This is called the **helium flash**.

When the *temperature* in the degenerate core of a star of $M_i \lesssim 2M_\odot$ shoots up and reaches a value of a few times 10^8 K, the degeneracy is lifted. This is because the limit between degenerate and ideal gas is set by $T_c \sim \rho_c^{2/3}$ (Figure 4.4). For any density, there is a T where the gas is no longer degenerate. The density of the degenerate He core of $0.5M_\odot$ is about 10^6 g cm^{-3}, so degeneracy is lifted when $T_c \approx 3 \times 10^8$ K. When that happens, the pressure increases when the temperature rises, so the core expands very quickly and the fusion rate drops. The star then settles into normal nondegenerate core He-fusion in hydrostatic and thermal equilibrium. The luminosity produced in the core during the He flash is very high, of the order of $10^{10}L_\odot$, but the He flash itself is extremely brief—it has never been observed! This is because the energy of the flash is used to expand the originally degenerate core by a factor of 100 in volume: from a degeneracy density of the order of 10^6 to nondegeneracy at 10^4 g cm^{-3}. Moreover, neutrinos remove a substantial fraction of the fusion energy.

Q (17.1) *Can you think of a way to estimate the duration of the He flash if it produces a luminosity of $10^{10}L_{\odot}$ and neutrinos carry off about 80% of the energy?*

17.2 Helium Fusion in the Core: Horizontal Branch Stars

Figure 17.1 shows the evolutionary track of a star of $M_i = 5M_{\odot}$ as a typical example. Phases B-C-F are similar to those in Figure 16.1 for a star of $1M_{\odot}$. We will discuss the core He-fusion phases of stars of $M_i = 1$ and $5M_{\odot}$ simultaneously.

The red giant branch ends at point F with the ignition of He in the core, when the core mass is $M_c \approx 0.50M_{\odot}$. At the beginning of core He-fusion in equilibrium, the star is at point G. A star of $M_i = 1M_{\odot}$ is clearly out of equilibrium during the He flash; therefore, a track in the HRD between points F and G in Figure 16.1 is not known and is not plotted. The ignition of the He core fusion in a star of $M_i \gtrsim 2M_{\odot}$ occurs in a nondegenerate core and gravity regulates the start of the fusion as described above. The star remains in equilibrium so the evolutionary track can be calculated: see phases F to G in Figure 17.1.

When core He-fusion has started, the star settles into a new equilibrium and we find the stars on the **horizontal branch (HB)** in the HRD. The core He-fusion (G-H) of a star of $M_i = 1M_{\odot}$ occurs at $L \sim 10^2 L_{\odot}$, with a core mass of $\sim 0.5M_{\odot}$. The core He-fusion (G-H) of a star of $M_i = 5M_{\odot}$ occurs at $L \sim 10^3\ L_{\odot}$, with a core mass gradually increasing from ~ 0.6 to $1.0M_{\odot}$.

The structure of an HB star is shown schematically in Figure 17.2. It can be compared with the Kippenhahn diagram of Figure 16.1 for a star of $1M_{\odot}$. We recognize the following structure from inside out.

- A **core with He-fusion:** He → C.
  Due to the strong T-dependence of the triple α process, the inner $\sim 1/3$ of the core is convective because $L/4\pi r^2$ is very high (Section 7.9).
- An **inert He region** around it without fusion.
- A **H-fusion shell**, which contributes significantly to the luminosity.
- A **H-rich envelope**; this envelope is mainly in radiative equilibrium, but with a convective outer region.

At the beginning of core He-fusion (phase G in Figures 16.1 and 17.1), the stars are still close to the Hayashi line because they are mainly convective, but they have a smaller radius than stars at the tip of the RGB. This is because of the virial theorem and the resulting mirror effect of the H-fusion shell. The core expands when core He-fusion starts, causing the envelope to contract. This results in a smaller R and L than stars at the tip of the RGB.

During the core He-fusion phase, the star traces a *blue loop* in the HRD. This loop is the mirror image of the small red loop that stars trace in the HRD during their core H-fusion. (see Figure 13.6). It is due to the expansion of the star during

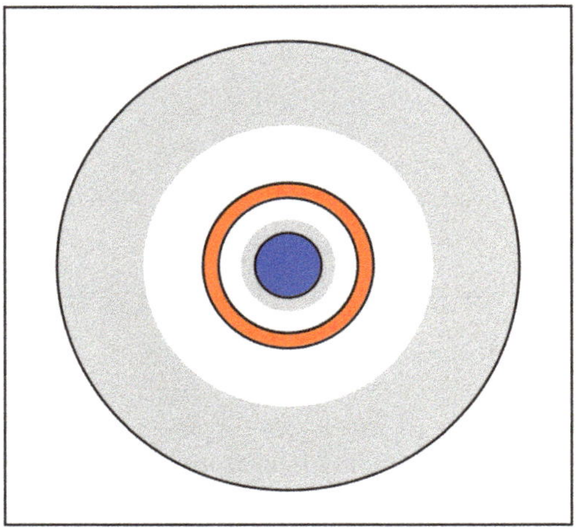

Figure 17.2. Schematic diagram of the structure of a horizontal branch star. Blue: He-fusion in a convective core. Red: shell H-fusion. Gray: convective zones. The figure is not to scale; the core is very small and the convective envelope is very extended.

core H-fusion and the contraction at the end. During core He-fusion, the core does the same, but the mirror action of the shell transforms it into a blue loop. This is shown by the loop G-H in the evolutionary track of a $5M_{\odot}$ star, whose $T_{\rm eff}$ increases to about 6000 K (Figure 17.1), and to a smaller extent by the track of the $1M_{\odot}$ star, which reaches 5000 K at its smallest radius (Figure 16.1). Figure 16.1 shows that the increase and decrease of the radius is produced by a decrease and later an increase of the mass and depth of the convective envelope during phase G-H.

The slight increase in luminosity during core He-fusion is due to the same μ-effect, $L \sim \overline{\mu}^4$, that is responsible for the brightening during the MS phase (Section 13.3.2). The fusion of He into C implies that μ increases from 4/3 to 12/7 in the core.

The duration of the HB phase is about 120 Myr for a $1M_{\odot}$ star and about 22 Myr for a $5M_{\odot}$ star. This is longer than the expected nuclear lifetimes of these stars (with a luminosity of about $10^2L_{\odot}$ and a core mass of $0.5M_{\odot}$, and a luminosity of $10^3L_{\odot}$ with a core mass of $0.6M_{\odot}$, respectively), considering the small mass deficiency of He-fusion of $\Delta m/m = 0.00065$. This is due to the large contribution of the energy production by the H-shell fusion (see Exercise 17.1).

The He-fusion occurs originally by the triple α-process (3He $\rightarrow$ C), but as the He abundance decreases and the C abundance increases the reaction He + C $\rightarrow$ O becomes more important. At this point the C abundance starts to decrease again and the O abundance increases in the core.

17.3 Evolution on the Horizontal Branch

The highest value of $T_{\rm eff}$ that a star can reach on the HB depends on the envelope mass at the time that core He-fusion starts. The smaller the envelope mass, the shallower the convective envelope, the smaller the radius, and the higher the $T_{\rm eff}$.

This is demonstrated in Figure 17.3, which shows the location in the HRD of low-mass HB stars with $Z = 0.001$ (characteristic of old globular clusters) during the core He-fusion phase. All models have the same core mass of $0.49 M_{\odot}$, but different envelope masses from 0.03 to $0.43 M_{\odot}$, so the total mass ranges from 0.52 to 0.92 $M_{\odot}$. Note that the smaller the envelope mass is, the bluer the star is during the HB phase. This is easy to understand: a more massive convective envelope will be deeper, which in turn implies a larger radius and a cooler star.

We have shown that, at the end of the core He-fusion phase, the core contracts and the star expands and moves back to the Hayashi line with increasing luminosity. At this phase, the evolution in the HRD is to the upper right. This is shown by the thin blue lines in Figure 17.2. However, if the envelope mass is too small (i.e., $M_{\rm env} < 0.03 M_{\odot}$ for a core of $0.5 M_{\odot}$), the mass of the envelope is insufficient to create an extended convection zone, so the radius of the star remains small. This could be the result of the increasing mass of the He core from the shell fusion, or by mass loss during the HB phase. Hot HB stars lose mass by a radiation-driven stellar wind.

HB stars with $M_{\rm env}/M \lesssim 0.1$ move up and *to the left* in the HRD. They become **hot subdwarfs**. Because they do not move back to the asymptotic giant branch, they are sometimes called **AGB-manqué stars**: stars that missed the AGB.

17.4 The Observed HB of Globular Clusters

Because the He-fusion phase is much shorter than the H-fusion phase, all HB stars in a cluster at any time come from a small range in initial masses. This implies that they

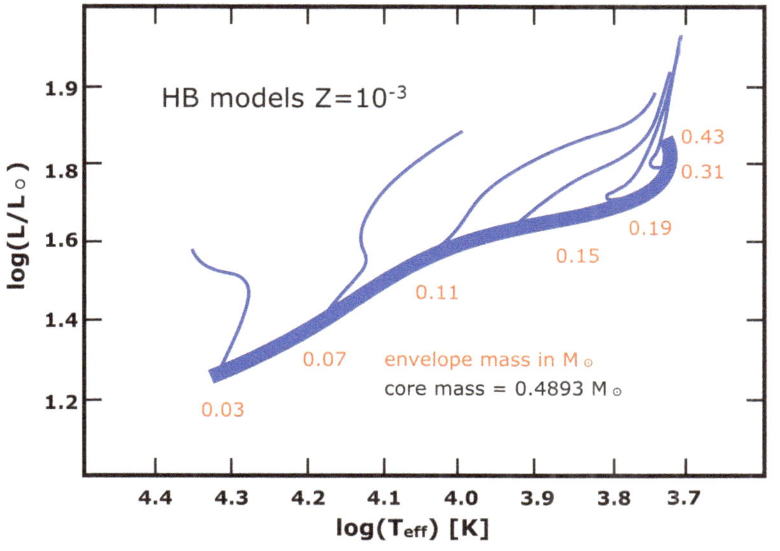

Figure 17.3. Blue lines show the predicted evolutionary tracks of HB stars with the same helium core mass of $0.49 M_{\odot}$ for different envelope masses ranging from 0.03 to $0.43 M_{\odot}$ (indicated by red numbers). The thick blue line shows the location at minimum radius and the thin lines show the track of the subsequent evolution during core He-fusion. Note the relation between $T_{\rm eff}$ and the envelope mass. (Reproduced from Maeder 2009. © Springer-Verlag Berlin Heidelberg 2012.)

all have about the same core mass and therefore also about the same luminosity L at the beginning of He-fusion. This is the reason that the HB is approximately horizontal in the HRD. From the evolutionary tracks of 1 and $5M_\odot$ stars (Figures 16.1 and 17.1), we can see that a cluster of 12 Gyr has an HB at $10^2 L_\odot$, and a cluster of 0.1 Gyr has an HB at $10^3 L_\odot$.

In an observational color–magnitude diagram the HB is not horizontal at all, but curves steeply down at high temperatures (Figure 2.7). This is due to the fact that the bolometric correction, $BC \equiv M_v - M_{\rm bol}$, increases steeply at high T because most radiation in hot stars is emitted in the UV. On the other hand, the HB is almost horizontal in the HRD. This is shown in Figure 17.4. The figure also shows the strong sensitivity to metal abundance mentioned in Section 16.5.

Observations show that *metal-poor clusters generally have an extended blue HB and metal-rich clusters generally have a short red HB*. Figure 17.5 shows the CMDs of the metal-rich globular cluster 47 Tuc with $Z = 0.17Z_\odot$, and the metal-poor cluster M15 with $Z = 0.006Z_\odot$. The conclusion is that stars in more metal-rich clusters have a larger envelope mass when they arrive at the HB than those in metal-poor clusters. This would imply that stars with higher Z have lost less mass during

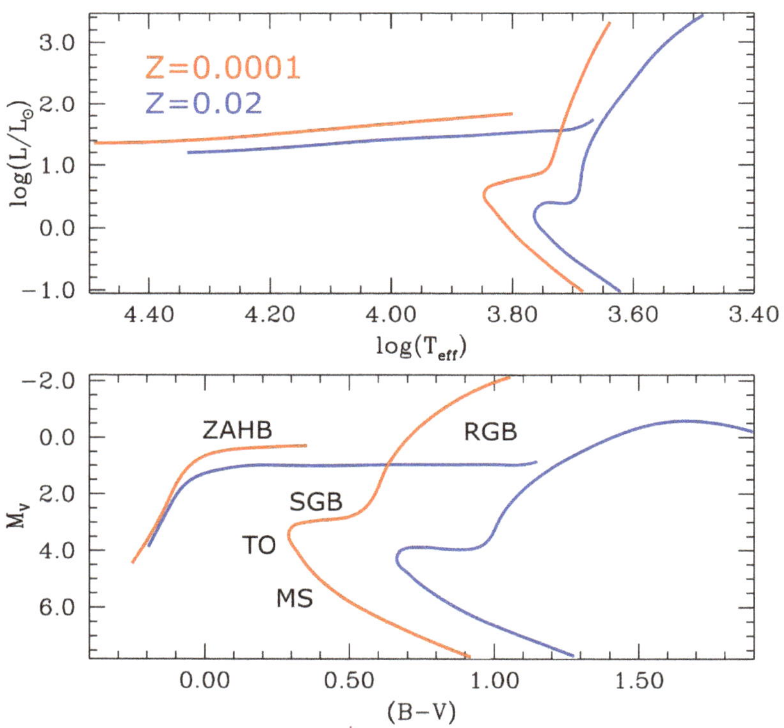

Figure 17.4. Top: the predicted locations of the main sequence (MS), the turn-off (TO), the subgiant branch (SGB), the red giant branch (RGB), and the horizontal branch (ZAHB) in the HRD of L versus $T_{\rm eff}$. The HB is indeed almost horizontal. Bottom: the same phases in the CMD of M_v versus B-V. The lines show predicted isochrones at $t = 10$ Gyr for models with $Z = 0.02$ (blue) and $Z = 0.001$ (red). (Reproduced from Salaris & Cassisi 2005. Copyright © 2005 by John Wiley & Sons, Inc. Reprinted by permission of John Wiley & Sons, Inc.)

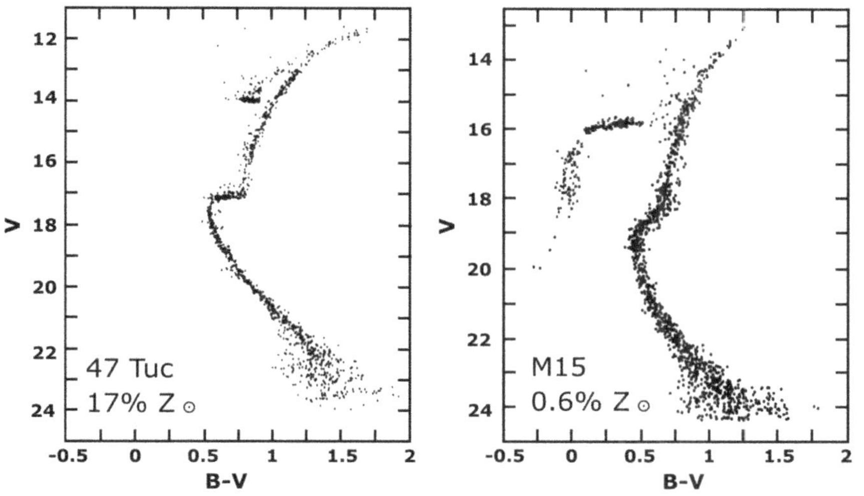

Figure 17.5. CMDs, in terms of M_V versus B-V, of two old globular clusters of different metallicities. The HB of the cluster with the lower Z (M15, right) extends to much bluer B-V colors (i.e., to higher $T_{\rm eff}$) than that of the high-Z cluster. (Reproduced from Carroll & Ostlie 1996. © Cambridge University Press. Reprinted with permission.)

the RGB phase. This contrasts expectations that the RGB mass loss is expected to increase with metallicity (because higher Z implies more dust and more molecules, resulting in stronger radiation pressure).

Even more curious is the fact that some *globular clusters with the same metallicity may have either a blue HB or a red HB!* This suggests that there must be another mechanism, apart from RGB mass loss and metallicity, affecting the extent of the convective envelope and $T_{\rm eff}$ of the HB stars. This problem is called the **second parameter problem**.

Some of the solutions that have been suggested involve the following.

- Differences in the initial He abundance. A higher He abundance reduces the opacity, $\sigma_e \sim (1 + X)$, which results in less convection in the envelope and a hotter HB star.
- Stars in some clusters may be rotating faster than those in other clusters. This would produce extra mixing, resulting in a more massive He core and a less massive envelope.
- Convective overshooting may have played a role. This would also result in a more massive core and a less massive envelope.

These last two effects could result in different envelope masses for stars of the same age and metallicity, and in different depths of the convection zone on the HB. See Dotter 2013 for a review.

Remark: this problem may be solved when the internal structure of stars can be determined empirically by asteroseismology (see Section 21).

17.5 Summary

1. Horizontal branch stars are low-mass stars with core He-fusion and shell H-fusion. HB stars are about 100 times more luminous than MS stars of the same mass.
2. In stars of $M_i \lesssim 2M_\odot$, He-fusion is ignited in a degenerate core, which produces a He flash. The He flash has not been observed.
3. In stars of $2 \lesssim M_i \lesssim 8M_\odot$, the ignition of He-fusion is regulated by gravity and the star remains in hydrostatic and thermal equilibrium.
4. During the phase of core He-fusion a low-mass star makes a leftward loop in the HRD, which mirrors the rightward loop during core H-fusion.
5. The highest $T_{\rm eff}$ and the bluest color that an HB star can reach depends on the mass of the envelope. The smaller the envelope mass, the higher $T_{\rm eff}$ can be.
6. In general, low-metallicity clusters have a long HB (extending far to the blue), whereas high-metallicity clusters have a short HB, close to the RGB; however, exceptions to this trend exist. This unexplained effect is called the second parameter problem.

Exercises

17.1 Calculate the potential energy needed to reduce the density in a core of $0.5M_\odot$ from 10^6 g cm^{-3} to 10^4 g cm^{-3} during the He flash.
Suppose that the peak energy production during the He flash is $10^{10}\ L_\odot$ and that neutrinos carry off 80% of this energy. Calculate the duration of the flash.

17.2 Study the Kippenhahn diagram for a star of $1M_\odot$.
  a. Derive the amount of He that is fused during the horizontal branch phase (G-H) and calculate the total amount of energy generated.
  b. Derive the amount of energy generated by the H-shell fusion during the same period.
  c. Which fraction of the energy is due to the H shell and which fraction is due to the He core?
  d. Compare your results with the luminosity and the duration of the HB phase.

References

Carroll, W. & Ostlie, D. A. 1996, An introduction to Modern Astrophysics (Reading, MA: Addison-Wesley)

Dotter, A. 2013, MmSAI, **84,** 97

Maeder, A. 2009, Physics, Formation and Evolution of Rotating Stars (Berlin: Springer)

Pols, O. 2011, in Stellar Structure and Evolution, Utrecht University Lecture Notes, 123, https://www.astro.ru.nl/~onnop/education/stev_utrecht_notes/

Salaris, M. & Cassisi, S. 2005, Evolution of Stars and Stellar Populations (Chichester: John Wiley and Sons)

Henny J.G.L.M. Lamers and Emily M. Levesque

Chapter 18

Double Shell Fusion: Asymptotic Giant Branch Stars

At the end of the core He-fusion phase, a star has a carbon–oxygen core (CO core). The luminosity is provided by a He-fusion shell and a H-fusion shell. The star has a very large radius and is located in the HRD on an even more luminous extension of the red giant branch, the **asymptotic giant branch (AGB)**. The AGB phase is one of the most interesting phases in the evolution of low- and intermediate-mass stars because of the various interacting effects that play a role: a degenerate core, double fusion shells, thermal pulses, and dredge-up phases that change the surface composition from O-rich to C-rich, and high mass-loss rates. The combination of all of these effects determines the end phase of low- and intermediate-mass stars with initial masses of $M_i < 8M_{\odot}$.

18.1 The Start of the AGB Phase

At the end of the HB phase, when He is exhausted in the core, the resulting CO core is without an energy source so it will contract. Because the star still has a H-fusing shell with mirror action, the core contraction results in envelope expansion and the star moves to the right in the HRD. Because the expanding envelope absorbs part of the energy, the luminosity decreases slightly during the expansion. This is seen in the track of the $5M_{\odot}$ star (Figure 17.1) near point H. The core contracts until He-fusion is ignited in a shell around the CO core. The CO core becomes degenerate because it is not hot enough to remain an ideal gas (Figure 8.10). The star is now at the bottom of the AGB (at point H in Figures 17.1 and 18.2).

An AGB star consists, from inside out, of

- a **degenerate CO core**,
- a **He-fusion shell:** He → C and later C → O,
- a **He-rich intershell zone**,

doi:10.1088/978-0-7503-1278-3ch18

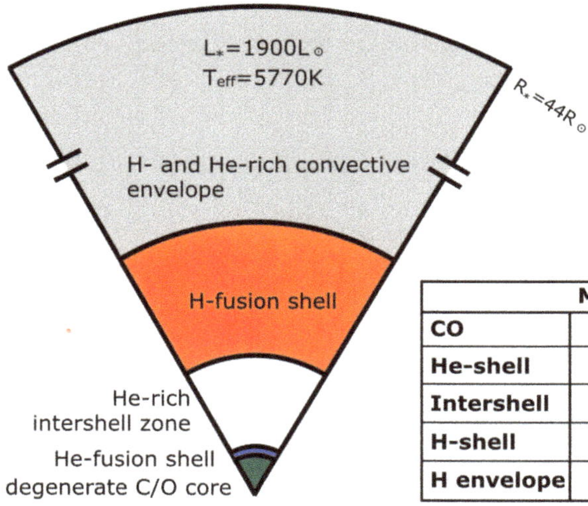

	M (M⊙)	ΔR/R
CO	0.4	2×10^{-5}
He-shell	0.1	2×10^{-6}
Intershell	0.5	7×10^{-5}
H-shell	0.05	4×10^{-5}
H envelope	3.95	>0.9999

Figure 18.1. Schematic figure of the internal structure of a star of $M_i = 5M_\odot$ at the start of the AGB phase: $L = 1.9 \times 10^3\ L_\odot$, $T_{\rm eff} = 5770$ K, $R = 44R_\odot$. The table shows the mass (in $M_\odot$) and thickness (in $R_\odot$) of each zone. (Reproduced from Carroll & Ostlie 1996. © Cambridge University Press. Reprinted with permission.)

– a **H-fusion shell:** H → He, and
– a **convective H-rich envelope**.

Figure 18.1 shows the structure of an AGB star of $5M_\odot$. The star has a radius of $R = 44R_\odot$. Note that the core, including the H-fusion shell, is very small: only 0.0056 $R_\odot$. It contains only 2×10^{-12} of the volume of the star, but 20% of the stellar mass!

18.2 The $M_{\rm core}$–L Relation of AGB Stars

The He-fusion shell is directly on top of the degenerate CO core. This implies that the pressure in the shell and its energy production are mainly set by the mass of the core. This is analogous to the case of the H-fusion shell on top of the degenerate He core in a low-mass RGB star (Section 16.4).

The M_c–L relation of AGB stars is called the **Paczynski relation** after the Polish-American astronomer Bohdan Paczýnski (1940–2007), who derived it in 1970 (Paczynski 1970). This relation for AGB stars with a metallicity of $Z = 0.03$ is

$$\boxed{\frac{L}{L_\odot} \approx 5.9 \times 10^4 \left(\frac{M_c}{M_\odot} - 0.522\right)} \quad \text{for } 0.57 < M_c/M_\odot < 1.39. \tag{18.1a}$$

Various alternative versions of the M_c–L relation exist for lower-metallicity stars in globular clusters (Boothroyd & Sackmann 1988). For instance, the relation for AGB stars with $Z = 0.001$ can be described by

$$\boxed{\frac{L}{L_\odot} \approx 5.2 \times 10^4\left(\frac{M_c}{M_\odot} - 0.456\right)} \quad \text{for } 0.52 < M_c/M_\odot < 0.70. \qquad (18.1b)$$

In both expressions, M_c is the total mass enclosed by the H-fusion shell. It includes the mass of the CO core, the He-fusion shell and the intershell zone, which all affect the pressure of the H-fusion shell.

Q (18.1) *Explain why the core mass determines the pressure of the H-fusing shell.*

The M_c–L relation of AGB stars has interesting consequences.

(a) If the luminosity of an AGB star is known, then its core mass is known (but the envelope mass and the total mass are not).

(b) For a given luminosity, the rate at which fusion occurs in the H and He shells is known. The fusion in these shells adds mass to the core, so the growth of the core mass and the increase of L is known. This implies that one can easily calculate the speed with which these stars will "climb" the AGB (we will show this later in Section 18.7).

18.3 The Second Dredge-up at the Beginning of the AGB Phase

The internal evolution and the evolutionary track in the HRD of a star with an initial mass of $5M_\odot$ during the AGB phase is shown in Figure 18.2. Shortly after He-shell fusion starts (point H), the H-shell fusion is switched off. This is because the He shell expands when the shell fusion starts. This forces the layers above it to expand. As a result, both T and ρ in the H shell decrease and the H-shell fusion

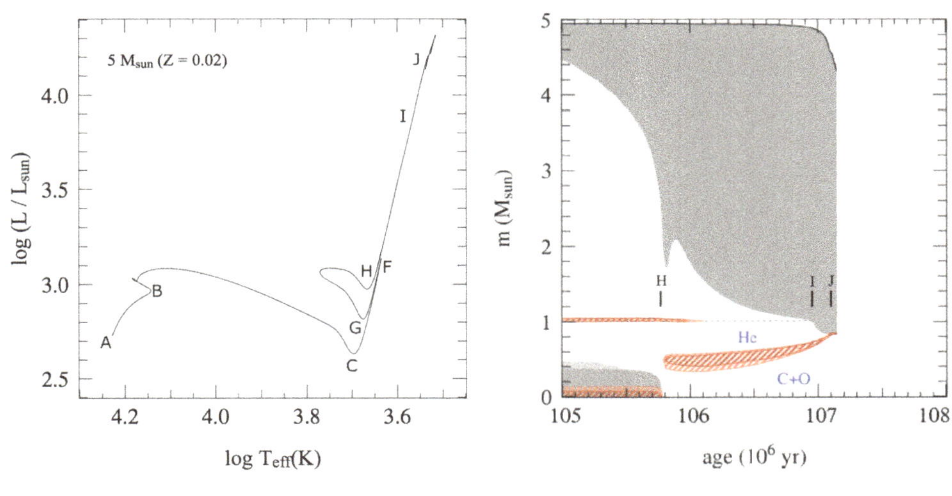

Figure 18.2. Evolutionary track (left) and Kippenhahn diagram (right) of the AGB phase, H to J, of a star of $M_i = 5M_\odot$. The letters correspond to points on the evolutionary track. Note the second dredge-up at point I when the convection zone reaches the He-rich intershell region. (© Pols 2009.)

almost stops (shortly after point H in the right part of Figure 18.2). In stars with $M_i \lesssim 4M_\odot$, the H-shell fusion remains active at a low level.

With He-shell and H-shell fusion going on, the increase in core mass results in an increasing luminosity, up to point J in the evolution track. Since the star is very extended and on the Hayashi track with an almost constant $T_{\rm eff}$, an increase in L implies an increase in radius. This is achieved by deepening the convection zone because an outer convection zone bloats a star: the deeper an outer convection layer reaches, the larger the star.

If the convection zone reaches the He-rich intershell zone, at I, it mixes the products of H-fusion (CNO cycle) into the outer envelope and convection brings them to the surface. This is called **the second dredge-up.** The intershell zone has no more H, a very low abundance of C and O, and a high abundance of He and N, so this second dredge-up brings the same CNO cycle products to the surface as the first dredge-up during the RGB phase (Section 16.2), but in much larger quantities. In stars of $M_i \lesssim 4M_\odot$, where H-shell fusion remains active, the convection does not reach the He-rich layer so these stars do not experience a second dredge-up.

18.4 The Thermal Pulsing AGB Phase (TP-AGB)

When an AGB star climbs the Hayashi track, it goes through a series of He flashes that are called **thermal pulses** at time intervals of $\sim 10^3$ to 10^5 yr (Do not confuse these thermal pulses with the He flash that started He core burning). The phase of the repeating thermal pulses is called **the thermal pulsing AGB phase (TP-AGB phase).** It usually starts near the end of the AGB phase when the luminosity is high and the CO core is sufficiently massive and compact to produce a high pressure in the He-fusion shell. A star may experience on the order of 10 thermal pulses near the tip of the AGB.

In order to understand the thermal pulses, we follow the evolution of the star's interior step by step, as shown in Figure 18.3.

1. For most of the time during the AGB phase, the H-fusion shell produces much more energy than the He-fusion shell, which is barely active.
2. The H-shell fusion leaves He behind, so the mass of the intershell zone (ISZ) increases with time.
3. As the ISZ grows in mass, the pressure at the bottom of that region steadily increases because more and more He is piled on top of the degenerate CO core. When the pressure reaches a critical value, He is reignited in a thin shell.
4. Ignition in a thin shell leads to a thermal instability of the same type as the ignition in a degenerate core (though here the gas is not degenerate). The reason is as follows. When the He-fusion is reignited, the gas is heated and it expands; however, if the shell is very thin, its expansion cannot lift the layers on top of it sufficiently to reduce the pressure that these layers exert on the shell, so the pressure hardly changes when He is ignited in a thin shell. With P remaining about constant, but ρ decreasing during the expansion, T must rise in the shell. This leads to runaway fusion that is called a **thermal pulse**; see Kippenhahn & Weigert (1990) for details. (Note that it looks like the

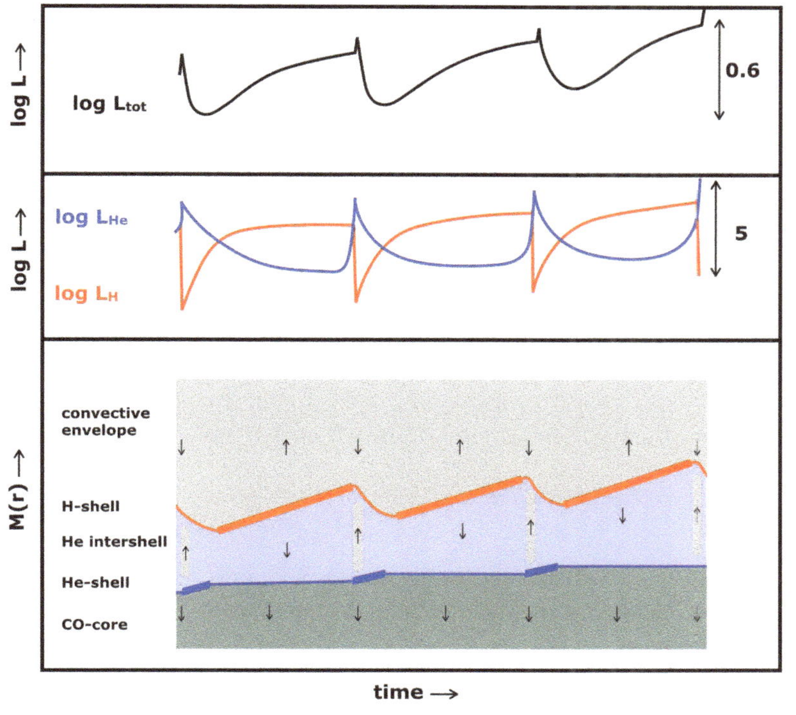

Figure 18.3. Schematic picture of thermal pulses in an AGB star. Top: variations in the radiative luminosity during several cycles. The total range in L is about a factor of 4 and about a factor of 2 in R, because $T_{\rm eff}$ remains almost constant. Middle: variations in the luminosity produced by the He shell and H-shell fusion. The total range in L is about a factor of 10^5. Bottom: variations in the internal structure in terms of the mass $M(r)$. The CO core is degenerate. Thick and thin lines of the H and He shells indicate which fusion shell is dominant. Gray regions indicate convection. Downward arrows indicate contraction and upward arrows indicate expansion of the regions.

flash is produced by the ignition of He-fusion in the degenerate core of an RGB star, but the physical reason is different.)

5. At the peak of the runaway He-shell fusion, it produces so much energy, up to $10^8\ L_\odot$, that the ISZ becomes convective (because $L/4\pi r^2$ is so large that it cannot be transported by radiation). The convective ISZ expands and the H shell on top of it is pushed upward. With the H shell moving outward its pressure drops to a value that cannot keep efficient H-fusion going, so when He-shell fusion starts the H-shell fusion is extinguished. When the H-fusion in the shell is extinguished, the envelope convection reaches deeper in mass and brings enriched gas to the surface. This is the **third dredge-up**.
6. While the He-shell fusion is going on and the ISZ is expanding, the pressure in the He-shell, mainly produced by the overlaying ISZ, reduces. Eventually, the pressure on the He-shell is so low that the energy production in the He-shell decreases and the layers on top of it contract again. The increase in T and ρ results in a reignition of H-shell fusion. When the H-shell fusion

reignites again it starts at the bottom of the convective envelope and moves outward in mass. At that point, the cycle starts again at 1.

Figure 18.3 shows a schematic sketch of the process of the thermal pulses. Note the short convective phase of the intershell zone when He-fusion is reignited in the shell.

When the intershell zone is convective during the thermal pulse, the mirror action of the H-fusion shell implies that the envelope contracts, so the radius of the star decreases. Because the star is on the Hayashi line, its $T_{\rm eff}$ hardly changes, so the luminosity decreases as well. The radius varies by about a factor of 2 and the luminosity varies by about a factor of 4. The very high peak luminosity of about 10^8 $L_\odot$ during the thermal pulse does not reach the photosphere; most of this energy is used for the expansion of the intershell zone.

Q (18.2) *Why is the H shell during the TP-AGB phase always at the bottom of the convective envelope?*

Q (18.3) *Why is the core contracting during the TP-AGB phase?*

The repetition time, t_p, of thermal pulses depends mainly on the core mass. From a large number of models, Paczynski (1975) showed that it follows approximately

$$\log t_{\rm p} \approx 3.05 - 4.5\left(\frac{M_c}{M_\odot} - 1\right) \text{yr}. \tag{18.2}$$

For a star with $M_i = 2M_\odot$ and a core mass of $0.6M_\odot$, the repetition time is $\sim 10^5$ yr, and for a star with $M_i = 8M_\odot$ and a core mass of 1 $M_\odot$ it is only $\sim 10^3$ yr.

During the thermal pulsing AGB phase, the general evolutionary trend of increasing radius and luminosity continues, due to steadily increasing core mass: the star keeps climbing along the Hayashi track. As the star gets brighter and brighter, the fusion goes faster and faster, so the period between thermal pulses gets shorter and shorter. This process continues until the star leaves the AGB due to the loss of the H envelope (Section 18.8).

Figure 18.4 shows the predicted thermal pulses in an AGB star of $1.8M_\odot$ (Salaris & Cassisi 2005). Note the general trends of increasing luminosity, decreasing $T_{\rm eff}$, and the increasing peak luminosity produced by the He shell during the thermal pulses. The pulses start with a periodicity of about 2×10^5 yr, decreasing to 1×10^5 years after 12 pulses. A thermal pulse lasts about $\sim 10^2$ yr. Due to the short duration of the peak luminosity, thermal pulses have not yet been observed!

During a thermal pulse, the mass-loss rate is higher than it is during the interpulse phase. This increased mass loss produces a circumstellar shell. Figure 18.5 shows the observations of such a shell around the AGB star R Sculptoris. The star is at a distance of 270 pc and has a luminosity of $4300L_\odot$. The radius of the shell is 2.6×10^{-2} pc and its expansion velocity is 14.3 km s^{-1}. This shows that the shell was ejected 1800 yr ago during a thermal pulse. The shell has a mass of 3×10^{-3} $M_\odot$. Modeling of the shell suggests that its ejection lasted ~ 200 yr at a rate of $\sim 1.5 \times 10^{-5}$ $M_\odot$/yr. This is 30 times higher than the present mass-loss rate (Maercker et al. 2014).

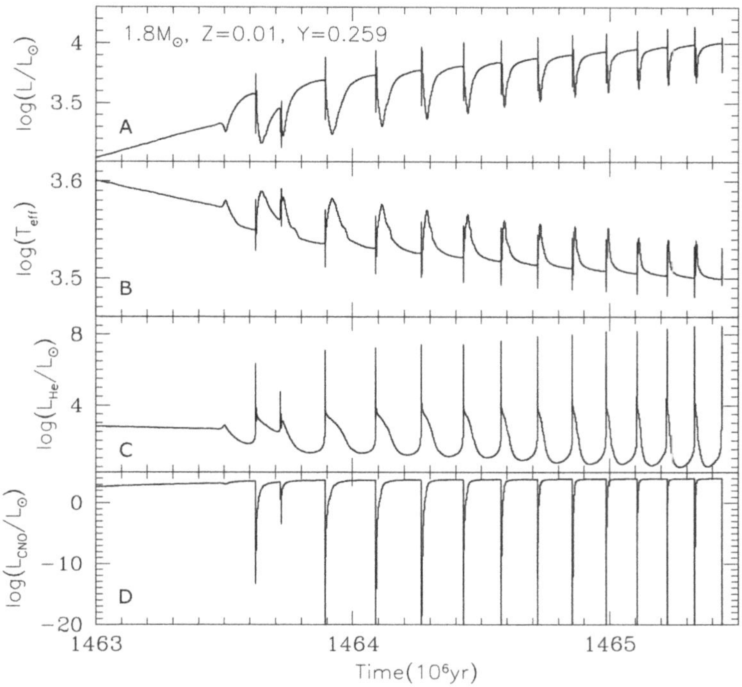

Figure 18.4. Predicted evolution of an AGB star of $1.8M_\odot$ of $Z = 0.01$ during the TP-AGB phase over a period of 2.5 Myr. A: the luminosity with the regular peaks and dips. The peak–peak variation is about a factor of 3. B: the variation in $T_{\rm eff}$. The peak–peak variations are about 15% superimposed on the general trends of decreasing $T_{\rm eff}$. C: the variation in the luminosity produced by the He-fusion shell, reaching peaks of 10^8 $L_\odot$. The energy of these peaks is absorbed by the expansion of the intershell zone and does not reach the surface. D: the energy produced in the H-fusion shell, which in this case is via the CNO cycle. Note the large difference in scale of panels C and D. (Reproduced from Salaris & Cassisi 2005. Copyright © 2005 by John Wiley & Sons, Inc. Reprinted by permission of John Wiley & Sons, Inc.)

18.5 The Third Dredge-up

The occurrence of thermal pulses with the alternate switching between He shell and H-shell fusion leads to efficient mixing and gradual changes in the atmospheric abundances.

This process occurs in two steps, as shown in Figure 18.6.

1. During the thermal pulse, when the ISZ is convective for a short time, products of the He-fusion (mainly C) are distributed throughout the ISZ.
2. When the He-shell fusion is stable, the intershell zone contracts and the envelope expands (due to the mirror action of the H shell) by deepening the convection in the envelope. When the convection reaches the layers with increased C abundance, it brings the products of He-fusion into the atmosphere.

This is called the **third dredge-up**.

Each cycle brings more and more C to the surface. Eventually, massive AGB stars near the tip of the AGB phase have an envelope and photosphere with a C/O ratio > 1,

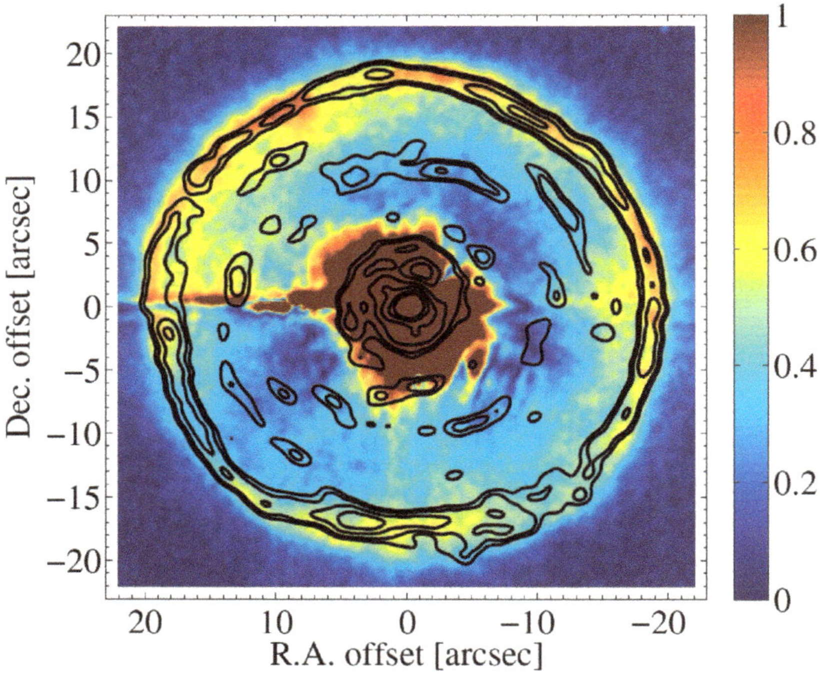

Figure 18.5. The observed shell around the AGB star R Scl was ejected during a thermal pulse 1800 yr ago. The colors show a polarized intensity image of dust-scattered light, observed in the *R*-band with the ESO 3.6-meter telescope. The black lines show the isophotes of ALMA observations of the CO emission. (Reproduced from Maercker et al. 2014, with permission. © ESO.)

instead of the normal case of O/C > 1. These stars are easily distinguished because instead of O-rich molecules and dust (i.e., silicates) they have C-rich molecules and dust (carbonaceous grains). The flip from an O-rich to C-rich atmosphere is drastic because the CO molecule is the most abundant and strongly bound molecule. When O/C > 1 all C is locked in CO and only the remaining O can form O-rich molecules (e.g., OH) and O-rich silicate dust. On the other hand, if C/O > 1, all O is locked in CO molecules and only the remaining C can form molecules (such as CH, etc.) and carbon dust.

In massive TP-AGB stars, as shown in figure 18.6, the H shell is active at such a high temperature that H-fusion occurs via the CNO cycle and not via the pp-chain. This implies that C (that was transported by convection of the ISZ during the time when the He shell was active) is converted into N. This process is called **hot-bottom burning (HBB)** and it may prevent massive TP-AGB stars from becoming C-rich; they instead become N-rich at the surface. HBB also produces nuclei like ^{23}Na, ^{25}Mg, and ^{26}Mg (Section 8.4.4), which are found to be overabundant in old globular clusters (Gratton et al. 2004).

18.6 Summary of the Dredge-up Phases

It is useful to summarize the dredge-up phases of stars of $0.8 \lesssim M_i \lesssim 8 M_\odot$. Figure 18.7 shows, schematically, the position in the HRD where they occur.

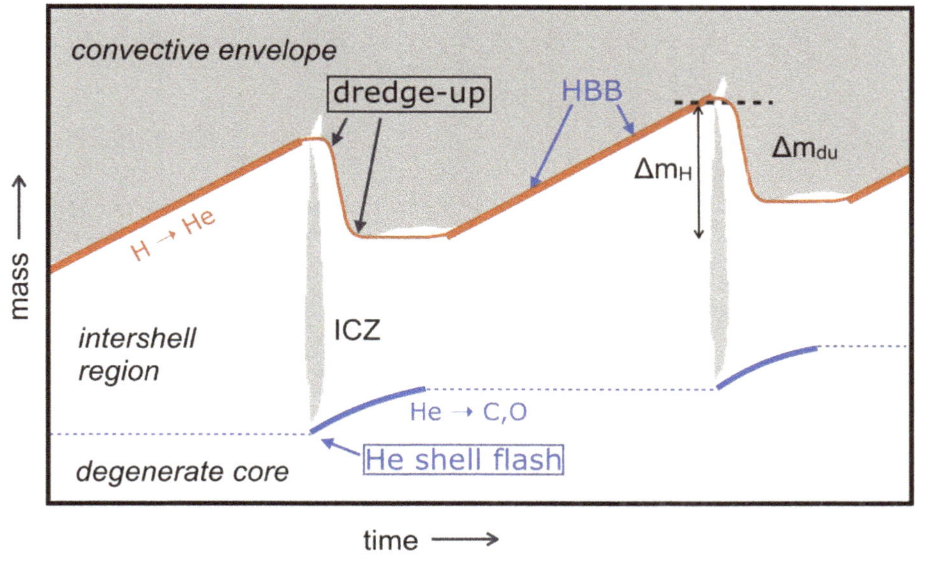

Figure 18.6. Internal evolution of an AGB star, explaining the dredge-up during two thermal pulses. The thick and thin red and blue lines indicate alternating H- and He-fusion shells. The gray areas indicate convective regions. The phases of dredge-up and hot-bottom burning (HBB in stars with $M_i \gtrsim 5M_{\odot}$) are indicated. The symbol Δm_{H} indicates the amount of mass converted into He between two thermal pulses, while Δm_{du} is the mass that is dredged up after each thermal pulse. (Figure is from Pols 2011.)

1. The **first dredge-up** occurs when the star is on the RGB and the envelope convection reaches the depth where He was produced during the MS phase. This results in an increase of the N abundance and a small increase of the He abundance at the surface (Figure 16.1).
2. The **second dredge-up** occurs in the early AGB phase for stars of $M_i \gtrsim 4M_{\odot}$ when the star expands after the HB phase. This expansion is produced by the growth of the convective envelope. The convection reaches even below the depth (in mass) of the H-shell of the HB phase. So it reaches into the He layer and brings He, N-rich, C-poor, and O-poor gas, mixed with the original envelope mass, to the surface (Figure 18.2).
3. The **third dredge-up**, or rather dredge-ups, occurs during the later TP-AGB phase in stars of $M_i \gtrsim 4M_{\odot}$ when the envelope convection reaches into the ISZ after a thermal pulse. The ISZ contains He and products of the He-fusion, such as C. The surface gradually becomes richer in C (Figure 18.7).
4. **C-stars**: In the massive AGB stars, the C/O ratio may change from C/O <1 to >1. These stars have a C-rich photosphere and a wind with C-rich dust.
5. **S-stars** are AGB stars in which the third dredge-up brings s-process elements, such as Technetium (Tc) to the surface. These elements are formed by slow neutron capture during thermal pulses.

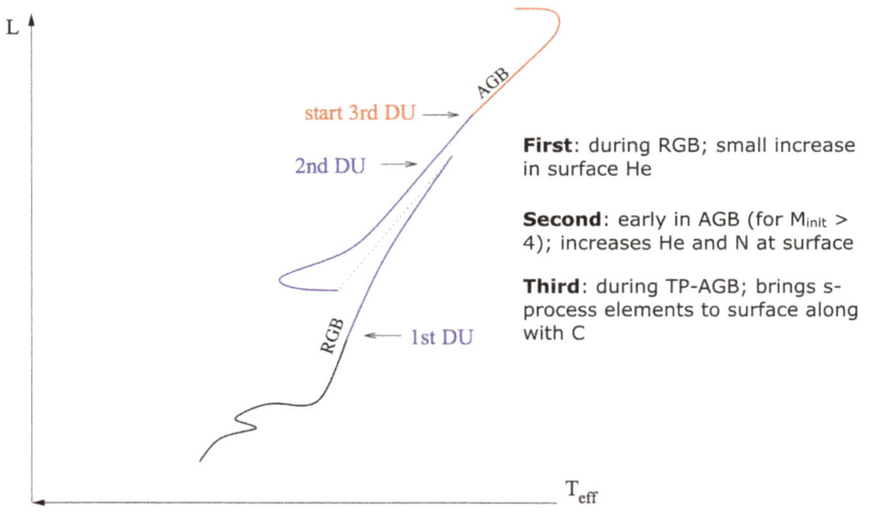

Figure 18.7. Dredge-up phases during the late evolution of low-mass stars. The changes in surface compositions are written next to the track. (Figure is based on Pols 2011).

All of the envelope mass of AGB stars is eventually expelled by mass loss during the late AGB phase. *AGB stars are the main producers of He, C, and s-process elements in the Universe.*

18.7 The Evolution Speed during the AGB Phase

Throughout the AGB phase, the luminosity of the star increases. Because the luminosity during the AGB phase is set by the core mass, while the growth of the core mass is set by the luminosity, there is a simple way to estimate the speed with which a star ascends the AGB-branch.

Let us assume that the luminosity of an AGB star is related to its core mass by the Paczynski relation (Equation (18.1a)). The core mass grows due to nuclear fusion. The fusion is dominated by H-fusion about 90% of the time and by He-fusion about 10% of the time. The energy production is 6×10^{18} ergs g^{-1} for H-fusion and 6×10^{17} ergs g^{-1} for He-fusion. We can then set the time-averaged energy production to approximately $\bar{\epsilon} = 5.5 \times 10^{18}$ erg g^{-1}.

The mass of the core grows as

$$\begin{aligned} \frac{dM_c/M_\odot}{dt} &= \left(\frac{L}{\bar{\epsilon}}\right)/M_\odot \\ &= \frac{(5.9 \times 10^4 \times 3.8 \times 10^{33})}{(5.5 \times 10^{18} \times 2.0 \times 10^{33})}\left(\frac{M_c}{M_\odot} - 0.52\right) \\ &= 2.1 \times 10^{-14}\left(\frac{M_c}{M_\odot} - 0.52\right)M_\odot s^{-1}. \end{aligned} \tag{18.2}$$

This can be expressed as

$$\frac{d\ln(M_c/M_\odot - 0.52)}{dt} = 2.1 \times 10^{-14}\ \mathrm{s}^{-1} = 1/1.4\ \mathrm{Myr}. \qquad (18.3)$$

So, the core mass grows as

$$\frac{M_c}{M_\odot} - 0.52 = \left(\frac{M_c}{M_\odot} - 0.52\right)_{t=0} e^{t/1.4\mathrm{Myr}}. \qquad (18.4)$$

Substitution of $M_c/M_\odot - 0.52 = (5.9 \times 10^4)^{-1} L/L_\odot$ yields

$$\boxed{L(t) = L(t=0)e^{+t/1.4\ Myr}} \rightarrow \text{an e-folding time of only} \sim 1.4\ \mathrm{Myr}, \qquad (18.5)$$

where $t = 0$ is the time when the star enters the AGB.

Suppose a star enters the AGB at $L \approx 1.2 \times 10^3\ L_\odot$ and $M_c \approx 0.54 M_\odot$. It will then reach

$$\begin{aligned} L &\approx 6 \times 10^3\ L_\odot \quad \text{at } t \approx 2.3\ \text{Myr with } M_c \approx 0.60 M_\odot, \\ L &\approx 1 \times 10^4\ L_\odot \quad \text{at } t \approx 3.2\ \text{Myr with } M_c \approx 0.70 M_\odot, \\ L &\approx 2 \times 10^4\ L_\odot \quad \text{at } t \approx 4.2\ \text{Myr with } M_c \approx 0.85 M_\odot, \\ L &\approx 3 \times 10^4 L_\odot \quad \text{at } t \approx 4.8\ \text{Myr with } M_c \approx 1.00 M_\odot. \end{aligned}$$

18.8 Mass Loss and the End of the AGB Evolution

White dwarfs (WDs) are the degenerate cores that remain after the end of the evolution of low-mass stars. Observations of WDs in clusters with different ages show that stars with initial masses of up to about 6–8$M_\odot$ end their lives as WDs. These studies also show that most WDs have a mass of about $M \sim 0.6$ to $0.7 M_\odot$ (Weidemann 1990). This implies that the majority of the low-mass stars, i.e., stars with $M_i \lesssim 3 M_\odot$, must terminate their AGB phase with $M_c \approx 0.6$ to $0.7 M_\odot$ and $L \approx 6 \times 10^3$ to $10^4\ L_\odot$, as shown in the calculations in the previous section. More massive stars of $M_i \approx 3$–$8 M_\odot$ produce a WD with $M \gtrsim 0.8 M_\odot$. A WD with $M \approx 1 M_\odot$ must have reached $L \approx 3 \times 10^4\ L_\odot$ at the tip of the AGB.

We have seen that stars climb the AGB in an exponential function of time and that the mass of the core also increases with time. What could stop the growth of the core? Mass loss! Observations show that all AGB stars suffer mass loss at a rate that increases from $\sim 10^{-7}\ M_\odot/\mathrm{yr}$ at the bottom of the AGB to very high values of $\sim 10^{-5}$ or even $10^{-4}\ M_\odot/\mathrm{yr}$ at the tip of the AGB (Sections 15.4.3 and 15.4.4). Let us consider a simple method for estimating the effect of mass loss on the AGB and predicting the mass of the resulting WDs.

Suppose we can describe the mass-loss rate $\dot{M}_w$ (subscript "w" for wind) as a function of L; we can then also describe it as a function of t (Equation (18.5)). Let $M_{env}(t = 0)$ be the envelope mass when the star enters the AGB. The envelope mass decreases by two effects.

- On the inside: due to nuclear fusion $(dM_{\rm env}/dt) = -dM_c/dt$.
- On the outside: due to mass loss by the stellar wind: $(dM_{\rm env}/dt)_{\rm w} = -\dot{M}_{\rm w}$.

The total mass loss of the envelope is

$$\frac{dM_{\rm env}}{dt} = -\frac{dM_c}{dt} - \dot{M}_{\rm w}. \tag{18.6}$$

Throughout most of the AGB phase, the wind is much more efficient in reducing the envelope mass than the growth of the core. So we can approximate

$$\frac{dM_{\rm env}}{dt} \simeq -\dot{M}_{\rm w} \rightarrow \Delta M_{\rm env} \simeq \int \dot{M}_{\rm w}\, dt. \tag{18.7}$$

Let us estimate how long it would take to remove the full envelope mass of a star that enters the AGB phase with $M = 3M_\odot$, consisting of a core of $M_c = 0.54M_\odot$ and an envelope of $M_{\rm env} = 2.46M_\odot$ at $t = 0$. The Paczynski relation predicts that $L_0 = 1.2 \times 10^3 L_\odot$ at $t = 0$.

We adopt van Loon's expression (Equation (15.18)) for the mass-loss rates of AGB stars in the LMC, increased by a factor of 2 to allow for the higher metallicity of Galactic AGB stars. Assuming $T_{\rm eff} = 4000$ K for Galactic AGB stars, we find

$$\dot{M}_{\rm w} \simeq 1.2 \times 10^{-10}\, (L/L_\odot)^{1.05} = 3.4 \times 10^{-7}\, e^{t/1.33{\rm Myr}}\, M_\odot/{\rm yr}. \tag{18.8}$$

Integration of this equation shows that the envelope has lost $2.4M_\odot$ after 2.3Myr. At that time, the luminosity was $6 \times 10^3 L_\odot$ and the core mass was $0.60M_\odot$.
This simple estimate is roughly in agreement with

- the observed maximum $L \sim 10^4 L_\odot$ of AGB stars in the globular clusters, and
- the derived mean mass of Galactic white dwarfs $\langle M_{\rm WD}\rangle \sim 0.6$ to $0.7M_\odot$.

This shows that

- the AGB phase terminates because mass loss has stripped (almost) the full envelope,
- the final mass of the WD is determined by the mass-loss rate $\dot{M}_{\rm w}$ during the AGB phase, and
- the maximum luminosity of the AGB stars is set by $\dot{M}_{\rm w}$ during the AGB phase.

The AGB mass loss prevents stars in the mass range of about 2 to $8M_\odot$ from becoming supernovae. If it were not for their high mass-loss rate, these stars would contract at the end of the AGB phase when they have a degenerate C-core. The ignition of C-fusion in a degenerate core produces a C-flash, similar to the He-flash at the ignition of He-fusion in a degenerate He core, as discussed in Section 17.1. However, a C-flash would destroy the star completely. This is called **C-detonation**, which produces a **thermonuclear supernova.**

Q (18.4) *How much higher would the supernova rate from single stars be if the lower limit of their initial mass was not* $8M_\odot$, *but* $4M_\odot$?

In our simple estimate, we have used van Loon's relation for $\dot{M}_w$ (AGB); however, observations show that mass loss on the AGB increases more drastically near the tip of the AGB and reaches a value of a few 10^{-5} or even 10^{-4} $M_\odot$/yr. This high mass-loss rate is due to pulsation (see Figure 15.5), because luminous AGB stars pulsate as Mira variables with periods of 100 to 600 days. The winds of these stars have a velocity of about 10 to 20 km s^{-1}. The very high mass loss at the end of the AGB phase is called the **superwind phase**. This is important for the formation of planetary nebulae, which will be discussed in Section 19.

18.9 Summary

The AGB phase is one of the most fascinating evolutionary phases because several interesting physical processes are happening.

1. AGB stars have a degenerate CO core, a He-fusion shell, a He-rich intershell zone, a H-fusion shell, and an extended outer convection zone.
2. The luminosity of AGB stars depends on the core mass and not on the total mass. As the core mass increases due to fusion, the star climbs the AGB in the HRD along a Hayashi track.
3. The two shell-burning phases alternate in producing the luminosity of the star, with a periodicity of about 10^3 to 10^5 years depending on the core mass and luminosity. The changes are triggered by shell flashes, called *thermal pulses.* Thermal pulses last about 10^2 years.
4. The very deep convection brings nuclear products of the H and He-fusion to the surface. These are called the *second and third dredge-up* (the first dredge-up occurred during the RGB phase).
5. The second dredge-up occurs early in the AGB phase, when the outer convection reaches into the He-rich intershell zone. It brings He and N to the surface.
6. The third dredge-ups occur after the shell flashes in a two-step process. During the thermal flash, the products of He-fusion are mixed in the ISZ during a short internal convection phase. Later, when the outer convection reaches into the ISZ, these products are brought to the surface.
7. The surface is gradually enriched by C and by s-products that were formed during the thermal pulse. This produces C-stars and s-stars near the tip of the AGB.
8. The AGB phase ends when the stellar wind has removed almost the entire envelope. This occurs on a timescale of a few Myr. During that time the AGB star reaches a luminosity of $\sim 10^4 L_\odot$ and a core mass of about 0.6–0.7 $M_\odot$.
9. Mass loss during the AGB phase sets the mass of white dwarfs and the supernova rate.

Exercises.

18.1 Figure 18.4 shows the luminosity produced by the H-fusion and the He-fusion shells of a star that started the AGB phase with a mass of $1.8M_\odot$. The peaks in luminosity last less than 100 yr and can be ignored.

(a) Estimate the mean luminosity produced by He-fusion from the beginning of the TP phase to the end.

(b) Do the same for the H-fusion.

(c) What is their ratio?

(d) What is the mass that took part in the He shell fusion and in the H-shell fusion during the full TP-AGB phase?

18.2 Sketch the changes in the surface abundances of He, C, and N expected during the three dredge-up phases of a star with $M_i = 3M_\odot$ and $5M_\odot$.

18.3 **Computer exercise**

Calculate the evolution of AGB stars numerically, using the M_c–L relation of Paczynski and Reimers mass-loss rates for different values of η_R.

(a) Write the differential equations that describe $L(t)$, $M_c(t)$, and $M_{env}(t)$.

(b) Solve them by computer for stars that start the AGB phase with

$$M_* = 2.0M_\odot \text{ and } M_c = 0.522M_\odot,$$
$$M_* = 5.0M_\odot \text{ and } M_c = 0.55M_\odot.$$

Assume that the AGB evolution ends when the envelope mass is smaller than $0.01M_\odot$.

(c) Calculate $L(t)$, $M_c(t)$, and $M_{env}(t)$

(d) What is the core mass M_c and L at the tip of the AGB? What is the mass of the resulting white dwarf?

(e) How long does the AGB phase last for these stars?

(f) How does the result depend on the adopted value of η_R in the Reimers relation? Explain the results.

(g) Which average value of η_R agrees best with the observed masses of white dwarfs?

18.4 **Computer exercise**

Calculate the evolution of two AGB stars that start the AGB with

$$M_* = 2.0M_\odot \text{ and } M_c = 0.522M_\odot,$$
$$M_* = 5.0M_\odot \text{ and } M_c = 0.55M_\odot.$$

Use the mass-loss rate derived from the IR flux of AGB stars and its relation with the period of pulsation (Section 16.4.4), or the Reimers relation with $\eta_R = 2$ (whichever is the larger of the two). Adopt the relation between T_{eff} and L from the track in Figure 18.2 and plot the evolution of the mass-loss rate, M_c, M_{env}, M, L, R, and T_{eff}. Explain the result.

References

Boothroyd, A. I., & Sackmann, I. J. 1988, ApJ, **328, 641**

Carroll, W., & Ostlie, D. A. 1996, An introduction to Modern Astrophysics (Reading, MA: Addison-Wesley)

Gratton, R., Sneden, C., & Carretta, E. 2004, ARA&A, **42, 385**

Kippenhahn, R., & Weigert, A. 1990, Stellar Structure and Evolution (Berlin: Springer)

Maercker, M., Ramstedt, S., Leal-Ferreira, M. L., Olofsson, G., & Floren, H. G. 2014, A&A, **570, 101**

Paczynski, B. 1970, AcA, **20, 47**

Paczynski, B. 1975, ApJ, **202, 558**

Pols, O. 2011, in Stellar Structure and Evolution, Utrecht University Lecture Notes, 124

Salaris, M. & Cassisi, S. 2005, in Evolution of Stars and Stellar Populations (Chichester: John Wiley and Sons), 192

Weidemann, V. 1990, ARA&A, **28, 103**

Henny J.G.L.M. Lamers and Emily M. Levesque

Chapter 19

Post-AGB Evolution and Planetary Nebulae

After the AGB phase, when a star has lost almost all of its very extended convective envelope, the remaining envelope mass shrinks. The luminosity remains constant because it is still produced by shell fusion on top of a degenerate core. The shrinking of the envelope implies a decreasing radius, so the star moves horizontally to the left in the HRD. For the first part of this leftward motion, the star is difficult to observe because it is enshrouded by a thick dusty wind that was ejected in the superwind phase at the end of the AGB phase with an outward expanding velocity of about 10 km s^{-1}. When the star reaches an effective temperature of ~30,000 K it develops a high-speed radiation-driven wind of $\sim 10^3$ km s^{-1}. The interaction between this fast wind and the slow AGB wind produces a planetary nebula. When the mass of the envelope is below a critical limit, the fusion stops and the star moves downward in the HRD toward the white dwarf region. Some stars experience a late thermal pulse after the planetary nebula phase. Those stars are observed to brighten and expand back to the Hayashi line in a few decades.

19.1 The Post-AGB Phase

Evolutionary calculations show that a star leaves the AGB when the mass in the H envelope has decreased to only about 10^{-2} to $10^{-3} M_{\odot}$, depending on the core mass. At that time, the reduced H-envelope mass can no longer sustain a fully developed convection zone. The convective envelope slowly shrinks and becomes radiative rather than convective. This transition occurs first in the deepest layer of the envelope, where κ is small.

The star still has double-shell fusion around the degenerate core and its luminosity is still set by the core mass. This post-AGB phase is short, about 10^3 to 10^4 years, so the core mass hardly increases during that time. This means that the luminosity remains constant and the star moves *horizontally to the left* in the HRD, eventually reaching a temperature as high as $T_{\rm eff} \sim 10^5$ K. The post-AGB evolutionary track of a star with a degenerate core of $0.6 M_{\odot}$ that leaves the AGB with an envelope mass of $0.003 M_{\odot}$ is shown in Figure 19.1.

doi:10.1088/978-0-7503-1278-3ch19

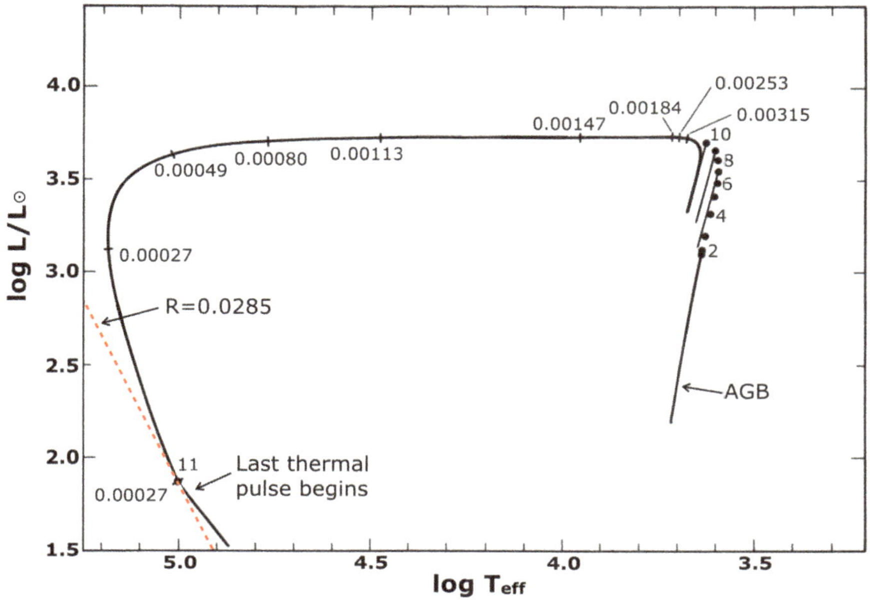

Figure 19.1. Post-AGB evolutionary track of a star with $M_i = 3M_\odot$ that has a core mass of $M_c = 0.60$ and an envelope mass of $M_{env} = 0.003M_\odot$ when it leaves the AGB. The numbers along the track indicate the decreasing envelope mass in $M_\odot$. Note the 10 thermal pulses at the tip of the AGB. A red dashed line of constant radius $R = 0.0285R_\odot$ is indicated in the lower left. (Reproduced from Iben 1982.)

Q (19.1) *Estimate the mean density in the convective envelope when the star leaves the AGB.*

The location of the star in the HRD during the horizontal crossing depends on the mass of the envelope. As long as $M_{env} \gtrsim 10^{-2}M_\odot$ for a high luminosity AGB star, or $10^{-3}M_\odot$ for a low luminosity AGB star, the star remains close to the AGB. But when mass loss has reduced M_{env} below this limit, the envelope goes into radiative equilibrium and contracts because it does not have enough mass to sustain an extended outer convection zone. The star then starts moving to the left in the HRD. This is called the **post-AGB phase**. During the first part of the post-AGB phase, the star is optically faint because it is surrounded by a thick dust shell that was ejected during the superwind phase at the end of the AGB phase.

When $T_{eff} \approx 30{,}000$ K and the radius has decreased from $\sim 200R_\odot$ to $\sim 3R_\odot$, two effects start to play a role.

- The star develops a line-driven stellar wind with a velocity of a few 10^3 km s^{-1} (similar to the ones that were discussed in Section 15.2 for hot luminous stars).
- The UV flux from the star destroys the dust and ionizes the gas of the AGB wind that surrounds the star.

The interaction of this fast wind with the slow AGB wind produces a **planetary nebula** and the star becomes a **central star of a planetary nebula (CSPN)**.

Phase	AGB	post-AGB	CSPN	end crossing
$M_{env}/M_\odot$	$\sim 3\times10^{-3}$	2×10^{-3}	10^{-3}	3×10^{-4}
$R/R_\odot$	150	25	2.7	0.25
t (yr)	0.0	1×10^{3}	2×10^{3}	7×10^{3}

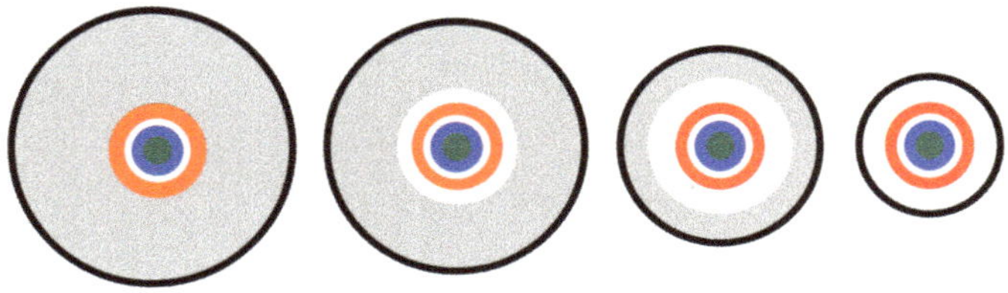

Figure 19.2. Schematic picture of the internal structure of a star of $M_i = 3M_\odot$ during the crossing of HRD. The H- and He-fusion shells are shown in red and blue, respectively, and the CO core is green. White zones are in radiative equilibrium. The gray area indicates the convective envelope, which decreases in mass and size. The envelope mass, the stellar radius, and the time since the star left the AGB are indicated. (Data are from Blöcker 1995.)

The mass-loss rates of CSPNs are $\sim 10^{-7}$ to 10^{-9} $M_\odot$ yr^{-1}, strongly dependent on L as $\dot{M} \sim L^{1.6}$, and the terminal wind velocity is 500 to 3000 km s^{-1} (Kudritzki & Puls 2000). From now on, the envelope loses mass even faster, not only to the core but also to the wind, so the speed with which the star moves to the left in the HRD increases. The speed of the crossing is determined by the continuing mass loss from the envelope. The crossing time is of the order of 10^3 to 10^4 years.

Figure 19.2 shows the change in the internal structure of a post-AGB star of $M_i = 3M_\odot$ during the leftward crossing in the HRD. During the first part of the post-AGB track, the star is invisible because it is hidden in the dust that was ejected at the end of the AGB phase, during the superwind phase when the mass-loss rate was as high as 10^{-4} $M_\odot$ yr^{-1}.

When the envelope mass has decreased to as little as $\sim 3 \times 10^{-4}$ or $3 \times 10^{-5} M_\odot$, depending on L, the envelope is almost fully in radiative equilibrium and the star is on the left of the HRD at $T_{\rm eff} \sim 10^5$ K with a small radius of only $\sim 0.3 R_\odot$. Soon thereafter, the fusion stops completely because the envelope is no longer producing sufficient pressure for the fusion to continue.

Q (19.2) *By what factor does the volume of the star decrease during the post-AGB crossing? How much did the core decrease in volume?*

19.2 Born-again AGB Stars

In some cases, a star experiences a last thermal pulse while it is moving to the left on the HRD. This is possible because the time between two thermal pulses is on the order of 10^3 to 10^4 yr and the HRD crossing time is of the same order. It is estimated that about one-fourth of all the AGB stars will have a late thermal pulse when they have already left the AGB.

The two most famous examples are **Sakurai's object** (V4334 Sgr), discovered in 1996 by the Japanese amateur astronomer Yukio Sakurai, and FG Sagittae (FG Sge).

FG Sge was discovered to be variable in 1943. In 1955, it was a blue B-type star that was slowly getting redder. In 1991, it was a yellow F-type star and continued moving to the red side of the HRD, where it stopped as a K-type star on the AGB. The presence of an ionized circumstellar shell shows that the star was a CSPN before it started to move back to the right on the HRD. This means that it had been an AGB star before and had already crossed the HRD to the left and produced a PN a few thousand years ago. When it was a hot star, the last thermal pulse produced so much energy that the thin envelope expanded again. This resulted in an increase of the radius and a decrease of $T_{\rm eff}$, until the star was on the AGB again. Therefore, these stars are called **born-again AGB stars** (Duerbeck & Benetti 1996).

19.3 Planetary Nebulae

For a long time, PNe were explained in terms of the central star ionizing the previous AGB wind. The problem with this idea was that the expansion speed of the PNe is typically $v_{\rm exp} \sim 30$ to 50 km s^{-1}, but the AGB winds are ejected with $v_{\rm AGB} \sim 10$–15 km s^{-1}. How could the AGB material have been accelerated? After the discovery of stellar winds from CSPN (Heap 1979; Patriarchi & Perinotto 1991) it became clear that PNe are the result of the interaction between the slow AGB wind and the fast CSPN wind (Kwok et al. 1978).

Figure 19.3 (left) shows an image of the planetary nebula NGC 7293, also known as the Helix nebula. The nebula, at a distance of 220 pc, has an inner radius of 0.26 pc, an outer radius of 0.4 pc, a surrounding outer shell with a radius of 0.9 pc, a total mass of the ionized gas of $0.1 M_\odot$, and an expansion velocity of 31 km s^{-1}. The ratio between the outer radius and expansion velocity indicates a kinematic age of about 12,000 yr. The right side of Figure 19.3 shows a schematic picture of the interaction between the fast CSPN wind and the slow AGB wind. The fast wind (white) from the central star runs into the slow wind ejected during the AGB phase. The interaction region (green) is shock-heated and ionized, producing the visual nebula. Outside the nebula is the unshocked cold but highly diluted AGB wind (dark red).

The velocity of the interaction region between the slow AGB wind and the fast CSPN wind can be derived from the conservation of momentum of the two interacting winds (Lamers & Cassinelli 1999)

$$V_{\rm PN}(t) \simeq V_{\rm AGB}\left\{1 + \sqrt{\frac{\dot{M}_{\rm CS} V_{\rm CS}}{\dot{M}_{\rm AGB} V_{\rm AGB}}}\right\}. \tag{19.1}$$

The mass of the interaction region is

$$M_{\rm PN}(t) \simeq t \times \dot{M}_{\rm CS}\sqrt{\frac{\dot{M}_{\rm AGB}/V_{\rm AGB}}{\dot{M}_{\rm CS}/V_{\rm CS}}}, \tag{19.2}$$

where t is the time since the fast wind of the CSPN has reached the slow AGB wind.

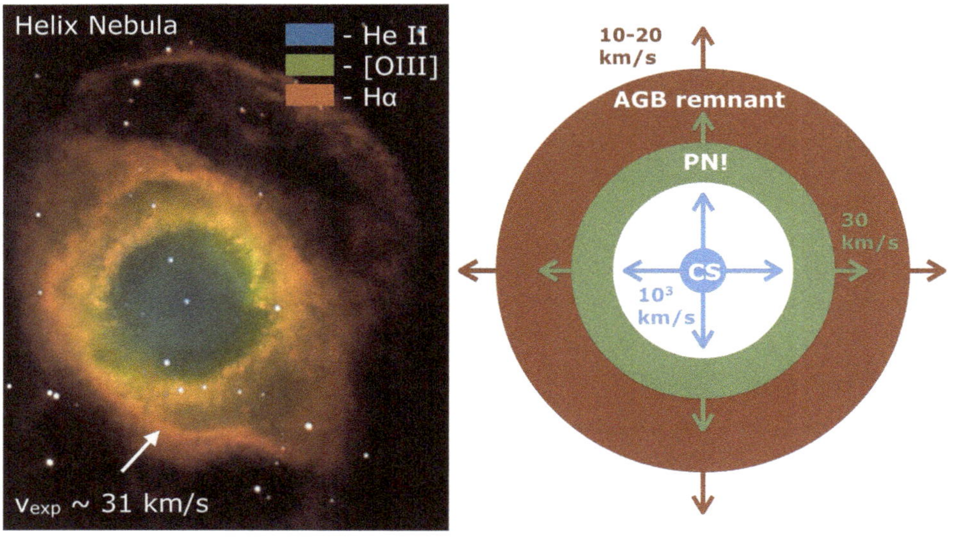

Figure 19.3. Left: *Hubble Space Telescope* image of the Helix nebula, NGC 7293, observed in different emission lines. Right: interacting wind model of planetary nebulae. (Courtesy of NASA;[1] model reproduced from Lamers & Cassinelli, 1999 © Cambridge University Press. Reprinted with permission.)

Adopting $\dot{M}_{\rm CSPN} \approx 10^{-7}\, M_\odot\, {\rm yr}^{-1}$, $\dot{M}_{\rm AGB} \approx 10^{-5}\, M_\odot {\rm yr}^{-1}$, $V_{\rm CS} = 3000$ km s^{-1}, and $V_{\rm AGB} = 20$ km s^{-1}, we find that $V_{\rm PN} \approx 30$ km s^{-1} and the mass of the PN at $t \sim 10^4$ yr is $\sim 0.1 M_\odot$, in reasonable agreement with the characteristic values of PNe.

Q (19.3) *Explain why the expansion velocity of a PN depends on the ratio* $\dot{M}_{\rm CS} V_{\rm CS} / \dot{M}_{\rm AGB} V_{\rm AGB}$.
Q (19.4) *Explain why the mass of a PN depends on the parameters as in* Equation (19.2).
Hint: $\dot{M} = 4\pi\rho v r^2$ *for a spherically symmetric wind.*

This simple description is reasonably successful in explaining the gross properties of PNe. However, observations with facilities such as the *Hubble Space Telescope* have shown that most PNe have complex morphologies. This suggests the presence of additional effects such as rotation and nonspherical winds of AGB stars, binary stars, and magnetic fields (Balick & Frank 2002).

19.4 Fading to the White Dwarf Phase

When the envelope mass has decreased to $M_{\rm env} < 10^{-4}$ to $10^{-6} M_\odot$, depending on luminosity, shell fusion stops and the luminosity decreases. The star moves down and to the right in the HRD along a *cooling track*. Figure 19.5 shows the cooling tracks from the post-AGB phase via the CSPN phase to the WD phase for stars of different initial masses.

[1] http://www.spacetelescope.org/images/opo0432d/

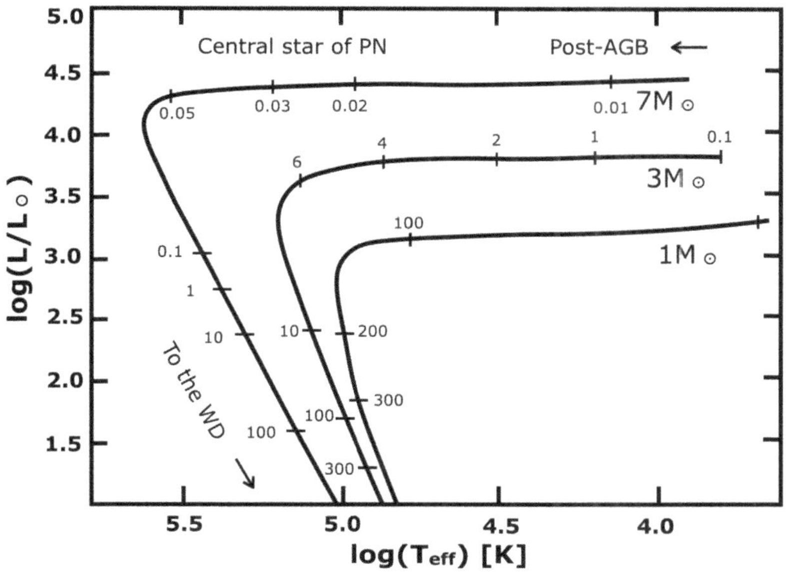

Figure 19.4. Evolutionary tracks of stars with initial masses of $M_i = 7$, 3, and $1M_\odot$. The ages since the stars left the AGB are given along the tracks in 10^3 yr. (Figure adapted from Blöcker 1995.)

Table 19.1. Evolution from AGB star to WD (Based on Blöcker 1995)

$M_i/M_\odot$	$M_c/M_\odot$	$\log(L/L_\odot)$ (AGB tip)	$\log(L/L_\odot)$ (turn)	$\log T_{\rm eff}$ (turn)	$t_{\rm crossing}$ yr	$t_{\rm fading}$ yr
7	0.94	4.30	4.42	5.62	6×10^1	1×10^6
3	0.60	3.75	3.26	5.20	7×10^3	3×10^6
1	0.52	3.20	2.90	5.00	2×10^5	4×10^6

Notes.

a. M_c is the core mass and the mass of the resulting WD.
b. The term "turn" is the hottest phase, i.e., the end of the leftward crossing.
c. $t_{\rm crossing}$ is the time from the tip of the AGB to the hottest phase.
d. $t_{\rm fading}$ is the time from the hottest phase to $L = L_\odot$.

Q (19.5) *What could be the reason that the H-shell fusion stops when the envelope mass is too small?*

Q (19.6) *Does the cooling track of a star of $M_i = 7M_\odot$ follow a line of constant radius? If not, does the radius increase or decrease? Explain the reason.*

Table 19.1 lists the characteristics of stars of initial masses 7, 3, and $1M_\odot$ in-between the AGB phase and the WD phase. Note the extreme sensitivity of the crossing time on M. The stars fade to the WD phase in a few megayears.

Because the radius of a star during the crossing and the fading depends strongly on the envelope mass, the predicted values of $t_{crossing}$ and t_{fading} are sensitive to the assumed mass-loss rate during the post-AGB phase, especially the mass-loss rate of the CSPN.

19.5 Summary

1. A star leaves the AGB when the envelope mass is below a critical limit of the order of 10^{-4} to $10^{-3} M_{\odot}$, because it can no longer sustain a deep convection zone. The envelope slowly goes into radiative equilibrium. The star shrinks and moves to the left in the HRD. The crossing time in the HRD is typically 10^4 yr, but depends strongly on core mass and luminosity. During the crossing, the energy is still produced by He-shell and H-shell fusion.
2. The luminosity is set by the mass of the degenerate core, which does not change anymore. The values of the radius and T_{eff} are set by the decreasing envelope mass.
3. Some post-AGB stars experience a late thermal pulse after they leave the AGB. These stars temporarily evolve back to the AGB. They are called **born-again AGB stars.**
4. When $T_{eff} \approx 30{,}000$ K the star develops a fast line-driven wind. This fast wind runs into the slow AGB wind. The interaction region is heated by the shock and ionized by the UV radiation of the star. This produces a planetary nebula **(PN).** The star is now a central star of a planetary nebula **(CSPN)** that continues to move leftward in the HRD.
5. When the envelope mass is below $\sim 10^{-5}$ to $10^{-4} M_{\odot}$, the pressure in the fusion shells is so low that fusion is slowly extinguished.
6. When fusion is extinguished, the star moves down in the HRD toward the white dwarf region. This fading takes a few 10^6 yr.

Exercises.

19.1 (a) Calculate the radius of the star of $M_i = 3M_{\odot}$ and the density in the convective envelope when it leaves the AGB. Realize how small it is compared to the mean density in the convective envelope of the Sun (Appendix C).

(b) Calculate the change in the mean density in the envelope (i.e., outside the H- fusion shell) during the transition from right to left in the HRD.

19.2 The location of a post-AGB star during the crossing of the HRD from right to left depends on the core mass and the envelope mass. Consider the evolutionary track shown in Figure 19.1.

(a) Calculate the evolution time during the horizontal crossing of the HRD from $T_{eff} = 5000$ K ($M_{env} = 0.00184 M_{\odot}$) and $T_{eff} = 100{,}000$ K if the star had no stellar wind.

(b) Before stellar winds from post-AGB stars were discovered from the study of their UV spectra in the 1980s, astronomers had already concluded that post-AGB stars must have stellar winds. Can you think of arguments for this conclusion?

(c) Calculate the crossing time of the same post-AGB stars if the mean mass-loss rate by a stellar wind was 10^{-9}, 10^{-8}, and 10^{-7} $M_{\odot}$ yr^{-1}. Realize how strongly the evolution timescale depends on the mass-loss rate. Describe the physical reason for this strong dependence.

References

Balick, B., & Frank, A. 2002, ARAA, **40, 439**

Blöcker, T. 1995, A&A, **299, 755**

Duerbeck, H. W., & Benetti, S. 1996, ApJL, **468, L111**

Heap, S. R. 1979, in IAU Symp. 83, Mass Loss and Evolution of O type Stars (Dordrecht: Reidel), 99

Iben, I. 1982, ApJ, **260, 821**

Kudritzki, R. P., & Puls, J. 2000, ARA&A, **38, 613**

Kwok, S., Purton, C. R., & Fitzgerald, P. M., 1978, ApJL, **219, 125**

Lamers, H. J. G. L. M., & Cassinelli, J. P. 1999, Introduction to Stellar Winds (Cambridge: Cambridge Univ. Press)

Patriarchi, P., & Perinotto, M. 1991, A&AS, **91, 325**

Henny J.G.L.M. Lamers and Emily M. Levesque

Chapter 20

White Dwarfs and Neutron Stars

All stars with initial masses smaller than about $8M_\odot$ end their lives as **white dwarfs (WD)**, degenerate stars with no nuclear energy source. They radiate as they lose thermal energy from their degenerate interior and we can describe their evolution due to this cooling. We will also show that there is a maximum mass for white dwarfs, which is different for H-rich and He- or C- and O-rich white dwarfs. Observationally, white dwarfs come in three spectral classes: **DA** (spectrum dominated by H-lines), **DB** (spectrum dominated by He-lines), and **DC** (continuum spectrum without absorption lines). This distinction is based only on the spectrum and does not give information on the *internal* composition. The gravity at the surface of a WD is high ($\sim 10^8$ cm s^{-2}) and the atmosphere is so stable that gravitational diffusion occurs: heavier elements sink to the bottom of the photosphere and lighter elements appear at the top. For instance, in He-rich or C- and O-rich WDs the He, C, and O may have settled at the bottom of the atmosphere, leaving an extremely thin H layer in the photosphere that produces the A-type spectrum. We show that the luminosity of white dwarfs is due to cooling of their ions and we derive an expression for the cooling and the decrease in luminosity. We also discuss the structure of **neutron stars (NSs)**, which are more massive degenerate stars. These stars are the end products of the evolution of more massive stars with initial masses of $8 \lesssim M_i \lesssim 25M_\odot$.

20.1 Stars That Become White Dwarfs

White dwarfs are the end points of the evolution of low-mass stars. Three scenarios can be distinguished.

(1) $0.8 \lesssim M_i \lesssim 8M_\odot$: these stars go through core H- and core He-fusion and end as C- and O-rich degenerate stars, **CO white dwarfs.**

(2) $0.5 \lesssim M_i \lesssim 0.8M_\odot$: these stars reach core H-fusion, but they are not massive enough to reach core He-fusion. They will end their lives as **He-rich white dwarfs.** The lifetime of a star with $M_i \lesssim 0.8M_\odot$ is longer than the age of the

doi:10.1088/978-0-7503-1278-3ch20

Universe. This implies that no He-rich WD could have formed yet by single star evolution. Therefore, He-rich WDs must have been formed via the stripping of an initially more massive star by a close companion. In this process, the star loses its envelope by mass transfer to the companion, leaving behind a degenerate core at the end of its truncated evolution.

(3) Binary-stripping of even lower-mass stars can result in the formation of **H-rich white dwarfs.**

20.2 The Structure of White Dwarfs

White dwarfs are stars that are electron nonrelativistic degenerate, so their EoS can be expressed as $P \sim \rho^{5/3}$. This means that they are polytropes with $\gamma = 5/3$ or $n = 1.5$. The structure of polytropic stars was discussed in Sections 11.2 and 11.3. In Section 11.3.1, in Equation (11.14), we have shown that there is a strict *mass–radius relation* for degenerate stars

$$R = 0.012\, M^{-1/3}\ (\mu_e/2)^{-5/3} \qquad \text{if } M \text{ and } R \text{ are in solar units.} \tag{20.1}$$

This relation implies that the volume of the star is inversely proportional to its mass! For fully nonrelativistic degenerate white dwarfs, the numerical values are

$$\begin{array}{lll} \text{H}-\text{WD} & \mu_e = 1 & R = 0.046 \times (M_{\text{WD}}/0.5M_\odot)^{-1/3} R_\odot \\ \text{He or CO}-\text{WD} & \mu_e = 2 & R = 0.014 \times (M_{\text{WD}}/0.5M_\odot)^{-1/3} R_\odot. \end{array} \tag{20.2}$$

Q (20.1) *What is the radius and density of a He-rich WD of* $0.5M_\odot$*? Compare it with the radius and mean density of Earth.*

Q (20.2) *How does the mean density of WDs vary with mass?*

20.3 The Chandrasekhar Mass Limit for White Dwarfs

The mass–radius relation for WD (Equation (20.2)) shows that R will decrease as M increases. Because the volume of a WD is inversely proportional to its mass, the mean density increases as $\rho \sim M^2$. This implies that the density in the center of a WD may be high enough for the electrons to become relativistic degenerate (RD). The more massive a WD, the larger the mass fraction of the relativistic degenerate core. If the WD is completely relativistic degenerate, the EoS is $P \sim K\rho^{4/3} \rightarrow n = 1/(\gamma - 1) = 3$ (Section 4.5.2). We have shown in Section 11.3.4 that polytropes with fixed K and $n = 3$ can only exist for one specific mass. This is the **Chandrasekhar mass limit.** This limit is named after the Indian–American Nobel laureate Subrahmanyan Chandrasekhar (1910–1995), who derived it at the age of 19 while traveling by boat from India to the U.K. (Chandrasekhar 1931). The modern value is

$$M_{\text{ch}} = 1.46(2/\mu_e)^2 M_\odot, \tag{20.3}$$

with $\mu_e = 1$ for H-WDs and $\mu_e = 2$ for He-WDs and CO-WDs. This implies that WDs with $M > M_{\text{ch}}$ cannot exist. If a WD in a binary system accretes enough mass from its companion to reach this limit, it will collapse into a neutron star; in practice,

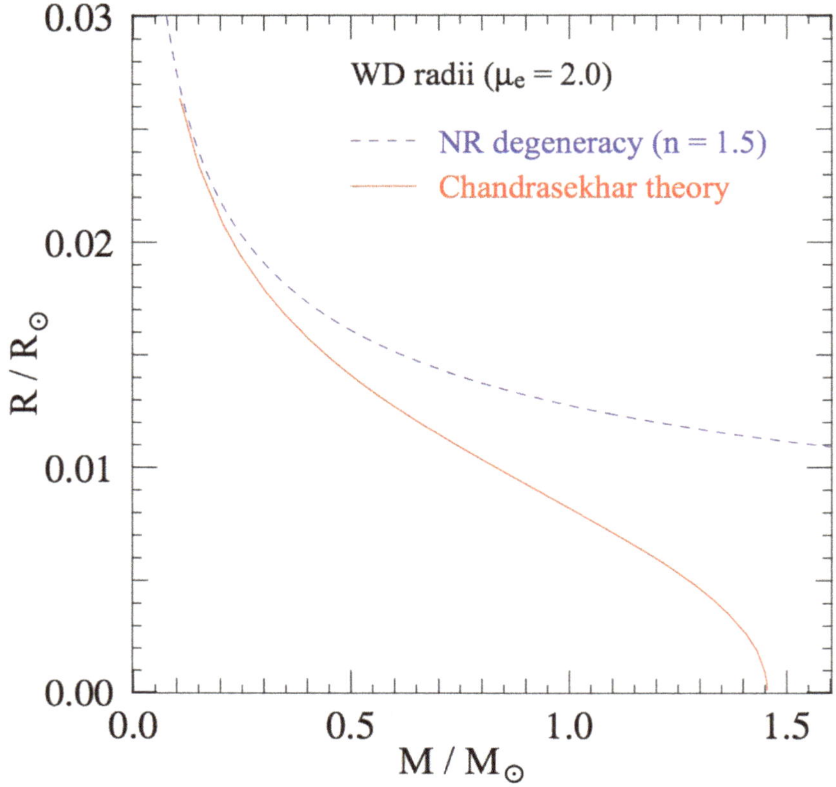

Figure 20.1. Mass–radius relation of WDs. The blue dotted line is for completely nonrelativistic degenerate He-rich or CO-rich WDs with $\mu_e = 2$. The red line is for partly relativistic degenerate WDs. The radius goes to zero at the Chandrasekhar limit of $1.46 M_{\odot}$. (© Pols 2011.)

CO-WDs accreting mass from a companion appear to undergo a runaway nuclear fusion reaction that leads to their explosion as a *supernova of type Ia* (discussed in Chapter 27).

The predicted mass–radius relation of WDs is shown in Figure 20.1. The blue lines are for nonrelativistic degenerate stars (Equation (20.2)). The red line is for WDs with a relativistic degenerate core. The radius goes to zero at the Chandrasekhar limit.

20.4 The Cooling of White Dwarfs

The luminosity of WDs comes from cooling. The electrons cannot cool because they are degenerate, so their energy distribution is set by the density, which does not change. Only the ions can cool and they contain almost all of the mass of a WD. Their thermal energy is transported outward by conduction and radiated by the very thin nondegenerate photosphere.

Initially, the ions have a temperature of the order of 10^8 K, which was the temperature of the He-fusion shell and the isothermal core. Such a young hot WD cools down fast, with a relatively high luminosity of $\sim 10^{-1} L_\odot$. However, as the ions cool, the luminosity decreases and so the cooling slows down and the luminosity decreases slowly over time.

The cooling of a WD depends on its energy loss and hence on its luminosity. The luminosity depends on the temperature of the interior. Because the core has no energy source, the temperature gradient must be about zero, so the core is isothermal (apart from a very small gradient due to heat conduction). The luminosity–temperature dependence can be derived if we assume for simplicity that the WD consists of a fully degenerate core with a thin photosphere made up of ideal gas with a Kramers law opacity: $\kappa = \kappa_0 \rho T^{-7/2} \sim PT^{-9/2}$ (Equation (5.8)). In that case, Eddington's equation for radiative transfer (Equation (6.1)) in the photosphere, combined with the HE equation $dP/dr = \rho GM/r^2$, can be written as

$$dT/dP = APT^{-7.5} \qquad \text{with } A = \frac{3\kappa_0}{16\,\pi ac} \frac{\mu}{\mathfrak{R}} \frac{L}{M}. \tag{20.4}$$

The solution of this equation with the approximate outer boundary conditions, $T \to 0$ and $P \to 0$, gives

$$T^{8.5} = 4.25A \times P^2 \qquad \text{and} \qquad \rho \sim P/T \sim T^{3.25} \tag{20.5}$$

in the photosphere.

Now consider the boundary (b) between the degenerate core and the photosphere. At that boundary the pressure at the bottom of the photosphere, $P_b \sim \rho_b T_b$, is equal to that at the top of the core, $P_b \sim \rho_b^{5/3}$. So

$$P_b \sim \rho_b^{5/3} \sim \rho_b T_b \to \rho_b^{2/3} \sim T_b \qquad \text{with } \rho_b \sim T_b^{3.25}. \tag{20.6}$$

Substitution of the temperature from Equation (20.5) gives an expression for T_b. Inserting the constants results in

$$T_b = 5.9 \times 10^7 \left\{ \frac{L/L_\odot}{M/M_\odot} \right\}^{2/7} \text{K} \tag{20.7}$$

for a CO or He WD. Adopting typical values of $L \sim 10^{-2} L_\odot$ and $M \sim 0.6 M_\odot$, we find $T_b \approx 1.8 \times 10^7$ K. Since the core is isothermal, its temperature is $T_c \approx T_b$.

The thermal energy of the ions that is available for cooling is $E_{th} = c_V\, M\, T_c$, where $c_V = 3k/2m_{ion}$ is the specific heat per gram · K and $m_{ion} = 4m_H$ for a He WD. The luminosity of a WD is due to cooling of the ions, so

$$L = -\frac{dE_{th}}{dt} = -c_V M \frac{dT_c}{dt} = -c_V M \frac{dT_b}{dt}. \tag{20.8}$$

We have shown in Equation (20.7) that $T_b \sim (L/M)^{2/7}$, so $L = AMT_b^{7/2}$, where A is a constant. Adopting $T = T_b = T_c$ results in an expression of the form

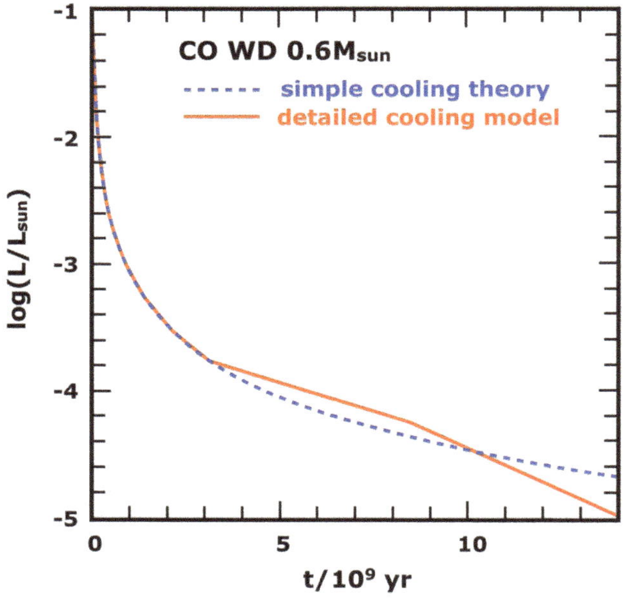

Figure 20.2. Cooling curve of a CO WD of $0.6M_{\odot}$. Blue line: simplified model described above. Red line: detailed model with crystallization. (© Pols 2011.)

$$\frac{dT}{dt} = -(A/c_V)T^{7/2} \rightarrow T^{-5/2} = T_0^{-5/2} + (5A/2c_V)t, \tag{20.9}$$

where t is the time since the cooling started. Substitution of the constant A and the resulting $T(t)$, from equation (20.9), into expression (20.7) for the luminosity, $L\sim MT^{7/2}$, results in the cooling curve shown in Figure 20.2. We see that the cooling is initially rapid. After this initial fast cooling, when $T \ll T_0$, the temperature decreases as $T \sim t^{-2/5}$. Substitution in equation (20.7) gives $L \sim Mt^{-7/5}$. Substitution of the constants yields

$$\boxed{\frac{L}{L_{\odot}} \approx 4.3 \times 10^{-2} \frac{M}{M_{\odot}} \mu_{\rm ion}^{-7/5} \left(\frac{t}{\rm Gyrs}\right)^{-7/5}} \tag{20.10}$$

with $\mu_{\rm ion} = 4$ for a He WD. This shows that an increase in t by a factor of 1.64 reduces L by a factor of 2. It also shows that a He-rich WD will be fainter than a H-rich WD of the same cooling age.

Figure 20.2 shows the calculated luminosity of a CO WD of $0.6M_{\odot}$. The dashed blue line is the prediction by the simple theory (Equation (20.10)). It shows that after a few gigayears the luminosity of a WD decreases slowly on a timescale of gigayears. The red line in Figure 20.2 is the prediction from a more detailed model in which partial degeneracy and crystallization are taken into account. Crystallization occurs in the center of a WD when the ions form a lattice structure. This structure grows as the white dwarf cools. Crystallization releases energy and produces the bump in L at $t \approx 5$ Gyr.

20.5 Neutron Stars

Neutron stars (NSs) are the descendants of single stars with initial masses between about 8 and $25M_\odot$. The evolution of these stars will be discussed in Chapter 23.

At densities in excess of about 1×10^7 g cm^{-3}, electrons are captured by protons, forming a neutron-rich gas at a density so high that even the neutrons are degenerate. Compact stars with $M > M_{\rm ch}$ collapse into neutron stars, which are supported by the pressure of degenerate neutrons. The equation of state for neutron degeneracy is similar to that of electron degeneracy (Equations (4.20) and (4.22)):

$$P_n = K_{n,1}(\rho/\mu_n)^{5/3} \qquad \text{if } \rho \lesssim 6 \times 10^{15}\ \text{g cm}^{-3}, \tag{20.11}$$

with $K_{n,1} = 5.5 \times 10^9 \frac{\text{dyne cm}^{-2}}{\left(\text{g cm}^{-3}\right)^{5/3}}$ and

$$P_n = K_{n,2}(\rho/\mu_n)^{4/3} \qquad \text{if } \rho \gtrsim 6 \times 10^{15}\ \text{g cm}^{-3}, \tag{20.12}$$

with $K_{n,2} = 1.24 \times 10^{15} \frac{\text{dyne cm}^{-2}}{\left(\text{g cm}^{-3}\right)^{4/3}}$ and $\mu_n = 1$ for a gas fully consisting of neutrons. Equation (20.11) is the EoS for extreme nonrelativistic degenerate neutrons. Equation (20.12) is the EoS for extreme relativistic degenerate neutrons. We see that nonrelativistic degenerate neutron stars are polytropes with $n = 1.5$, so they have a similar mass–radius relation to white dwarfs:

$$R_{\rm ns} \approx 4.9(M_{\rm ns}/M_\odot)^{-1/3}\ \text{km}. \tag{20.13}$$

This equation predicts $R_{\rm ns} \approx 4$ km for a typical NS mass of $M_{\rm ns} \approx 1.5M_\odot$. In reality, the radius of a neutron star is about 10 km, with a large uncertainty due to the uncertainty in the EoS (Özel & Freire 2016). With such a small radius, the density of a neutron star is on the order of 10^{15} g cm^{-3}! (For comparison, the density at the center of the Sun is $\sim 10^2$ g cm^{-3}.)

The structure of a neutron star is sketched in Figure 20.3. The crust consists of three layers: an outside layer of nonrelativistic electron degenerate gas, a middle layer of heavy nuclei (e.g., ^{56}Fe) and relativistic degenerate electrons, and an inner layer of even heavier neutron-rich nuclei, free neutrons, and relativistic degenerate electrons. The outer core consists of superfluid neutrons and protons and relativistic degenerate electrons. The inner core consists of hyperons, which are nuclei consisting of three quarks, including at least one "strange quark." The quarks interact via the strong nuclear force.

Complete relativistic degenerate neutron stars are $n = 3$ polytropes and exist for only one value of the mass, so neutron stars can only exist if their mass is smaller than some maximum value (analogous to the Chandrasekhar mass limit of white dwarfs). This maximum mass is about 2 to $3M_\odot$, depending on their uncertain EoS. This limit is called the Oppenheimer–Volkoff limit (Oppenheimer & Volkoff 1939). If the mass of a neutron star exceeds this limit, it collapses into a black hole. This happens at the ends of the lives of massive stars with $M_i \gtrsim 25M_\odot$ when they explode as core-collapse supernovae.

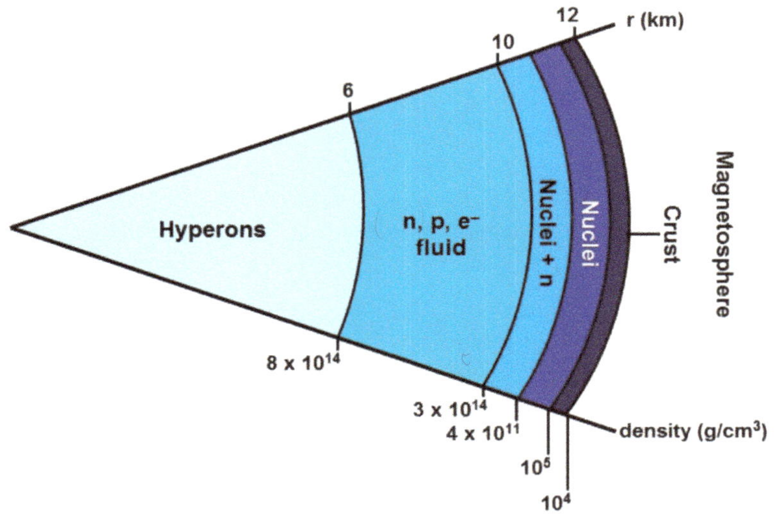

Figure 20.3. Schematic representation of the structure of a neutron star. The different regions, described in the text, are indicated by different colors. The density scale and the distance scale are indicated. (Reproduced from Bowers & Deeming 1984, courtesy of Boston: Jones and Bartlett.)

20.6 Summary

1. White dwarfs are electron degenerate stars.
2. CO white dwarfs are the remnants of stars with $0.8 \lesssim M_{\rm i} \lesssim 8M_\odot$. He- and H-rich WDs are formed by binary-stripping.
3. The mass–radius relation of nonrelativistic degenerate WDs is $R \sim M^{-1/3}$. WDs have a radius on the order of the Earth radius and a density of $\sim 10^5$ to 10^6 g cm^{-3}. WDs with $M \gtrsim 0.5M_\odot$ are partly relativistic degenerate in their core and the mass–radius relation becomes steeper.
4. The Chandrasekhar upper limit for He and CO WDs is $1.46M_\odot$.
5. The luminosity of WDs is due to the loss of thermal energy of the ions. The cooling time is on the order of a few gigayears and $L \sim t^{-7/5}$.
6. Neutron stars are the degenerate end products of stars with $M_i > 8M_\odot$. They have a mass in the range of 1.46 to $\sim 3M_\odot$ and a density of the order of 10^{15} to 10^{17} g cm^{-3}. The upper mass limit is uncertain due to uncertainties in the EoS.

Exercises

20.1 The lower limit for the initial mass of single stars that become neutron stars is equal to the upper limit for the initial mass of single stars that become WDs. This last limit can be derived empirically from the study of a large number of clusters. Can you invent a way to do this?

20.2 (a) How much fainter or brighter is a H-rich WD compared to a He-rich WD and a C-rich WD of the same cooling age in the slow cooling phase? (b) Explain the basic reason for this difference in physical terms.

20.3 What is the expected lower luminosity limit of CO-rich WDs resulting from stars formed shortly after the Big Bang? (Assume that the stellar evolution of the first generation stars was about the same as that of stars that are formed now.)

References

Bowers, R. L., & Deeming, T. 1984 in Astrophysics I: Stars (Boston: Jones and Bartlett), 287

Chandrasekhar, S. 1931, ApJ, 74, L81

Oppenheimer, J. R., & Volkoff, G. M. 1939, PhRv, 55, 374

Özel, F. Freire, P. 2016, ARA&A, 54, 401

Pols, O. 2011, Stellar Structure and Evolution, Utrecht University Lecture Notes, 146 https://www.astro.ru.nl/~onnop/education/stev_utrecht_notes/

Winget, D. E., Hansen, C. J., Liebert, J., et al. 1987, ApJ, 315, L77

AAS | IOP Astronomy

Understanding Stellar Evolution

Henny J.G.L.M. Lamers and Emily M. Levesque

Chapter 21

Pulsating Stars

A surprisingly large fraction of stars pulsate, from the low-mass solar-type stars to the intermediate-mass δ Scuti stars, to the high-mass long period variables and Cepheids. Pulsation can occur during various evolutionary stages: the main-sequence phase, the horizontal branch phase, the AGB phase, and when stars cross the Hertzsprung gap in the HRD. Several types of stars are radial pulsators that are located in the so-called "classical" instability strip in the HRD. We discuss their pulsation mechanisms and the timescale of the radial pulsations. In recent years, with the advent of accurate photometry measured from space, many nonradial pulsators have also been discovered. We briefly discuss some of their properties.

21.1 Classical Radial Pulsators

Radially pulsating stars periodically change in size. Their pulsation is due to the existence of a partial ionization zone in the envelope. The location of the classical radial pulsators in the HRD is shown in Figure 21.1. Most of these radial pulsators are located in the **instability strip**, in-between the main sequence on the hot side and the Hayashi line for red giants and AGB stars on the cool side. This location corresponds to a specific depth of the ionization zone of H or He, as will be discussed below.

Low-mass and high-mass stars cross the instability strip when they evolve from the MS to the Hayashi track during the H-shell fusion phase (for low-mass stars) or during the core He-fusion phase (for high-mass stars). Because the crossing time during H-shell fusion is much shorter than that during core He-fusion the number of classical radial pulsators in the core He phase is higher.

The properties of the most important radially pulsating stars are listed in Table 21.1. The **pulsation constant** $\bar{Q}$ is a characteristic of the radial pulsation period and will be discussed below.

- **δ Cepheids,** or **Type I Cepheids** (or simply **Cepheids)**, are massive metal-rich (Pop I) stars that cross the instability strip when they describe a loop in the

doi:10.1088/978-0-7503-1278-3ch21

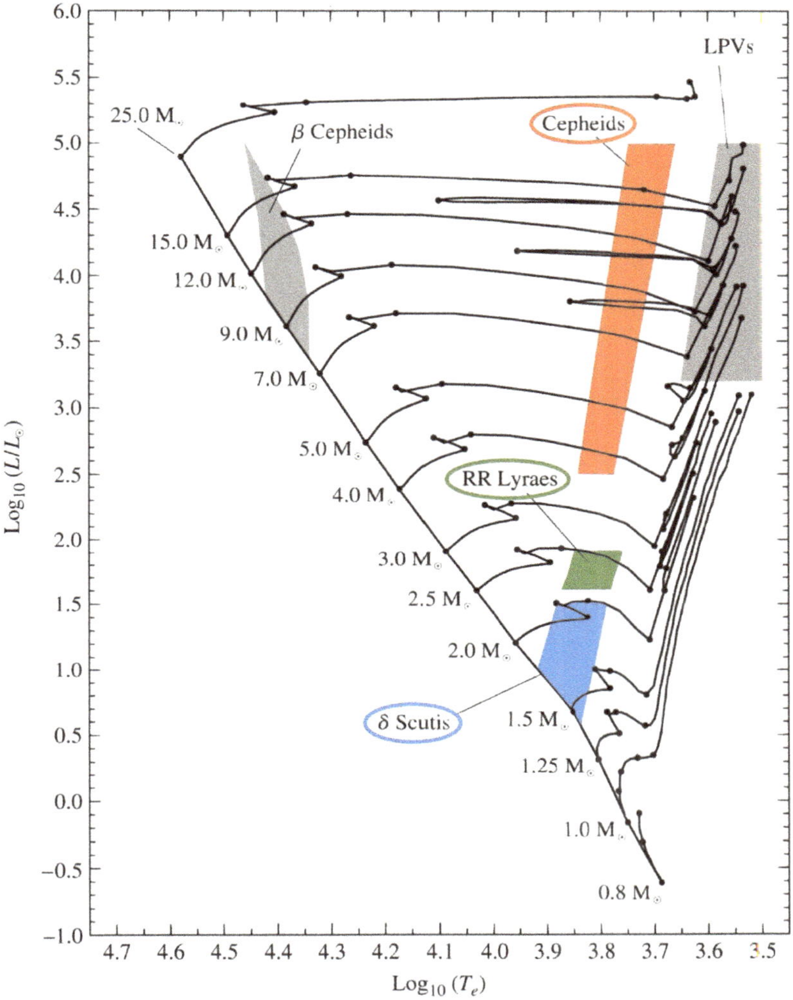

Figure 21.1. Location in the HRD of various types of radially pulsating stars. Cepheids (red), RR Lyrae stars (green), and δ Scuti stars (blue) are located in a nearly vertical instability strip. The long period variables (LPVs) and β Cepheids are outside of this strip. The gap between the red, green and blue regions are due to the smaller number of stars in this region. (Reproduced from Iben 1967, with permission)

Table 21.1. Properties of the Fundamental Mode of Radially Pulsating Stars.

Type	Pop	Period (days)	log $(L/L_{\odot})$	Spectral Type	$\bar{Q}$(days)
δ Cepheids	I	2–60	2.8–4.6	F–G	0.04
δ Scuti	I	0.04–0.2	0.9–1.6	A–F	0.04
β Cepheids	I	0.1–0.2	4.0–4.7	B1–B2	0.03
W Virginis	II	1–20	2.0–2.9	F–G	0.06
RR Lyrae	II	0.3–1	1.5–1.6	A–F	0.04

Note. Data are from Cox (2000).

HRD during the core He fusion phase. Their pulsation is driven by a partial ionization zone in the star.
- **δ Scuti** stars, or **dwarf Cepheids**, are on an extension of the instability strip at lower luminosity down to the main sequence.
- **β Cepheids**, or **β Canis Majoris stars**, are massive main-sequence stars or subgiants of about 7 to $20M_\odot$. They are not located in the main instability strip because their pulsation is not driven by partial ionization of H or He, but by the high opacity in a layer of $T \sim 2 \times 10^5$ K below the photosphere, due to iron-like chemical elements.
- **W Virginis stars,** or **Type II Cepheids** are the metal-poor (Pop II) equivalents of Type I Cepheids.
- **RR Lyrae stars** are low-mass metal-poor stars in globular clusters that cross the main instability strip during core He-fusion on the HB.

21.2 Pulsation Periods of Classical Radial Pulsators

The pulsation period, *P*, of a star depends on the timescale for restoring equilibrium. For stars that pulsate radially, the restoring force is gas pressure. In Section 9.1, we have shown that the characteristic time for a sound wave to travel from the center to the surface of a star is the dynamical time that is inversely proportional to the square root of G × $\bar{\rho}$ (Equation 9.2), so we can expect a relation of the type

$$\boxed{P = C \times \tau_{\rm dyn} = \frac{C}{\sqrt{G\bar{\rho}}} = Q \times \left(\overline{\rho_\odot}/\bar{\rho}\right)^{1/2}}, \qquad (21.1)$$

where *C* is a constant of the order of unity and *Q* is the **pulsation constant.** The mean value of *Q* is given for the different types of variables in Table 21.1. It is about 0.05 days for most radial pulsators. The pulsation is in fact a standing pressure wave, driven in the ionization zone in the envelope and traveling between the stellar center and the surface of the star. Because the temperature structure in the different types of stars is not the same, the sound speed crossing time depends on the evolutionary phase, so the constants *C* and *Q* are slightly different for the various types of variables. In fact, the value of *Q* for any type of variable changes slightly with the period. For instance, $Q = 0.036$ for Cepheids with $P = 2.5$ days and 0.045 for Cepheids with $P = 60$ days (Cox 2000).

21.2.1 Period–Luminosity Relations

Radial pulsators have periods that strictly follow period–luminosity relations. For a given type of radial pulsator, a higher luminosity or larger radius corresponds to a lower mean density and a longer period. The *P-L* relations depend on the observed photometric band. The most accurate *P-L* relations of Cepheids are those in the infrared, because the IR is in the Rayleigh–Jeans part of the spectrum, where the magnitudes are less sensitive to the details of the changes in spectral type.

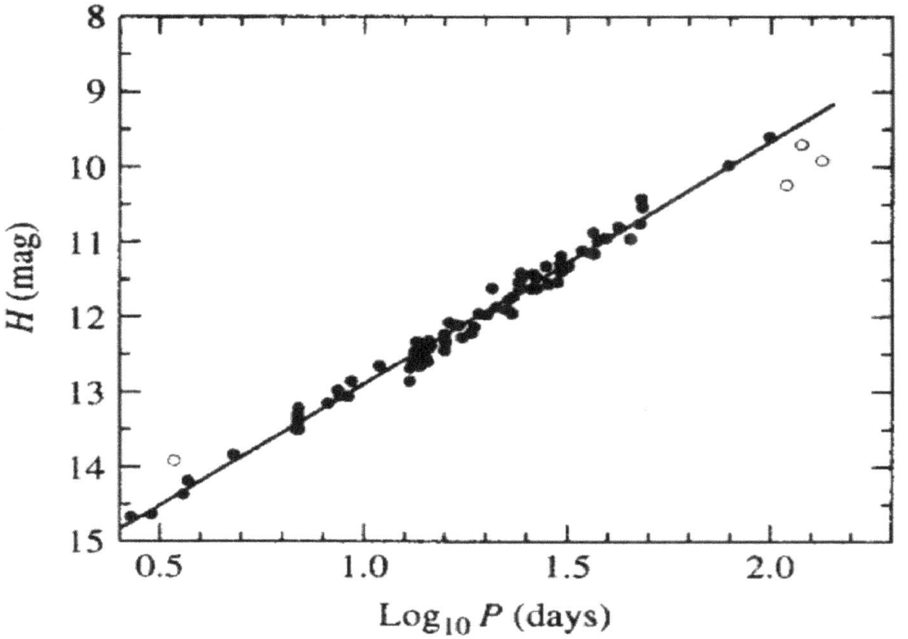

Figure 21.2. Period versus the H-band magnitude relation of Cepheids in the LMC. The mean relation is H = −3.14 × log (*P*/days) + 16.00. (Reproduced from Testa et al. 2007, with permission. © ESO.)

The existence of a *P-L* relation for stars in a small range of $T_{\rm eff}$ can easily be understood. We have seen that $P \sim \bar{\rho}^{-1/2} \sim M^{-1/2} \times R^{3/2}$ with $R \sim L^{1/2}$ for constant $T_{\rm eff}$ and $M \sim L^{1/x}$ if L depends on M as $L \sim M^x$. This results in $P \sim L^{(3x-2)/4x}$. The exponent is approximately 1/2 for a mass–luminosity relation with $2 < x < 4$.

An accurate *P-L* relation for Galactic Cepheids has been derived from trigonometric parallaxes measured with the Hubble Space Telescope:

$$M_v = -2.43 \log (P/\text{day}) - 1.62, \tag{21.2}$$

where M_v is the mean value during the pulsation (Benedict et al. 2007). The relation is slightly different for Type II Cepheids in the LMC, due to their lower metallicities (Gieren et al. 1998):

$$M_v = -2.77 \log (P/\text{day}) - 1.29. \tag{21.3}$$

Figure 21.2 shows the very tight relation between the period and the mean H-band magnitude of Cepheids in the LMC. Cepheids are therefore ideally suited for determining their absolute magnitude from the observed period and, thus, determining the distance of their host star cluster or galaxy from the comparison between the absolute and apparent magnitude.

Q (21.1) *Give at least two reasons why the observed P-L relation is tighter in the near-IR than in the visual.*

21.3 The κ-mechanism of Classical Radial Pulsators

RR Lyrae stars, Cepheids, δ Scutis, and their Pop II equivalents are variable because the κ-mechanism in the H and He ionization zone excites a radial pulsation mode, where κ refers to the absorption coefficient. The **κ-mechanism** works in layers, which are partly ionized because, in ionization zones, the degree of ionization and the opacity can change during compression and expansion. The κ-mechanism is also called the **Eddington valve mechanism,** because Eddington described it in 1926 in his influential book "The internal constitution of the stars" (Eddington 1926).

To explain this, let us first look at the working of a normal combustion motor in a car. A car is driven by the motion of pistons in cylinders. For an engine to work, it needs heat input in the cylinder at the right moment (i.e., when the gas is compressed). This is done by producing a spark during compression. This spark heats the gas, which expands and drives the cylinder upward. If you ignited the spark at the moment of minimum compression (i.e., when the cylinder is at its highest position), the motor would not work; the crucial aspect of the engine is the addition of heat when the gas is compressed. In the expansion phase, the exhaust valve opens so that the hot gas escapes through the exhaust pipe and the piston can move inward to begin a new compression phase.

In the κ-mechanism for pulsating stars, the heat input during compression is provided by an increase in opacity during the compression of a partially ionized zone in a star. This blocks the transfer of radiation and heats that layer whereby energy is delivered to the oscillation, hence the name "κ-mechanism." During expansion the gas cools, κ decreases with ρ, and excess heat is released so that the gas can start a new compression phase, hence the name "valve mechanism."

Let us compare the effect of compression in a fully and a partly ionized layer where the absorption coefficient is described by the Kramers relation (Equation (5.8)) $\kappa \sim \rho T^{-7/2}$.

In a fully ionized or fully neutral layer:

Compression: $\rho\uparrow \rightarrow T$ increases adiabatically (Equation (4.33)) $T \sim \rho^{2/3}\uparrow$

$\rightarrow \kappa \sim \rho^{-4/3} \downarrow$ so radiation escapes more easily if a layer is compressed.

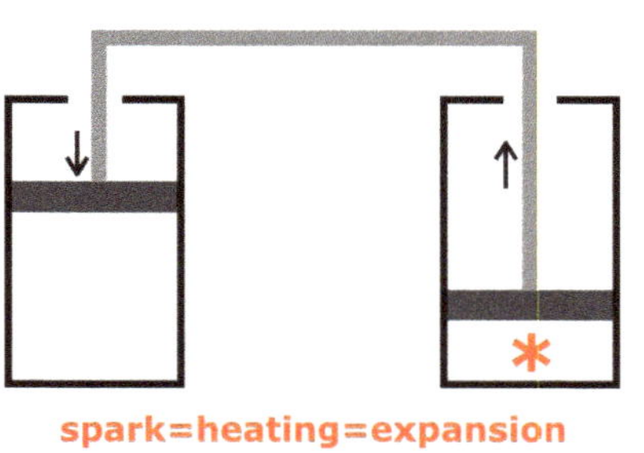

Figure 21.3. Basic principle of a combustion motor. A spark ignites the fuel (gas) when the gas is compressed. This heats the gas, so its expands and drives one piston up and the other one down. The production of a spark during compression in the alternating cylinders keeps the motor running.

In a partly ionized layer:

Compression: $\rho\uparrow \rightarrow T$ rises marginally because the heat goes into ionizing the gas:
$P \sim \rho^{\gamma}$ and $T \sim \rho^{\gamma-1} \rightarrow \kappa \sim \rho T^{-7/2} \sim \rho^{(9-7\gamma)\,/\,2} \rightarrow \boldsymbol{\kappa \uparrow}$ **if** $\boldsymbol{\gamma < 9/7}$.
$\rightarrow \kappa\uparrow$, so the radiation flow is blocked (trapped) in the compressed layer
$\rightarrow$ energy input during compression (like in an engine).

Expansion: $\rho\downarrow$ but T is about constant as gas recombines and releases energy:
$P \sim \rho^{\gamma}$ and $T \sim \rho^{\gamma-1}$ with $\gamma < 9/7 \rightarrow \kappa \sim \rho T^{-7/2} \sim \rho^{(9-7\gamma)/2} \downarrow \rightarrow \kappa\downarrow$
and the energy escapes by radiation.

This works like a motor because the ionization layer stores energy during compression and releases it during expansion. The condition that $\gamma < 9/7$ is crucial for the κ–mechanism! In fully ionized gas $\gamma = 5/3$, so the κ-mechanism does not work. However, γ is reduced in partly ionized gas and reaches a value of $1.19 < \gamma < 1.29 \approx 9/7$ (see Figure 4.5). So *the partial ionization zones are the motors that drive the pulsations in the classical instability strip.*

There are three ionization zones in a star, starting from outside in: the $H \rightarrow H^+$ zone, the $He \rightarrow He^+$ zone, and the $He^+ \rightarrow He^{++}$ zone. Detailed models show that the $He^+ \rightarrow He^{++}$ zone, which is the deepest of the three and occurs at a temperature of about 4×10^4 K, is mainly responsible for the observed radial pulsations.

In principle, the κ-mechanism could work in the partial ionization zone of any star, but it only produces an efficient pulsation if two conditions are fulfilled.

1. The ionization layer cannot be too deep; otherwise the layers above it may quench the pulsation. This happens in cool stars where the ionization zone lies deep and the convection in the envelope on top of it produces efficient quenching.
2. The ionization layer cannot be too close to the surface; otherwise, there is not enough gas on top of the ionization zone to produce efficient pulsation. This happens in hot stars.

Stars only pulsate via the κ*-mechanism if the* $He^+ \rightarrow He^{++}$ *ionization zone has the right depth* (i.e., the star has the right surface temperature), slightly depending on gravity and pressure. This is the reason why there is a nearly vertical instability strip in the HRD!

Q (21.2) *Explain why the instability strip in the HRD runs to the lower left and not vertically or to the lower right.*

Figure 21.4 shows the depth of the ionization zone in stars of different $T_{\rm eff}$. The left model refers to the coolest radial pulsator in the instability strip. If $T_{\rm eff}$ is less than about 5500 K, the He ionization zone is too deep and the layers above it quench the pulsation. The right model refers to the hottest radial pulsator in the instability strip. If $T_{\rm eff}$ is higher than about 7500 K, the star is too hot and the ionization zone is not deep enough: there is not enough mass above the ionization zone to push the ionization zone down again. The ionization zone is at the right depth for pulsation in the model of $T_{\rm eff}$ = 6500 K. Notice that in these radially pulsating stars the $He^+ \rightarrow He^{++}$ ionization zone has only about 10^{-7} to 10^{-4} of the stellar mass above it.

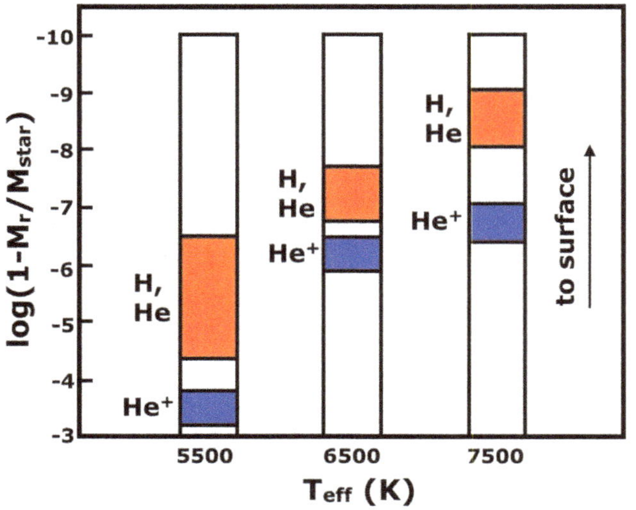

Figure 21.4. Schematic picture of the depth of the H and He ionization zones (IZ) in stars of different $T_{\rm eff}$ in the classical instability strip. Red indicates the location of the combined H→H$^+$ and He→He$^+$ zone and blue shows the location of the He$^+$→He^{++} zone. The radial pulsation is driven by the He$^+$→He^{++} zone. The vertical axis gives the logarithm of the fraction of the stellar mass above the zones. Left column: maximum depth of IZ for radial pulsation. Right column: minimum depth of IZ for radial pulsation. (Reproduced from Carroll & Ostlie 1996. © Cambridge University Press. Reprinted with permission.)

Q (21.3) *Why are the H→H$^+$ and He→He$^+$zones partly overlapping in mass, and why is the He$^+$→He^{++} ionization zone separated in mass from the other zones?*

21.4 An Example: The Pulsation of δ Cephei

Let us consider the observed variation of a Cepheid as an example of a radially pulsating star. When a star pulsates radially, it not only changes R but also $T_{\rm eff}$ and L. The bolometric magnitude changes with $R^2 {T_{\rm eff}}^4$. In general, the star is hotter when it is smaller. The photometric variation at a given wavelength depends on how the monochromatic flux per unit surface depends on $T_{\rm eff}$. At short wavelengths, i.e., shorter than the flux maximum, an increase in $T_{\rm eff}$ results in a large increase in monochromatic flux per unit surface. At wavelengths longer than the flux maximum, i.e., in the Rayleigh–Jeans tail of the Planck function, the monochromatic flux per unit surface scales with $T_{\rm eff}$, so the changes in magnitude will be smaller.

Figure 21.5 shows the variation in V-magnitude, $T_{\rm eff}$, R, and radial velocity $v_{\rm rad}$ (negative for expansion and positive for contraction) of the photosphere of the prototype Cepheid, δ Cephei. This star has a mass of $4.5 M_\odot$, a mean radius of 44.5 $R_\odot$, and pulsation period of 5.37 days. The average radial velocity of the star is -13.5km s^{-1}. The peak-to-peak change is 11.5% in radius and a factor of 1.7 in luminosity. Note that the velocity curve and $M_{\rm bol}$ have a sawtooth shape, with

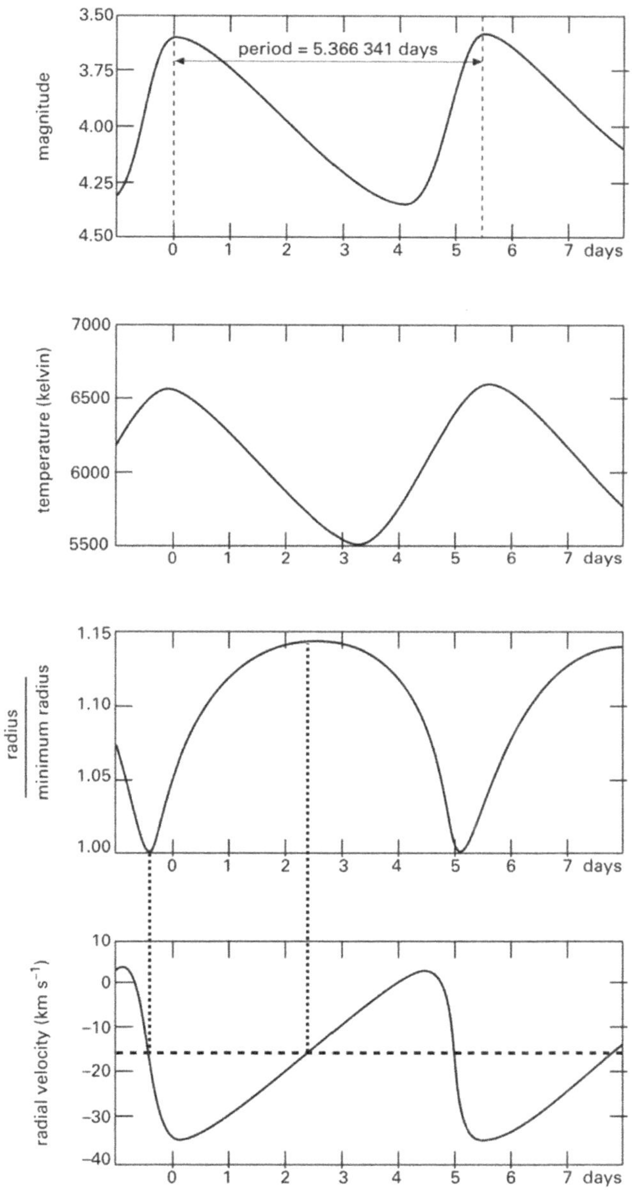

Figure 21.5. Variation in visual magnitude, $T_{\rm eff}$, $R/R_{\rm min}$, and $v_{\rm rad} = -v_{\rm expansion}$ during the pulsation of δ Cephei, which has a period of 5.37 days. The phase of $v_{\rm rad} = 0$ corresponds to either maximum or minimum radius. The maximum brightness is reached during maximum expansion velocity. (Reproduced form Collins Dictionary of Astronomy, 2006. © Market House Books. Ltd.)

maximum brightness corresponding to maximum outward velocity. This agrees with the explanation of the κ-mechanism, which requires that the radiation that is trapped in the ionization zone is released during expansion. The variations in $T_{\rm eff}$ and R are almost anti-correlated, which indicates that the minimum radius occurs just before maximum T_{eff} and vice-versa.

Most radial pulsators vary in the fundamental mode. Throughout the star, the gas moves in the same direction, alternatingly inward or outward. In such radially pulsating stars, driven by the κ-mechanism, the motions are restricted to the outer layers and are quenched at larger depths. Stars can also pulsate in radial overtones. In that case, there are more $n > 0$ nodes in the stellar interior. The period of an overtone pulsation is shorter than the fundamental pulsation period by about a factor of $1/n$. This makes it easy to identify overtones.

21.5 Nonradial Pulsations and Asteroseismology

The stars discussed above pulsate in a dominant radial mode; however, stars can also pulsate in many radial and nonradial modes simultaneously. In that case, the star not only changes its radius but it changes its shape as well, deviating from spherical symmetry. These changes can be described by spherical harmonic functions. Contrary to the fundamental radial mode in Cepheids, which has a period on the order of days, the nonradial pulsations in solar-like stars or red giants have periods on the order of minutes or hours, respectively. Nonradial pulsations are often called **oscillations.** The study of stellar structure by means of detected oscillations is called **asteroseismology.**

The oscillation pattern of a nonradial pulsator, NRP, is characterized by two quantum numbers: (m, ℓ).

- ***m*** describes the number of *meridional nodes*: $m = 0$ is symmetric around the rotation axis.
- ***ℓ*-*m*** is the number of nodes in the latitude direction. If $\ell = m$ there are no latitude nodes.

A pulsation with $\ell = m = 0$ is a radial pulsation. Pulsation modes with $m \neq 0$ describe waves traveling in the azimuthal direction. An $m = 3$ and ℓ-$m = 0$ pulsation mode is like a peeled orange with six parts alternatingly expanding and contracting. An $m = 0$ and ℓ-$m = 3$ mode is like a tomato with three horizontal node lines that separate the star into four latitudinal zones, alternatingly expanding and contracting. Pulsations with $m > 0$ represent traveling waves that move around the star parallel to the equator. At the surface of a star these waves are analogous to water waves in a river, but they also occur in deeper layers. They produce horizontal motions at the surface that can be detected as spectral line profile variations.

Figure 21.6 shows the modes for $\ell = 3$ and different values of ℓ-m.

Nonradial pulsation modes are more difficult to detect than radial modes because different parts of the star may pulsate out of phase: while one part contracts and gets hotter, the other part expands and gets cooler, resulting in opposite effects and partial canceling of observable characteristics. The higher the quantum numbers m and ℓ-m (i.e., the larger the number of surface nodes), the smaller the variations in

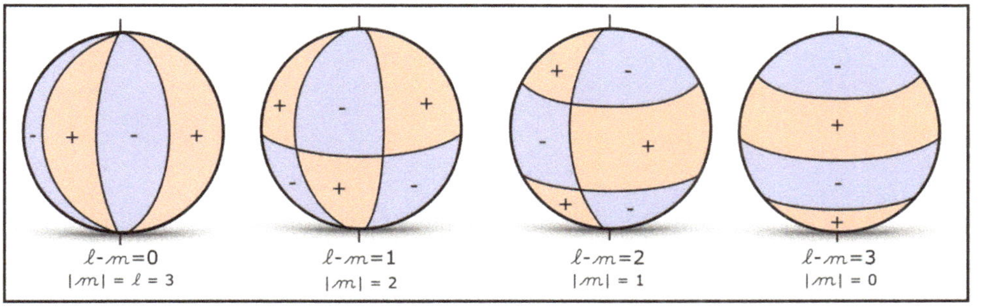

Figure 21.6. The topology of nonradial pulsations (NRPs) for $\ell = 3$ modes and different m modes. From left to right: (m, ℓ-m) = (3,0), (2,1), (1,2), (0,3). Notice that ℓ-m is the number of latitudinal nodes and m is the number of longitudinal nodes. The number of longitudinal and latitudinal zones is one higher than the number of nodes, e.g. the right model has 3 latitudinal nodes and 4 zones. The blue sections are expanding and the red sections are contracting with the + and − signs indicating positive or negative Doppler shifts.

brightness and in line profiles. Brightness variations of NRP stars are therefore generally on the order of 1% or less (i.e., more than 10 times smaller than those of radial pulsators). See Coen Schrijvers' personal website[1] for movies with examples of line profile variations for different nonradial pulsations.

Nonradial pulsations can be due to

- **pressure modes** or **p-modes,** in which the pressure force is the dominant restoring force. The pressure force can cause both horizontal and vertical motions. P-mode pulsations are trapped in the upper layers of the star. Therefore, the frequencies of p-mode pulsations provide information about the T, ρ, and chemical structure of the outer envelope.
- **gravity modes** or **g-modes,** in which gravity (buoyancy) is the dominant restoring force. G-modes represent internal gravity waves that travel deep into the interior of the star. Therefore, the frequencies of the g-mode pulsations provide information about the T, ρ, and chemical structure deep inside the star. The analysis of g-mode oscillations provides a great way to study internal mixing by convection, overshooting, and meridional circulation.

See Christensen–Dalsgaard (2014) for a detailed description of theory and observations and Aerts et al. (2010) for a review.

Figure 21.7 shows the propagation of sound waves in the Sun. The paths of the waves are bent because the sound speed increases with depth. This effect is analogous to the bending of light rays when they enter a medium with a higher speed of light. At a certain depth, depending on wavelength and on the local temperature and density conditions, the waves will be reflected.

At the surface the waves are reflected by the rapid outward decrease in density, so the waves are trapped in the layer between the surface and the inner turning point.

[1] http://staff.not.iac.es/~jht/science/

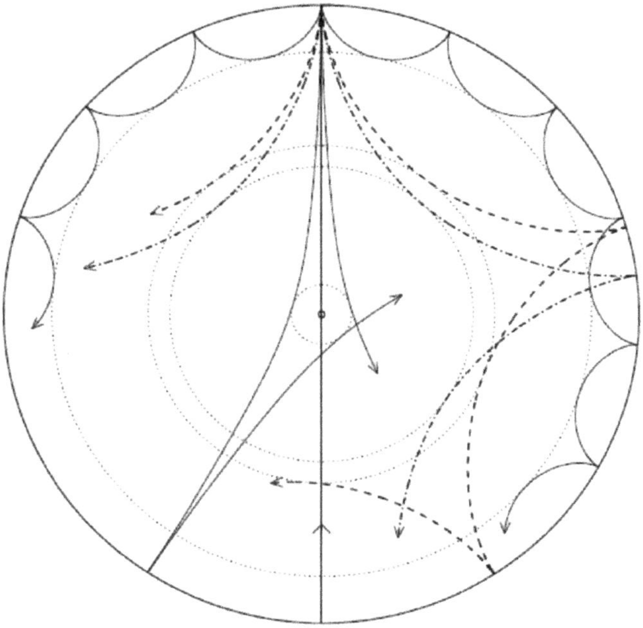

Figure 21.7. Path of sound waves of several frequencies that produce oscillations in the Sun. The wavelength λ of the sound waves is specified by the degree ℓ with $\lambda = 2\pi R/\sqrt{(\ell(\ell+1)}$. The dotted circles show the depth where the waves are reflected. The other lines show the path of waves of degree $\ell = 0$ (a straight path through the center), 2 (passing close to the center), 20, 25 (dashed lines), and 75 (trapped in a thin surface layer). (Reproduced figure with permission from Christensen–Dalsgaard 2014. Copyright 2002 by the American Physical Society)

From all possible wavelengths, only those that fit with an integer number around the circumference of the star will produce an oscillation. The figure shows the path of a few waves of wavelength $\lambda = 2\pi R/\sqrt{(\ell(\ell+1)}$ that produce an oscillation. The pulsations of degree ℓ can be in the fundamental mode (i.e., with $n = 0$ nodes), or in overtones with $n > 0$.

High-accuracy photometric observations of NRP stars usually show a series of frequencies. Most of the oscillations are of mixed modes, with both pressure and gravity as restoring forces. In the limiting case of single modes, there is a significant difference between the periodicities of p-mode and g-mode pulsations. High-order *p-mode* pulsations (i.e., with overtones $n \gg 0$), *are equidistant in frequency,* whereas high-order pulsations in the *g-mode are equidistant in period.* This characteristic makes it easy to determine the types of observed modes.

The observed oscillations in the Sun have frequencies of about 1200 to 5000 μHz, with a mean period of about 5 minutes. They are equidistant in frequency, which indicates that they are p-modes. The waves are triggered by convection in the envelope of the Sun. The oscillation of B-type stars is due to low-order g-modes, with periods on the order of hours.

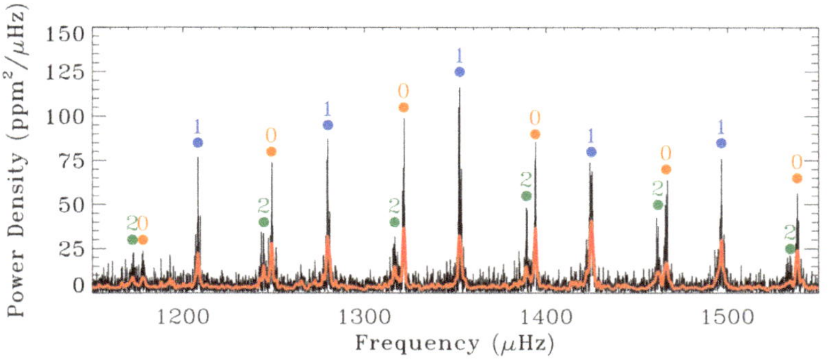

Figure 21.8. Observed power spectrum (black) of the G-type star KIC 6933899. The red line is the smoothed power spectrum. The pulsation periods are around 10 to 15 minutes. The equal spacing in frequency of the overtones shows that the peaks are due to nonradial p-mode pulsations. The peaks are identified with their value of ℓ. (Reproduced from White et al. 2012.)

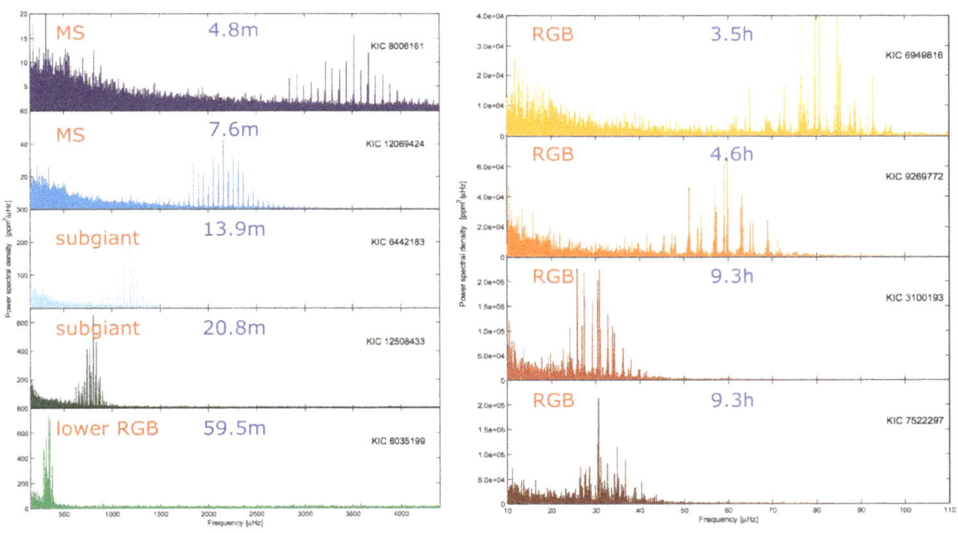

Figure 21.9. Frequency spectrum (in μHz) of nonradial pulsations of nine stars with a mass of $\sim 1 M_{\odot}$, observed with *Kepler*. The stars are arranged in order of decreasing surface gravity and mean density. The left panel shows, from top to bottom, two main-sequence stars, two subgiants, and an RGB star. The right panel shows four RGB stars with increasing radius. The mean nonradial pulsation period of each star is indicated in blue. (Reproduced from Chaplin & Miglio 2013, with permission)

Figure 21.8 shows the power spectrum of the nonradial pulsations in the solar type G-type star KIC 693 3899, observed with the Kepler satellite, as an example. The observed overtone pulsations of p-modes with $\ell = 0$, 1, and 2 are indicated. Note that they are equidistant in frequency.

Figure 21.9 shows the frequency spectrum of nine stars of about $1 M_{\odot}$ in different evolutionary phases (i.e., with different surface gravities and mean density). The mean nonradial pulsation period runs from 4.8 minutes to 9.3 hours. There is a clear trend from high-frequency oscillations in stars with high surface gravity to low frequencies at low surface gravity. This shows that the surface gravity can be derived accurately from asteroseismology.

Just like the fundamental radial pulsations, which are driven by the κ-mechanism in classical pulsators, nonradial pulsations must also be driven by some mechanism that triggers instability. Multiperiodic radial and nonradial p-mode pulsations, as observed in cool stars like the Sun, are due to the motions produced by the convective cells in the outer stellar envelope. All low-mass stars with convective envelopes are expected to be multiperiodic nonradial pulsators. On the other hand, the g-mode pulsations of B stars are most likely driven by the κ-mechanism, due to the iron opacity peak.

21.6 Summary

1. Classical radial pulsators are located in a nearly vertical band in the HRD, called the **instability strip,** centered around $T_{\rm eff} \approx 6500$ K.
2. The fundamental radial pulsation period P depends on the mean density of the star. It is characterized by the pulsation constant $Q = P/\left(\bar{\rho}_{\odot}/\bar{\rho}\right)^{1/2}$, with $Q \approx 0.05$ days.
3. The radial pulsators in the instability strip are driven by the κ mechanism. This mechanism operates in the ionization zones inside stars, mainly the $He^{+} \rightarrow He^{++}$ zone. In these zones, the opacity increases during compression and decrease during expansion.
4. In stars with $T_{\rm eff} \lesssim 5500$ K, the $He^{+} \rightarrow He^{++}$ zone is too deep and the pulsation is quenched in the layers above. In stars with $T_{\rm eff} \gtrsim 7500$ K, this zone is too shallow and there is not material on top of it to drive the expansion back. This is the reason that the instability strip is nearly vertical in the HRD.
5. The light curve of Cepheids shows a maximum when the expansion velocity reaches a maximum and the trapped radiation is released.
6. Stars can also pulsate nonradially in modes that are excited by convective motions in the envelope. The dominant restoring force can be the pressure force (p-mode pulsations) or gravity (g-mode pulsations). The high-order peaks in the power spectrum of p-mode pulsations are equidistant in frequency, while they are equidistant in period for the g-modes. P-mode pulsations give information about the structure of the star's outer layers and g-mode pulsations depend on the structure deep inside the star. Asteroseismology provides a method to determine empirically the depth of convection zones, internal mixing, internal rotation curves, etc.

Exercises

21.1 a. In chapter 9, we have predicted that the dynamical timescale of a star is of the order of $\tau_{\rm dyn} = (G\,\bar{\rho})^{-1/2}$. Observations show that the fundamental pulsation period is $P = C\tau_{\rm dyn}$.

Calculate the value of C from the data in Table 21.1.

b. If this value of C also holds for other types of stars, predict the pulsation period of the Sun, a red supergiant, a star at the tip of the AGB, a WD, and a neutron star.
c. Explain why these last estimates do not predict the pulsation periods accurately?

21.2 Explain the trend in the mean pulsation periods of the stars shown in Figure 21.9.

21.3 Stars on the RGB and the AGB are located in about the same region of the HRD. Suppose that an RGB star and an AGB star have the same values of $T_{\rm eff}$ and L and the same mass. Which one of the two would have the shorter fundamental period? Explain this.

21.4 Suppose that the RGB star and the AGB star of Exercise 21.3 with the same $T_{\rm eff}$ and L both show nonradial p-mode and g-mode pulsations. Which of the two modes provides the best diagnostic to distinguish between the RGB phase and the AGB phase?

References

Aerts, C., Christensen-Dalsgaard, J., & Kurtz, D. W. 2010, Asteroseismology (Dordrecht: Springer)

Benedict, G. F., McArthur, B. E., Feast, M. W., et al. 2007, AJ, **133, 1810**

Carroll, W., & Ostlie, D. A. 1996, An introduction to Modern Astrophysics (Reading, MA: Addison-Wesley)

Chaplin, W. J., & Miglio, A. 2013, ARA&A, **51, 353**

Christensen–Dalsgaard, J. 2014, Lecture Notes on Stellar Oscillations, http://astro.phys.au.dk/~jcd/oscilnotes/

Cox, A. N., 2000, Allen's Astrophysical Quantities (Berlin: Springer)

Eddington, A. S. 1926, The Internal Constitution of Stars (Cambridge: Cambridge Univ. Press)

Gieren, W. P., Fouqué, P. & Gómez, M. 1998, ApJ, **496, 17**

Iben, I. 1967, ARAA, **5, 571**

Testa, V., Marconi, M., Musella, I., et al. 2007, A&A, **462, 599**

White, T. R., Bedding, T. R., Gruberbauer, M., et al. 2012, ApJL, **751, L36**

Henny J.G.L.M. Lamers and Emily M. Levesque

Chapter 22

Observations of Massive Stars: Evidence for Evolution with Mass Loss

The distribution of massive stars in the Hertzsprung–Russell diagram is an important source of information about their evolution. The observed luminosity upper limit is not the horizontal line in the HRD, predicted by the Eddington limit, but a line that slopes downward from about $3 \times 10^6\ L_\odot$ at ~40 000 K to $5 \times 10^5\ L_\odot$ at ~8000 K. We will show that this is the result of the atmospheric Eddington limit. This indicates that the most massive stars do not evolve into red supergiants (RSGs); instead they lose a substantial fraction of their mass shortly after the main-sequence phase as luminous blue variables. Not surprisingly, the upper part of the HRD contains various types of stars in which the nuclear fusion products appear at the surface: N-rich O- and B-type stars, and N-rich or C-rich (and H-poor) Wolf–Rayet stars (WR). In this section, we will discuss the observational clues to their evolution. The number ratios of WR stars to RSGs and the ratio of C-rich to N-rich WR stars, observed in several galaxies, depends on the metallicity of that galaxy. This shows the importance of metallicity-dependent mass loss in the evolution of massive stars.

22.1 The Observed Upper Limit in the HRD

Figure 22.1 shows the observed distribution of luminous stars in the LMC with the empirical upper limit. This empirical limit is called the **Humphreys–Davidson limit or HD-limit,** after the U.S. couple Roberta Humphreys and Kris Davidson, who studied it in 1979 (Humphreys et al. 1979).

The observed distribution of massive stars shows a conspicuous absence of cool stars with $T_{\rm eff} < 8000$ K above $M_{\rm bol} = -9.5$ or $L = 5 \times 10^5\ L_\odot$. This corresponds to stars with an initial mass of about $M_i > 40 M_\odot$. This is surprising because luminous stars above this mass limit exist on the MS, and the evolutionary tracks of massive stars are approximately horizontal in the HRD; the lack of bright red supergiants—cool massive stars with a very extended convective envelope—indicates that stars more

doi:10.1088/978-0-7503-1278-3ch22

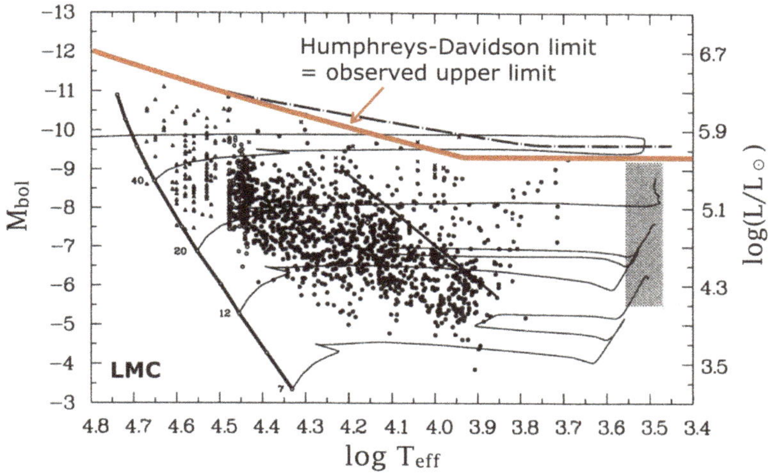

Figure 22.1. Distribution of the most luminous stars in the LMC. Two versions of the possible Humphreys–Davidson upper limit are shown (red line or black dashed–dotted line), as well as several predicted evolutionary tracks. The gray region is the location of the brightest red supergiants. (Reproduced from Massey 2003, with permission.)

massive than about $40M_\odot$ do not evolve to the far right of the HRD. Massive stars can be prevented from becoming red supergiants if they lose a significant fraction of their envelope before they reach the Hayashi line. It turns out that they lose this mass during a brief phase as luminous blue variables after the main sequence.

22.2 The Atmospheric Eddington Limit

The luminosity of massive stars is so high that they are close to their Eddington limit for radiation pressure (Equation (6.12)). For massive stars, with electron scattering as the dominant opacity in their interior, $L_E = 4\pi cGM/\sigma_e \approx 3$ to $4 \times 10^6 L_\odot$ and $M_{max} \approx 150$ to $200M_\odot$; however, the absorption coefficient is higher than σ_e in the atmosphere (see the peak in κ at $10^4 < T < 10^5$ K for $\rho < 10^{-6}$ g cm^{-3} in Figure 5.1). At atmospheric densities of the order of 10^{-10} g cm^{-3} the opacity has a peak around 10,000 to 20,000 K. A peak in κ implies a drop in the Eddington limit. We can define the **atmospheric Eddington limit** as

$$L_E^{atm} = 4\pi cGM/\kappa_{atm} = L_E(\sigma_e/\kappa_{atm}) < L_E, \tag{22.1}$$

where κ_{atm} is the maximum value of κ in the atmosphere at $10^{-2} < \tau < 10^3$. Because κ_{atm} has a peak around $T_{eff} \sim 10{,}000$ K, the atmospheric Eddington limit in the HRD will show a dip in luminosity, called the **Eddington trough**. Stars that reach that limit when they evolve to the right in the HRD after the MS phase will become unstable and suffer severe mass loss. The situation is sketched in Figure 22.2.

The predicted evolutionary tracks of massive post-MS stars are almost horizontal in the HRD, going from left to right. If a star reaches L_E^{atm}, its atmosphere becomes loosely bound, which may lead to instabilities and high mass-loss rates. This will continue until the star has lost so much mass that it can no longer sustain an

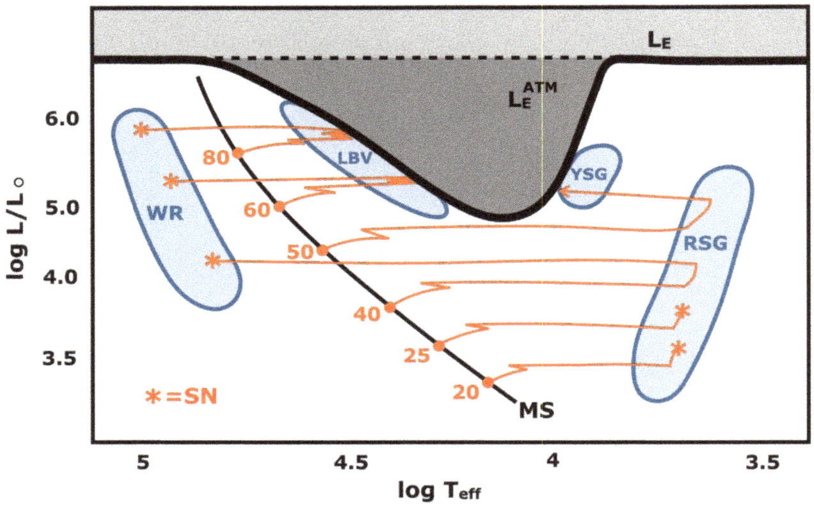

Figure 22.2. Schematic description of the effect of the dip in the atmospheric Eddington limit L_E^{atm} on the evolution of massive stars. Schematic evolutionary tracks are indicated, as well as the location of WR stars, luminous blue variables (LBVs), and red (RSG) and yellow (YSG) supergiants. A star at the end of a track indicates a supernova explosion.

extended convective envelope. It will shrink and move to the left in the HRD. Stars with a mass below the critical mass or luminosity may safely evolve "under the atmospheric Eddington trough" and reach the RSG phase. In principle, stars could exist and be stable on the cool side of the Eddington trough; however, because the tracks are nearly horizontal they cannot reach this area. The exceptions are stars with a mass just below the critical mass. They may become RSGs and move slightly upward along the Hayashi line. When they have lost enough mass, and can no longer sustain their large convective envelope, they will shrink and move back to the left of the HRD; however, during their leftward motion in the HRD, they will hit the Eddington dip on the cool side and become unstable at T_{eff} in the range of about 5000 and 10,000 K. These are the highly variable YSGs.

Figure 22.3 shows the location of the predicted atmospheric Eddington limit in the HRD for different metallicities from $Z/Z_\odot = 0.1$ to 2.0 (Ulmer & Fitzpatrick 1998). In this study, the atmospheric Eddington limit is defined as the luminosity where the force due to radiation pressure reaches a maximum of 90% of the gravity somewhere in the atmosphere between $10^{-2} < \tau < 10^3$. This means that the effective gravity is reduced by a factor of 10 compared to the values predicted by stellar evolutionary theory. Note the downward slope between about 50,000 K and 11,000 K, the minimum near 12,000 K, and the upturn at $T_{eff} < 10{,}000$ K. This atmospheric Eddington limit is sensitive to metallicity, with a L_E^{atm} being smaller for higher metallicities, because κ increases with Z. The predicted decrease of L with T_{eff} agrees qualitatively with the observed Humphreys–Davidson limit. This suggests that the HD-limit is due to high mass loss as stars become unstable and lose mass because of their loosely bound atmospheres.

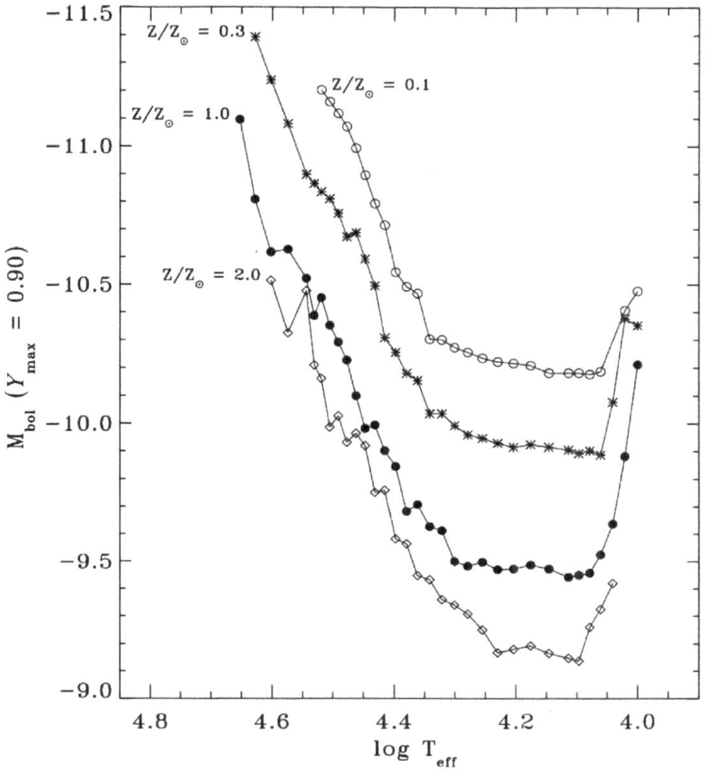

Figure 22.3. Predicted atmospheric Eddington limit, defined by $Y_{\max} = g_{\rm rad}/g_{\rm grav} = 0.90$ in the atmosphere between $10^{-2} < \tau < 10^{3}$. The metallicities approximately correspond to M31 ($Z/Z_{\odot} = 2$), Milky Way ($Z/Z_{\odot} = 1$), LMC ($Z/Z_{\odot} = 0.3$), and SMC ($Z/Z_{\odot} = 0.1$). (Reproduced from Ulmer & Fitzpatrick 1998.)

22.3 Luminous Blue Variables and the Atmospheric Eddington Limit

Luminous Blue Variables (LBVs) are very luminous blue supergiants with $L/L_{\odot} > 3 \times 10^{5}$ that show large and irregular variations in their V magnitudes (Humphreys & Davidson 1994). These variations occur on timescales from weeks to years, with occasional large eruptions.

- On timescales of yrs or decades, their visual brightness temporarily increases by a factor of about 2 to 5. After a few years to a decade the brightness drops again.
- Every 10^{2} to 10^{3} yr, they experience a large eruption when they eject $\sim$1 to $10 M_{\odot}$ of gas. The most famous of these eruptions are those of P Cygni in AD 1600 and η Carinae around AD 1843, when that star temporarily became the second brightest star in the sky.

Figure 22.4 shows an image of the nebula around η Car, which is a result of a series of outbursts in 1838–1856. The amount of mass in the nebula is estimated to

Figure 22.4. HST image of the Carina nebula that was ejected in a series of eruptions between 1838 and 1856 by the LBV η Carinae. The nebula, at a distance of 3000 pc, contains about $20M_{\odot}$ of gas and dust. (Figure is from J. Morse, University of Colorado and NASA HST. Courtesy of NASA.)

be $20M_{\odot}$. The nebula is highly aspherical, which suggests that either binarity or fast rotation has played a role in shaping the nebula.

When LBVs are in a phase of faint visual luminosity, their spectral type is O or B and $T_{\rm eff}$ is between 15,000 and 35,000 K. At maximum visual light $T_{\rm eff}$ is between 7000 and 10,000 K. Observations over the full spectrum have shown that the luminosities of LBVs remain "approximately" constant during these variations (but see Figure 22.6): the radiation is just redistributed more to the UV in the hot phase. This means that the variations in V are mainly due to large changes in radius. In the hot phase, LBVs have a typical radius of 30 to $100R_{\odot}$, but in the bright phase this can increase to about 100 or $300R_{\odot}$. The more luminous the LBV, the larger the variations in R and $T_{\rm eff}$.

Figure 22.5 shows the V-band light curves of S Doradus, an LBV in the LMC, between 1974 and 1992. The variations are irregular, with a timescale of several years, and large, with $\Delta V \approx 1.2$ mag.

Figure 22.6 shows the stellar parameters of S Dor during the large variations in 1984–1991, derived by comparing their energy distribution with model atmospheres. In this period, the radius varied between 100 to $380R_{\odot}$ and $T_{\rm eff}$ varied between 9000 and 20,000 K. The luminosity varies only slightly and has a minimum when R is at maximum. This suggests that part of the energy generated by fusion is used for the expansion.

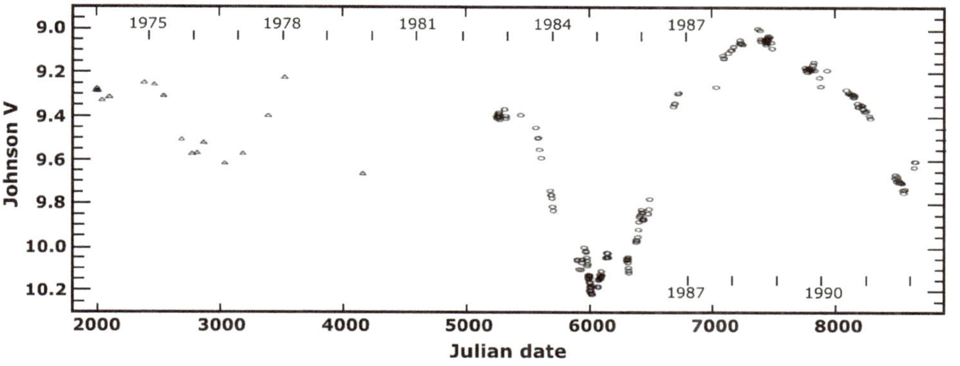

Figure 22.5. Visual light curve of the LBV S Dor in the LMC in 1975–1992. (Reproduced from Spoon et al. 1994, with permission. © ESO.)

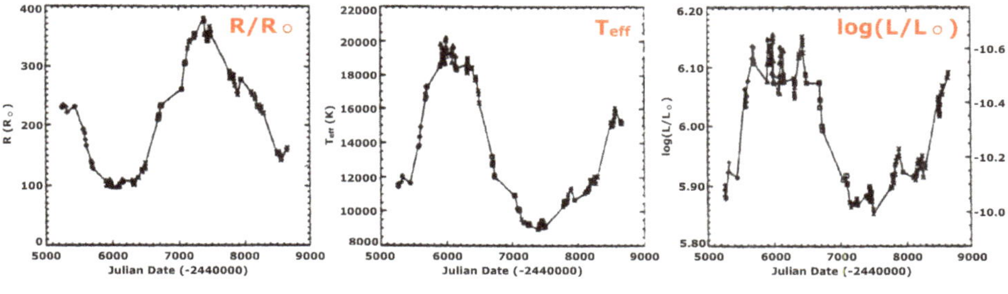

Figure 22.6. Variation of the stellar parameters of S Dor in the period 1984–1991. The variations in R and $T_{\rm eff}$ are approximately anticorrelated: $T_{\rm eff}$ reaches a minimum of about 9000 K when the radius reaches its maximum of about $380R_{\odot}$. The luminosity varies only slightly, by about 0.2 dex, and has a minimum when R reaches its maximum. (Figure is from Lamers 1995.)

Q (22.1) *Give at least two arguments for why the variations of LBVs cannot be due to some kind of pulsation.*

Figure 22.7 shows the location of LBVs in the HRD. For each star, the location is shown at two occasions, connected by dotted lines. The luminosity at the two occasions is about the same. This indicates that changes in $T_{\rm eff}$ are due to changes in R. During the hot phase (left in the HRD), the LBVs are located in a band of L decreasing with $T_{\rm eff}$. In the cool phase, they are located in an almost vertical band at $T_{\rm eff} \approx 9000$ K. This location is qualitatively remarkably similar to the predicted shape of the atmospheric Eddington limit!

There are only a handful of LBVs known in our Galaxy and only a few dozen in total. This indicates that it must be a short-lived phase. The origin of the instability is not yet known; however, it is significant that in the hot phase, when the star is faint in the visual, its location in the HRD is very close to the HD-limit and to the atmospheric Eddington limit. The instability of these stars is somehow related to the fact that their

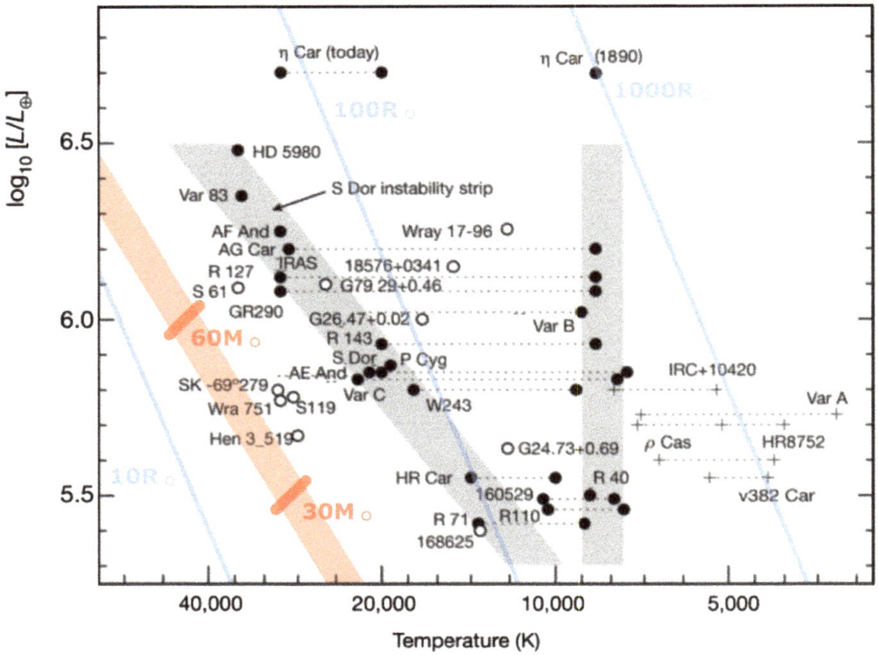

Figure 22.7. Location of LBVs (black dots) in the HRD. For each star, the location is shown twice, connected by a dashed line: at the optical maximum when the star is cool and R is large and at optical minimum when the star is hotter and R is smaller. The crosses indicate highly variable yellow supergiants. The location of the main sequence is indicated in red. The blue lines are lines of constant radius. (Data are from Humphreys & Davidson 1994 and Smith et al. 2004.)

atmospheres are loosely bound and a small internal disturbance may have a large effect. We will show later in Section 25 that critical rotation may play a role.

The mass-loss rates of LBVs are high and on the order of 10^{-6} to 10^{-4} $M_\odot\,\mathrm{yr}^{-1}$. In addition, the stars suffer large eruptions. From the statistics of LBVs with nebulae, the occurrence of these eruptions is estimated to be once in about 10^2 to 10^3 years. Because of the highly variable mass loss and the eruptions, the predicted post-main-sequence evolutionary tracks of stars encountering the atmospheric Eddington limit is uncertain. In most stellar evolution calculations, a high average mass-loss rate of about $10^{-4} M_\odot\,\mathrm{yr}^{-1}$ is adopted. For reviews, see Nota & Lamers (1997).

22.4 Wolf–Rayet Stars

Wolf–Rayet stars (WR stars) are luminous stars, $L \gtrsim 10^5 L_\odot$, with T_{eff} ranging from about ~30,000 to 100,000 K; see Crowther (2007) for a review. They are named after the French astronomers Charles Wolf (1827–1918) and Georges Rayet (1839–1906), who discovered in 1867 a new class of hot stars with strong and broad emission lines. The emission lines indicate that the stars have a strong stellar wind with a mass-loss rate of 4×10^{-6} to 4×10^{-5} $M_\odot\,\mathrm{yr}^{-1}$ and terminal velocities of ~1000 to 3000 km s^{-1}. The wind is so optically thick that the photosphere cannot be observed. The spectrum is therefore dominated by wind features (i.e., the strong and broad emission lines). The spectrum shows that the atmospheres are strongly enriched by fusion products.

Wolf–Rayet stars are classified according to their observed chemical compositions:

- **WNL stars** (W for Wolf–Rayet, N for N-rich, L for late spectral type, with $T_{\rm eff} \sim 30{,}000$ to 45,000 K) are N-rich but C- and O-poor. They are also He-rich but still have some H left. These abundances are clearly the products of H-fusion via the CNO cycle (Section 8.4.3).
- **WNE stars** (E for early spectral type, with $T_{\rm eff} > 45{,}000$ K) are N- and He-rich and C-poor, but, contrary to the WNL stars, they have no H left. These are also the products of H-fusion via the CNO cycle, but from a later stage when all H was converted into He.
- **WC stars** are C-rich and He-rich, but lack N and H. They are hot with $T_{\rm eff} >$ 45,000 K. They clearly show the products of He-fusion via the triple-α process (Section 8.5).
- **WO stars** are O-rich, C-rich, and He-rich stars whose spectrum is dominated by O lines. This is partly an abundance effect and partly an ionization effect. These stars are in fact so hot that the OIV lines become stronger than the CIV lines, even if the ratio C/O > 1.

Figure 22.8 shows the optical spectrum of WN and WC stars in the wavelength range of 3500 to 6000 angstroms. The subclasses 4 to 9 indicate different values of $T_{\rm eff}$, decreasing from subclass 4 to 9. Notice the difference in the spectra due to their different chemical compositions. The WN spectra show emission lines of HeI–HeII and NIII–NV, with HeII decreasing toward later subtypes. The WC spectra show emission lines of HeII, CIII–CIV, and OIV, with HeII decreasing and CIII increasing toward later subtypes.

The presence of the strong optical emission lines is evidence for the presence of dense extended winds around WR stars. The structure of the winds is shown schematically in Figure 22.9. The optically thick part of the star, $\tau > 20$, is shown in orange and the atmosphere at $2/3 < \tau < 20$ is shown in red. The blue extended region is the wind, with $n_e > 10^{12}$ cm^{-3} in dark blue and $n_e < 10^{12}$ cm^{-3} in light blue. The size of the WN8 star, the WC9 star, and the O star are plotted to scale. Note the large extension of the winds of the WR stars that hide the photospheres.

Figure 22.10 shows the location of WR stars in the HRD. The stars are located in-between the zero-age main sequence (ZAMS) of H-rich stars and He-rich stars. The location of the Humphreys–Davidson limit is also shown, as well as the evolutionary track of a star with an initial mass of $40M_\odot$. Single WR stars are the descendants of massive stars with $M_i \gtrsim 40M_\odot$, but their present mass is lower.

Although WR stars have lost almost all of their H, they are located in the HRD between the H and He MS. This is due to their extended envelope. Their luminosity is so high that the radiation pressure produced by the Fe opacity peak around 10^5 K in the envelope almost balances gravity. This means that the gas pressure is very low in the envelope, so the density is also low. This results in an extended low-density envelope, an increased radius, and a drop in $T_{\rm eff}$.

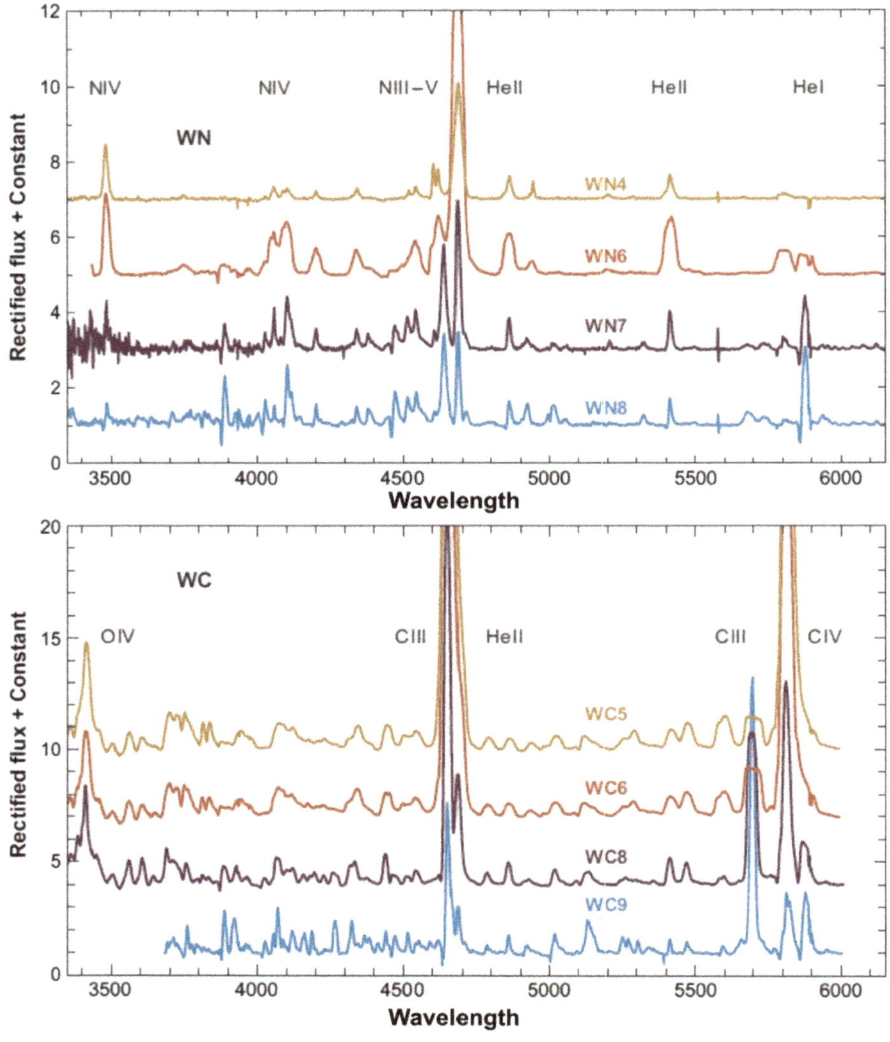

Figure 22.8. Optical spectra of WR stars. Top: WN stars; bottom: WC stars. (Reproduced from Crowther 2007, with permission.)

The strong enrichment patterns of the WR stars show that they are "peeled" stars. During their evolution after the MS phase, they have lost so much mass that the products of nuclear fusion are exposed at the surface. Because N-rich layers are the result of H-fusion and C-rich layers are the result of deeper He-fusion, WC stars have been peeled more severely than WN stars. In other words, if a massive post-MS star is peeled, the N-rich layers will appear first at the surface and the C-rich layers will appear later. This implies that WR stars will evolve from WN to WC due to their high mass-loss rates.

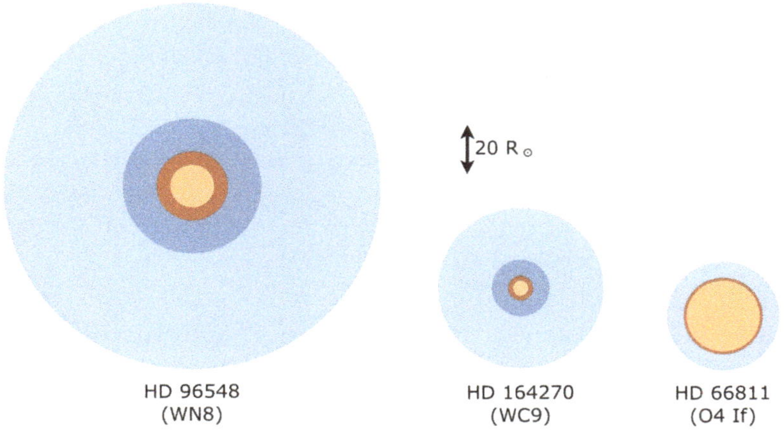

Figure 22.9. Sizes of a WN8 and a WC9 star are compared with that of an O4 If star. The sizes are plotted to scale. The optically thick regions of the stars ($\tau > 20$) are shown in orange and the atmospheres ($2/3 < \tau < 20$) are shown in red. The wind is shown in blue, with $n_e > 10^{12}$ cm^{-3} in dark blue and $n_e < 10^{12}$ cm^{-3} in light blue. Note the small radii and the extended winds of the Wolf–Rayet stars compared to that of the O star. (Reproduced from Crowther 2007, with permission)

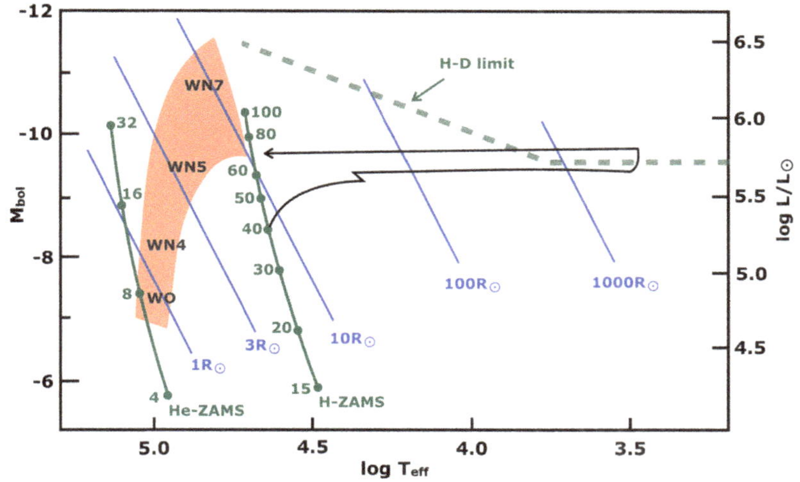

Figure 22.10. The locations of galactic WR stars in the HRD is indicated by a red region, with a few spectral types. The ZAMSs for He and H stars are indicated by green lines, with the location of stars of various masses marked as points. Lines of constant radii are shown in blue and the HD limit is shown as a dashed green line. The evolutionary track of a $40M_\odot$ star is shown in black. Note that the WR stars are located in-between the He-ZAMS and the H-ZAMS. (Adapted from Moffat et al. 1989. © Kluwer Academic Publishers 1989. With permission of Springer.)

Q (22.2) *Evolution calculations suggest that stars with initial masses less than about* $30M_\odot$ *explode as supernovae during the red supergiant phase. How can the presence of WR stars with* $L \lesssim 2 \times 10^5\ L_\odot$ *be explained?*

22.5 The Dependence of Massive Star Evolution on Metallicity

Observations of massive stars in the Galaxy, the LMC, and the SMC show that metallicity plays a crucial role in the evolution of massive stars. This is not surprising considering two effects that we have discussed before.

1. The high mass-loss rates of luminous stars are driven by radiation pressure in spectral lines of C-, N-, O-, and Fe-group elements (Section 15.2).
2. The atmospheric opacity, and hence the luminosity of the atmospheric Eddington limit, depends on metallicity (Figure 22.3).

22.5.1 The Observed Metallicity Dependence of Red Supergiants

Red supergiants are massive stars of $10 \lesssim M_i \lesssim 40 M_\odot$ with extended convective envelopes. They are located close to the Hayashi limit in the HRD. We have seen in Figures 16.3 and 17.4 that the minimum $T_{\rm eff}$ of a convective star depends on metallicity. The opacity in the atmospheres of very cool stars is due to H^- with the electrons provided by the low ionization stages of metals such as Fe, Cr, Ti, etc. The higher the metallicity, the larger the κ (as $\kappa \sim Z$) and the cooler the Hayashi limit.

This predicted trend is confirmed for massive stars by a study of the spectral types of RSGs. The spectral type distribution of RSGs in different galaxies shows a trend with metallicity. This is shown in Figure 22.11, where the distributions of RSGs over K and M subtypes in five galaxies with $Z = 1.0$ to $0.10 Z_\odot$ are plotted. There is a clear trend demonstrating that the spectral types are earlier (corresponding to a higher $T_{\rm eff}$ in cool supergiants) at lower Z. This shift in spectral type exceeds what would be expected from a simple Z effect on the molecular band strengths in the spectrum. It shows that the $T_{\rm eff}$ of RSGs is higher at low Z.

22.5.2 The Observed Metallicity Dependence of Wolf–Rayet Stars

Realizing that WR stars are peeled-off stars, we expect that the degree of peeling (and hence the relative number of WR stars and WR spectral types) depends on the mass-loss rate during previous evolutionary stages, which in turn depends on metallicity.

Figure 22.12 (top panel) shows the number ratio of RSG/WR stars for several Local Group galaxies with different metallicities. The metallicities, expressed here as the logarithm of the O/H ratio, are between 0.25 and 2 times that of the Galaxy. There is a clear trend with the WR/RSG ratio increasing toward higher metallicity. This can be explained by the fact that single stars become WR stars if their evolution tracks hit the atmospheric Eddington limit. This limit reaches lower luminosity at higher metallicity. So the number of WR stars will increase with metallicity compared to the number of stars that reach the RSG phase.

Metallicity also has an effect on the number ratio of WC/WN stars. Since WC stars are more peeled-off than WN stars, the number ratio of the WC/WN stars in galaxies is expected to increase with mass loss and metallicity. Figure 22.12 (bottom panel) shows that this is indeed the case.

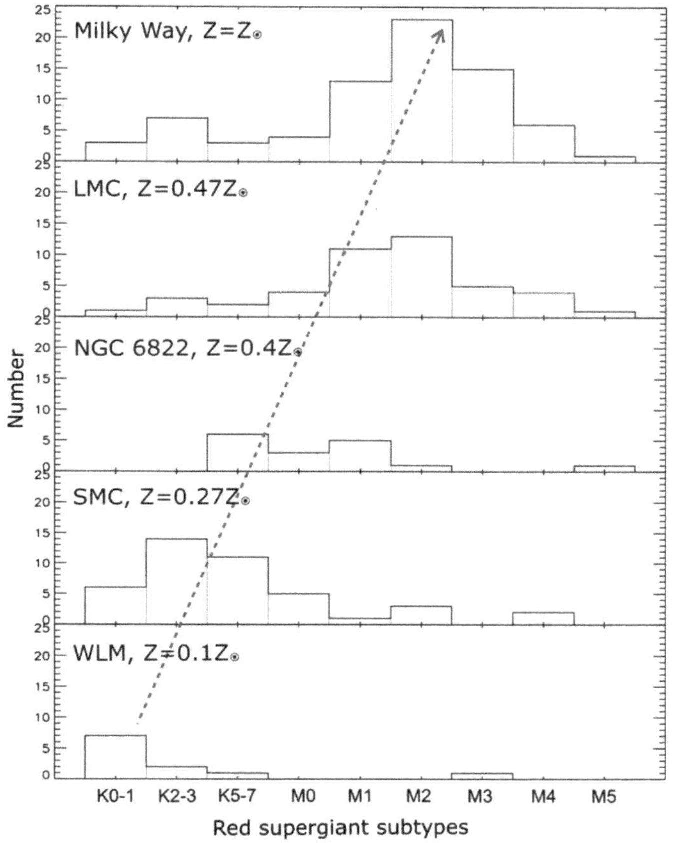

Figure 22.11. Observed distribution of red supergiant spectral subtypes in galaxies of decreasing metallicity. The RSGs have later spectral types, corresponding to cooler temperatures, at higher metallicity. (Reproduced from Levesque & Massey 2012.)

22.6 Summary

1. The luminosity upper limit of massive stars, called the **Humphreys–Davidson (HD) limit**, is a line in the HRD that drops between $50{,}000 \gtrsim T_{\rm eff} \gtrsim 9000$ K from $\log L/L_{\odot} \approx 6.3$ to 5.7 and is about constant at $T_{\rm eff} \lesssim 9000$ K.
2. The **atmospheric Eddington limit** in the HRD is lower than the classical Eddington limit because the opacity in the atmosphere is higher than $\sigma_{\rm e}$ in the interior. It has a dip (Eddington trough) at $T_{\rm eff} \approx$ 11,000 K. The hot side of the Eddington trough coincides with the HD-limit.
3. Stars more luminous than the deepest point of the Eddington trough cannot evolve into RSGs because they lose a large amount of mass as LBVs.
4. **Luminous blue variables (LBVs)** are massive stars that show large irregular variations in visual brightness on timescales of years to decades and large eruptions on timescales of centuries. The optical brightness variations are due to dramatic variations in radius at approximately constant luminosity.

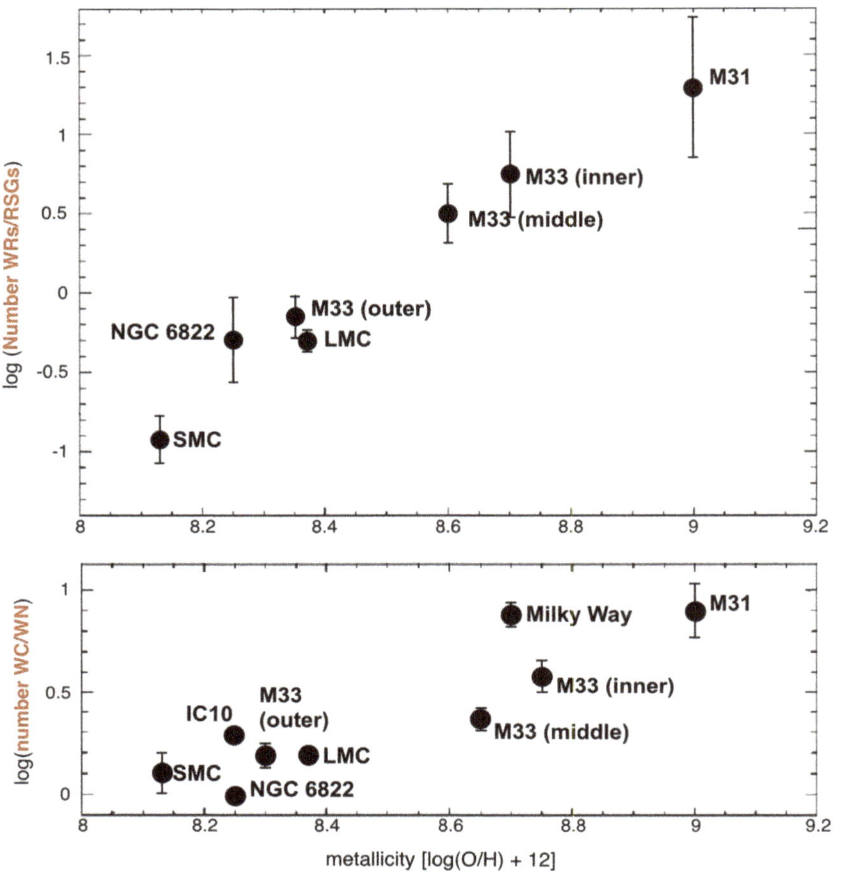

Figure 22.12. Top: the increase of the number ratio of WR/RSG stars in several Local Group galaxies with increasing metallicity, expressed as log (O/H) + 12. (N.B: log (O/H) + 12 ~ 8.7 for massive Galactic stars in the solar vicinity.) The trend is due to the metallicity dependence of the atmospheric Eddington limit. Bottom: the increase of the number ratio of WC/WN stars in the same galaxies. The trend is due to the metallicity dependence of the mass-loss rate. (Reproduced from Massey 2003, with permission)

At maximum radius, the stars are cool, $T_{\rm eff} \approx 9000$ K, while at minimum radius they are much hotter, close to the hot side of the Eddington trough in the HRD. The time-averaged mass-loss rate of LBVs is high, possibly as high as $10^{-4}\ M_\odot\ {\rm yr}^{-1}$.

5. **Wolf–Rayet (WR) stars** are luminous hot stars, located in the HRD between the H and the He ZAMS in the HRD. Their spectrum is formed in the stellar wind and shows broad emission lines that indicate high mass-loss rates. There are three classes: **WN stars** are N-rich, **WC stars** are C-rich and He-rich, and **WO stars** are O-rich and He-rich. WR stars are stripped massive stars with the products of H-fusion via the CNO cycle (WN) and He-fusion via the triple-α process (WC) in their atmospheres.

6. The number ratios of WR/RSG and WC/WN stars in a stellar population increase with metallicity. This is due to the metallicity dependence of both the atmospheric Eddington limit and the mass-loss rate.
7. All of these effects demonstrate that the evolution of massive stars is strongly influenced by mass loss and that the effect is stronger at higher metallicity.

Exercises

22.1 Calculate the maximum initial mass of stars that can evolve into red supergiants in galaxies with $Z/Z_{\odot} = 2$, 1, 0.3, and 0.1.

22.2 Estimate the duration of the LBV phase by comparing the number of LBVs in the LMC with $3 \times 10^5 \lesssim L \lesssim 2 \times 10^6\ L_{\odot}$, with the number of single WR stars in the same luminosity range, assuming a mean predicted lifetime for WR stars of about 0.5 Myr.
Hint: use the list of LBVs by Humphreys and Davidson (1994).
and use the lists of WR stars from Hainich et al. (2014) or later compilations.

22.3 Evolutionary calculations show that a massive post-main-sequence star must lose between 5 and $20 M_{\odot}$ during the LBV phase to prevent evolution into a red supergiant.
(a) What is the average mass-loss rate during the LBV phase?
(b) In-between eruptions, the mean mass-loss rate of LBVs is on the order of a few times $10^{-5}\ M_{\odot}\ \text{yr}^{-1}$. Which fraction of the mass that is lost during the LBV phase is due to quiescent mass loss and which fraction is lost in eruptions?
Remark: eruptions are not included in stellar evolution models, so the average mass loss during the post-MS phase of massive stars is adjusted to about $10^{-4}\ M_{\odot}\ \text{yr}^{-1}$ to match the observed lack of red supergiants above the Humphreys–Davidson limit.

22.4 **Computer exercise**
Use the MESA code to calculate the evolution of a star of $M_i = 40 M_{\odot}$, and derive the amount of mass in the convective envelope when the star leaves the red supergiant phase.

References

Crowther, P. A. 2007, ARAA, **45, 177**

Hainich, R., Rühling, U., & Todt, H. et al. 2014, A&A, **565, 27**

Humphreys, R. M., & & Davidson, K. 1979, ApJ, **232, 409**

Humphreys, R. M., & Davidson, K. 1994, PASP, **106, 1025**

Lamers, H. J. G. L. M. 1995, in ASP Conf. Ser 83, Astrophysical Applications of Stellar Pulsations, ed. R. S. Stobie, & P. A. Whitelock (San Francisco, CA: ASP)

Levesque, E. M., & Massey, P. 2012, AJ, **144, 2**

Massey, P. 2003, ARA&A, **41, 15**

Moffat, A. F. J., Drissen, L., & Robert, C. 1989, ASSL, **157, 229**

Nota, A., & Lamers, H. J. G. L. M. (ed.) 1997, ASP Conf. Ser. 120, Luminous Blue Variables: Massive Stars in Transition (San Francisco, CA: ASP)

Smith, N., Vink, J. S., & de Koter, A. 2004, ApJ, **615, 475**

Spoon, H. W. W., de Koter, A., Sterken, S., Lamers, H. J. G. L. M. & Stahl, O. 1994, A&AS, **106, 141**

Ulmer, A., & Fitzpatrick, E. L. 1998, ApJ, **504, 200**

Henny J.G.L.M. Lamers and Emily M. Levesque

Chapter 23

Evolution of Massive Stars of 8–25$M_\odot$

Stars with an initial mass above $8M_\odot$ evolve differently from stars with $M < 8M_\odot$ because they are massive enough to go through all nuclear fusion phases (Section 8.11). They do not develop a degenerate He or CO core. By core contraction they can therefore reach the high central temperatures that are needed for the ignition of the late nuclear fusion phases. Stars less massive than $M_i \lesssim 25M_\odot$ spend their last evolutionary phases as red supergiants, contrary to their more massive counterparts, which lose so much mass that they become Wolf–Rayet stars. Stars with $8 \lesssim M_i \lesssim 25M_\odot$ expand after the main-sequence phase and reach the Hayashi line and become red supergiants. Stars in the range of $8 \lesssim M_i \lesssim 12M_\odot$ temporarily leave the red supergiant region in the HRD as they describe a leftward loop (blue loop) and become yellow supergiants, It is during this loop that they cross the instability strip and pulsate as Cepheids. At the end of their life, stars with $8 \lesssim M_i \lesssim 25M_\odot$ explode in the red supergiant phase as core-collapse supernovae.

23.1 Predicted Evolutionary Tracks

Figure 23.1 shows the predicted evolutionary tracks of stars of $M_i = 1$ to $120M_\odot$, modeled with moderate convective overshooting of $\ell_{\rm os} = 0.25\ H_{\rm p}$ and mass loss. The left panel is for stars with solar composition ($Z = 0.02$), and the right panel is for low-metallicity stars ($Z = 0.001$). The hatched areas show the locations of slow nuclear fusion. The red band is the location of core H-fusion, the blue band is the location of core He-fusion, and the green band indicates core He-fusion and later fusion phases in red supergiants.

In this section, we concentrate the discussion on stars in the range of 8 to $25M_\odot$. Several tracks show **blue loops** (i.e., loops to the left in the HRD) during core He-fusion. These are due to the same effect that produces the horizontal branch during core He-fusion of low-mass stars: due to the mirror action of the H-fusion shell, the envelope contracts when the core expands. The presence of blue loops therefore depends on the expansion of the core.

doi:10.1088/978-0-7503-1278-3ch23

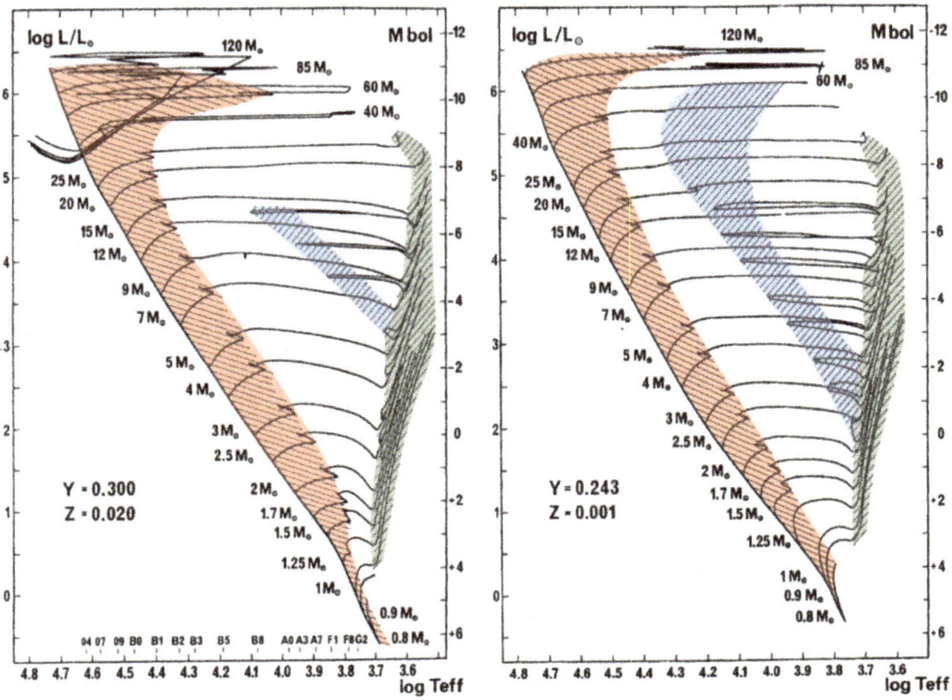

Figure 23.1. Evolutionary tracks of stars of about 1 to $120 M_\odot$ for metallicities of $Z = 0.02$ (left) and $Z = 0.001$ (right). The red band is the location of core H-fusion, the blue band is the location of core He-fusion, and the green band indicates core He-fusion and later fusion phases when the star is on the Hayashi line. Notice that core He-fusion in solar metallicity stars with $M_i \lesssim 12 M_\odot$ occurs during blue loops in evolution tracks. (Reproduced from Schaller et al. 1992, with permission. © ESO.)

In low-mass stars of $M_i < 5 M_\odot$ with solar composition, the loops are very short and close to the Hayashi track, whereas they reach further to the left in stars with $Z = 0.001$. These are the loops during the core He-fusion when the stars are on the horizontal branch that were discussed in Section 17. In more massive stars of $M_i \approx 8$ to $12 M_\odot$, the tracks for both metallicities show loops during core He-fusion, but the loops are more extended in the lower metallicity models. Solar metallicity stars with $M_i \gtrsim 12 M_\odot$ do not make loops during core He-fusion, despite having a H-fusion shell just like the lower-mass stars.

Detailed studies show that the presence or absence of blue loops depends on the surface gravitational potential, $\varphi_c \sim M_c/R_c$, of the core inside the fusion shell (Kippenhahn & Weigert 1990). If φ_c is smaller than some critical value, the star will describe a leftward loop. If φ_c is larger than this critical value, the star will not make a loop. Let us try to understand why stars with $M_i \gtrsim 12 M_\odot$ do not make a loop. Remember that the central temperature of a star in HE scales as $T_c \sim M/R$ (Equation 3.9). If the star is centrally concentrated, most of the central pressure is due to the mass of the core, so we may expect that $T_c \sim M_c/R_c \sim \varphi_c$. So, higher $T_c \rightarrow$ higher φ_c. The central temperature after the core H-fusion phase increases with stellar mass (see Figure 8.10), so φ_c increases with mass. Models show that in stars with $M_i \gtrsim 12M_\odot$ the potential of the core φ_c exceeds the maximum value for the occurrence of loops: these stars will not make blue loops in the HRD.

There are several additional subtle effects that play a role in setting the potential of the core and the occurrence of blue loops.

1. Convective overshooting increases the core mass and increases the potential, thus preventing blue loops.
2. Low-metallicity reduces the opacity and the convection in the core (according to the Schwarzschild criterion) during core H-fusion, so massive low-metallicity stars have less massive cores than solar metallicity stars at the end of core H-fusion. A smaller core has a lower potential, which facilitates the formation of loops. The tracks in Figure 23.1 show this effect.
3. The chemical structure above the core, which is sensitive to mixing and rotation, also plays a role.

These arguments show that the presence or absence of blue loops during the evolution of massive stars of $M_i \approx 8$ to $12M_\odot$ stars is uncertain. This is the reason that predicted evolutionary tracks of intermediate-mass stars by different groups may show different blue loops.

Q (23.1) *We will show in Chapter 25 that fast rotation induces extra mixing inside a star. What is the influence of fast rotation on the occurrence of blue loops?*

23.2 The Internal Evolution during the Post-MS Phase of Stars of 8 to 25$M_\odot$

Figure 23.2 shows the Kippenhahn diagram for the evolution of a star of $M_i = 15M_\odot$ and metallicity $Z = 0.02$. The evolutionary track is shown in the left panel of Figure 23.1. After the star has arrived on the Hayashi track with an extended convective envelope, it remains an RSG. During the RSG phase, the star climbs the Hayashi track from log $L/L_\odot = 4.5$ to 5.1. Notice in Figure 23.2 that each subsequent fusion phase is shorter in duration. This is partly due to the small mass defect of fusion of heavy ions and the increasing neutrino loss at high temperatures (Section 8.8).

The successive nuclear fusion phases in stars of $M_i = 15$ and $25M_\odot$ are listed in Table 23.1, together with the central temperature and density and the duration of the phases. Notice the rapid increase in ρ_c at each new phase. This is because at the end of each phase the core contracts and T_c increases until the next fusion reaction starts. The evolution accelerates very quickly in the latest stages. Whereas the core H- and core He-fusion phases take megayears, core C-fusion takes centuries, core Ne- and core O-fusion take years, and Si-fusion takes only days.

Q (23.2) *Why are the last evolutionary phases so short?*

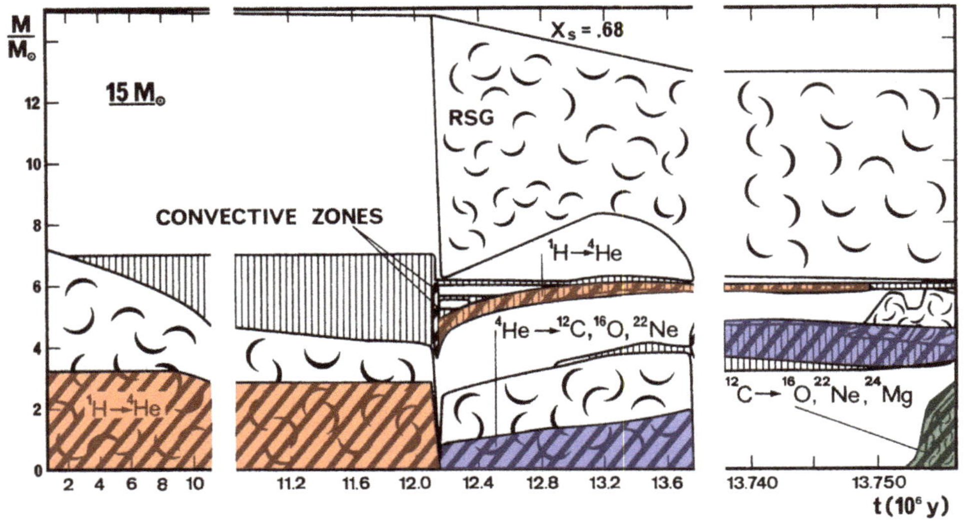

Figure 23.2. Kippenhahn diagram of a star of $M_i = 15M_{\odot}$ with $Z = 0.02$. Thick hatched regions indicate efficient fusion. At $t < 12.2$ Myr, when the star is on the main sequence, this is core H-fusion. At $12.2 < t < 13.6$ Myr, this is core He-fusion and H-shell fusion, followed after $t > 13.74$ Myr by He-shell and H-shell fusion. At $t = 13.753$ Myr, C-fusion starts in the core. Curly regions indicate convection. In vertically hatched regions, the abundances are changed by previous convection. During core He-fusion and later, the star is a red supergiant with an extended convective envelope. (Reproduced from Maeder & Meynet 1987, with permission. © ESO.)

Table 23.1. The Fusion Phases in Stars of $M_i = 15$ and $25M_{\odot}$. (Data are from Maeder & Meynet 1987)

Process			$15M_{\odot}$			$25M_{\odot}$		
Fusion	Fuel	Products	log T_c	log ρ_c	Duration	log T_c	log ρ_c	Duration
Hydrogen	H	He	7.48	0.76	11 Myr	7.58	0.28	6.7 Myr
Helium	He	C, O	8.25	3.14	2.0 Myr	8.29	2.88	0.8 Myr
Carbon	C	O, Ne	8.92	5.38	2000 yr	8.92	5.11	520 yr
Neon	Ne	O, Mg	9.21	6.86	0.73 yr	9.20	6.60	0.89 yr
Oxygen	O, Mg	Si, S	9.29	6.82	2.58 yr	9.30	6.56	0.40 yr
Silicon	Si, S	Fe, Ni	9.52	7.63	18 d	9.56	7.48	0.73 d

Note. T_c is in units of K, and ρ_c is in units of g cm^{-3}.

Figure 23.3 shows the variation of T_c and ρ_c during the evolution of stars of $M_i =$ 15 and $25M_{\odot}$. The arrow shows the overall trend of $T_c \sim \rho_c^{1/3}$ for a contracting core that was predicted on the basis of simple HE conditions (Equation (8.22)). During core fusion phases, the star adjusts itself slightly to a new equilibrium; this produces the little curls at each fusion phase. The contraction phases are fast, typically on the Kelvin–Helmholtz timescale of the core, and the core fusion phases last relatively longer, on the nuclear fusion timescale (see also Figure 8.10).

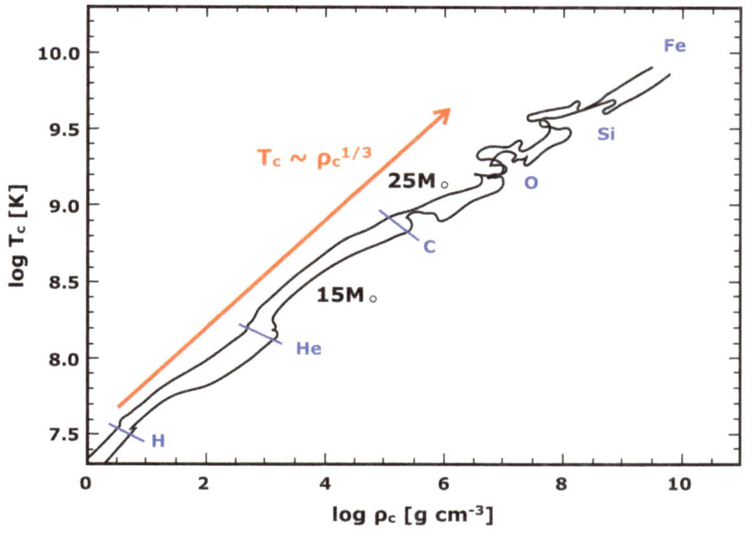

Figure 23.3. Internal evolution of T_c (in K) and ρ_c (in g cm^{-3}) during the evolution of stars of $M_i = 15$ and $25M_{\odot}$. The general trend of $T_c \sim M_c^{2/3}\rho_c^{1/3}$ is halted during the core fusion phases, indicated by the blue lines and labels. (Reproduced from Woosley et al. 2002. Copyright 2002 by the American Physical Society.)

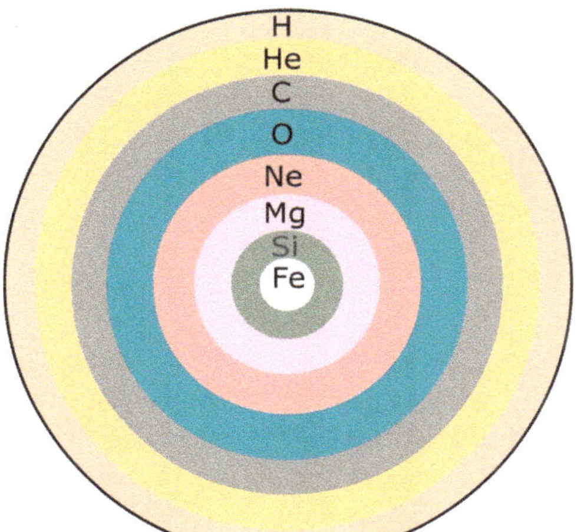

Figure 23.4. Schematic drawing of the onion shell structure of massive stars of $8 \lesssim M_i \lesssim 25M_{\odot}$ at the end of their evolution, with the dominant elements indicated. The outer envelopes of stars of $M_i > 25M_{\odot}$ are stripped before they explode as supernovae. Note that the figure is not to scale.

Each successive process requires a higher T_c and occurs in a smaller fraction of the stellar mass. The star therefore develops an "onion-skin" structure with the most advanced products in the core, surrounded by shells of less advanced fusion products. During these processes, the star remains a red supergiant until it explodes as a core-collapse supernova.

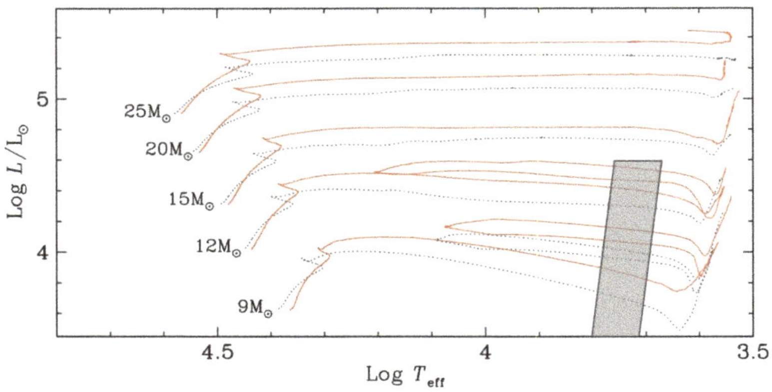

Figure 23.5. Location of Cepheids (gray region) in the HRD compared with predicted tracks of stars with $9 \lesssim M_i \lesssim 25 M_\odot$. Full red lines are tracks with initial rotational velocity of 300 km/s; dotted black lines are for nonrotating stars. The Cepheids are located in the blue loop. (Adapted from Meynet & Maeder 2000. Reproduced with permission © ESO.

23.3 Stellar Pulsation during Blue Loops

We have discussed radially pulsating Cepheids in Section 21. The large number of Cepheids indicates that these stars are in an evolution phase that lasts relatively long, much longer than the time that it takes stars to cross the Hertzsprung gap after the MS phase. This, together with their luminosity of $2.8 < \log L/L_\odot < 4.6$ and spectral type F–G, indicates that Cepheids are in the blue loop phase and core He-fusion phase of stars of $6 \lesssim M_i \lesssim 12 M_\odot$. The lack of Cepheids with $\log L/L_\odot > 4.6$ coincides with the lack of blue loops for more massive stars of solar metallicity. This is shown in Figure 23.5.

23.4 Summary

1. Intermediate-mass stars of $8 \lesssim M_i \lesssim 25 M_\odot$ evolve to the Hayashi line in the post-MS phase. The evolutionary tracks of stars with $M_i \lesssim 12 M_\odot$ describe **blue loops** in the HRD during core He-fusion. This is due to the contraction of the core and the mirror action of the H-fusion shell.
2. The occurrence of blue loops depends on the surface gravitational potential at the edge of the He core. The loops are longer and more pronounced in low-metallicity stars.
3. Cepheids are in the blue loop phase of stars of $M_i \lesssim 12 M_\odot$. The luminosity upper limit of Cepheids coincides with the luminosity upper limit of blue loops.
4. Stars with $12 \lesssim M_i \lesssim 25 M_\odot$ do not show blue loops but remain red supergiants throughout all successive nuclear fusion phases before they explode as core-collapse supernovae.

5. The core H- and core He-fusion phases last millions of years. The core C-fusion phase last thousands of years. The core Ne- and core O-fusion phases take years, and the core Si-fusion phase takes only days.

Exercises.

23.1 (a) Study the Kippenhahn diagram of a star of $M_i = 15M_{\odot}$ and relate the different phases to their locations on the evolutionary track.

(b) What was the adopted mass-loss rate during core H- and core He-fusion?

(c) Compare that with the mass-loss formulae in Chapter 15.

23.2 What could be the reason that the evolution of a star in the T_c–ρ_c diagram of Figure 23.3 becomes less steep at late evolution phases?

23.3 **Computer exercise**

Use the MESA code to follow the evolution of a star of $M_i = 25M_{\odot}$ and determine the amount of mass that takes part in each successive core fusion phase.

References

Kippenhahn, R., & Weigert, A. 1990, Stellar Structure and Evolution (Berlin: Springer)
Maeder, A., & Meynet, G. 1987, A&A, **182, 243**
Maeder, A. 2009, Physics, Formation and Evolution of Rotating Stars (Berlin: Springer)
Meynet, G, & Maeder, A. 2000, A&A, **361, 101**
Schaller, G., Schaerer, D., Meynet, G., & Maeder, A. 1992, A&AS, **96, 269**
Woosley, S. E., Heger, A., & Weaver, T. A. 2002, RvMP, **74, 1015**

Henny J.G.L.M. Lamers and Emily M. Levesque

Chapter 24

The Evolution of Massive Stars of 25–120$M_\odot$: Dominated by Mass Loss

Mass loss plays a dominant role in the evolution of stars more massive than about 25$M_\odot$. During the main-sequence phase, the stars have a convective core that diminishes as the He abundance increases. After the main-sequence phase, stars of $25 \lesssim M_\mathrm{i} \lesssim 50M_\odot$ evolve into red supergiants just like the stars in the range of 8 to 25$M_\odot$; however, massive red supergiants are so luminous and have such a low surface gravity that the convective envelope is almost completely stripped by mass loss. When the envelope mass is below a critical value, the stars contract and move to the left in the HRD where they appear as Wolf–Rayet stars with strongly enriched atmospheres. They remain Wolf–Rayet stars until they explode as supernovae. Stars with $M_i \gtrsim$ 40-50$M_\odot$ do not even reach the red-supergiant phase; shortly after the main-sequence phase they go through a phase of rapid and strongly variable mass loss as a luminous blue variable. They lose so much mass that they evolve straight into a Wolf–Rayet star and remain in that phase until they explode as supernovae.

24.1 The Effect of Mass Loss during the Main-sequence Phase

We have seen in Section 15.2.2 that the mass-loss rate of hot stars increases with luminosity as $\dot{M} \sim L^{1.5}$. At the same time, the mass–luminosity relation indicates that $L \sim M^y$ with $y \approx 2.9$ so $\dot{M} \sim M^{4.4}$. This shows that the influence of mass loss on the evolution will increase with stellar mass.

Figure 24.1 shows the first calculations of the influence of mass loss on the evolution of massive stars. The mass-loss rate, due to radiation-driven winds, is expressed in terms of N, where $\dot{M} = NL/c^2$(see Equation (15.4)). The figure shows the tracks during the main-sequence phase of a star of 30$M_\odot$ for strong mass loss (N = 300), moderate mass loss (N = 100), and conservative evolution (N = 0). The moderate mass loss is 7×10^{-7} $M_\odot$ yr^{-1} on the ZAMS. This is slightly more than the observed value of $\sim 3 \times 10^{-7}$ $M_\odot$ yr^{-1}.

doi:10.1088/978-0-7503-1278-3ch24

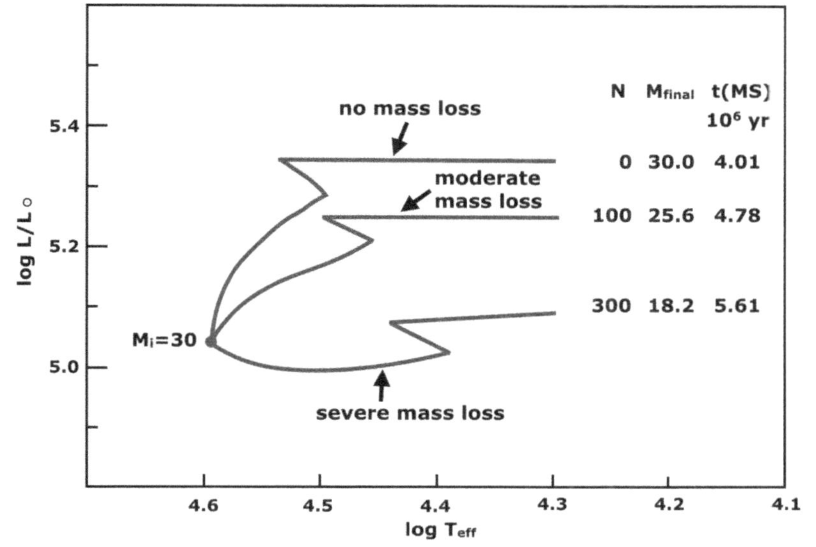

Figure 24.1. Effect of mass loss during the MS phase of a $30M_\odot$ star with a mass-loss rate specified by $\dot{M} = N \times L/c^2$. The mass at the end of the MS, $M_{\rm final}$, and the MS lifetimes are indicated. (Reproduced from Lamers & Cassinelli 1999. © Cambridge University Press. Reprinted with permission.)

The figure shows four effects.

1. The luminosity increases less in models with increased mass loss. This is because the convective core is smaller than in conservative evolution. As a consequence, μ increases less and thus L increases less (Equation (6.7)).
2. At the end of the MS phase, the star is less luminous than it would be in the case of conservative mass loss, but still more luminous than it would have been had the star started with $M_i = M_{\rm final}$ and evolved conservatively. This is because the star has a more massive He core than it would if it started with a lower mass but experienced no mass loss.
3. The MS phase lasts longer in the case of mass loss because L is lower than it would be in the case of conservative evolution.
4. The end of the MS phase for mass-losing stars is at lower $T_{\rm eff}$ than it would be in the case of conservative evolution. This is because the core H-fusion phase lasts longer, resulting in a widening of the main sequence for stars with substantial mass loss during core H-fusion.

Q 24.1 *What could be the reason that the mass-loss rate, adopted for calculating the evolution of the stars shown in* Figure 24.1, *is expressed as* $\dot{M} = N \times L/c^2$?

24.2 Predicted Evolution Tracks with Mass Loss

Figure 24.2 shows the evolutionary tracks of stars from 0.8 to $120M_\odot$, with mass loss and convective overshooting for solar metallicity ($Z = 0.014$) and SMC metallicity

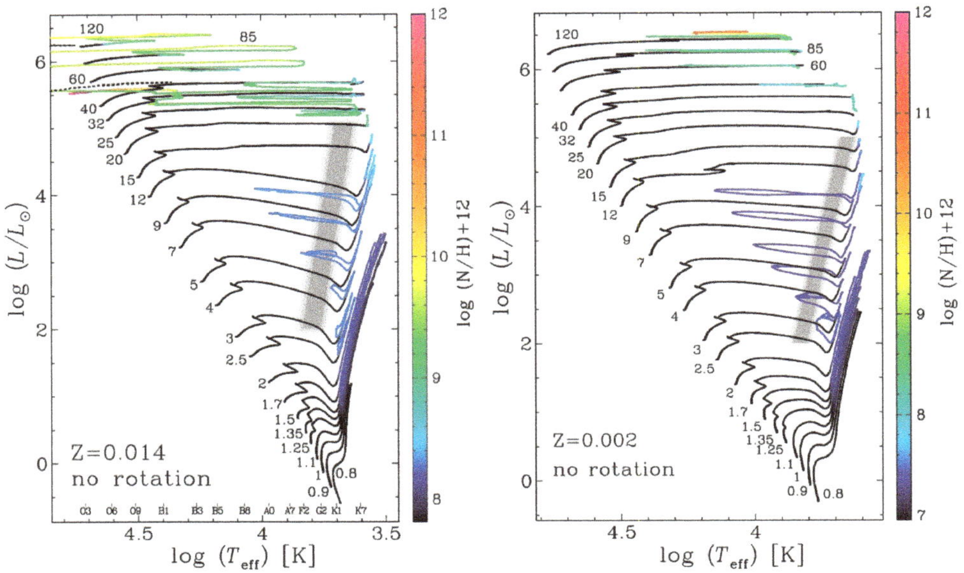

Figure 24.2. Evolutionary tracks of stars with solar metallicity ($Z = 0.014$) and SMC metallicity ($Z = 0.002$). The surface composition of N/H in terms of the number ratio $A_N = \log(N/H) + 12$ is shown in color. The initial composition of $A_N = 7.78$ is shown in black. The gray region shows the location of the pulsating Cepheids. (Reproduced from Ekstrom et al. 2012, with permission; Georgy et al. 2013, with permission).

($Z = 0.002$), calculated by the Geneva group (Ekstrom, S. et al. 2012, Georgy et al. 2013).

The predicted tracks show the main-sequence widening for stars of $25 < M_i < 120 M_\odot$. These stars lose between about 5% and 30% of their initial mass during the MS phase (Table 15.1). Stars of $M_i \gtrsim 50 M_\odot$ do not reach the Hayashi line as a red supergiant, because they have lost a substantial fraction of their mass during the MS phase and later as LBVs. The tracks return to the left before the stars reach the Hayashi line because their envelope does not have sufficient mass to sustain an extended convection zone. This agrees qualitatively with the observed lack of very massive red supergiants shown in Figure 22.1. The location of the predicted turnaround is uncertain because the highly variable mass loss during the LBV phase, which is still not well understood, cannot be properly modeled. The adopted mass loss rate in the LBV phase is usually adjusted to predict a turn around at approximately the observed Humphreys–Davidson limit. The tracks with solar metallicity turn around earlier (i.e., at higher $T_{\rm eff}$) than the tracks with SMC metallicity. This is because the radiation-driven mass-loss rate of massive stars is metallicity-dependent, with $\dot{M} \sim Z^{0.85}$ (Equation (15.14)).

In Section 24.2.2 we will discuss the evolution of a $60 M_\odot$ star, showing different processes that affect the evolution of massive stars with mass loss.

24.2.1 The N Surface Abundance of Peeled Stars

Mass loss produces a change in surface abundance of massive stars, because stripping of the outer layers exposes nuclear products that were formed earlier by fusion in deeper layers. At the end of the MS phase, the N abundance increases at the surface: the star becomes an **ON-star** (O for spectral type O, and N for nitrogen). This is because the overshooting from the convective core has brought products of the CNO cycle, in particular, N, into the radiative envelope, up to the point where $M(r) \sim 0.9M$. Once the outer $\sim 0.1M$ has been peeled off by mass loss during the MS phase, enriched layers appear at the surface (this is different from the dredge-ups of low-mass stars, where the enriched matter is dragged all the way to the surface by envelope convection). A comparison between observed and predicted surface abundances in various evolutionary phases provides a sensitive check to the accuracy of the evolutionary models. In particular, the role of rotation-induced mixing can be derived from the study of surface N abundance (Brott et al. 2011). In this section, we discuss the predicted changes in the N abundance at the surface.

The color-coding in Figure 24.2 indicates the abundance of N at the stellar surface. The original N/H number ratio is $\log(\mathrm{N/H}) + 12 = 7.98$ if $Z = 0.014$ and 7.13 if $Z = 0.001$. We can distinguish two mass regions.

- For stars of $M_i \lesssim 25M_\odot$, red giants and supergiants, horizontal branch stars, AGB stars, and blue loop stars show N enhancement at their surface. In all of these phases, the increase in N abundance is moderate because N is dredged up by convection into the envelope from the CNO fusion region below. The large envelope mass results in a severe dilution of the N abundance compared to that produced by the CNO cycle.
- In stars of $M_i \gtrsim 25M_\odot$ the increase in surface N abundance is much higher because shortly after the MS phase the stars are peeled off to the layers where CNO fusion has occurred. The higher the mass of the star, the stronger the N enhancement.

Q (24.2) *Explain why the N enhancement at the surface of massive stars, $M_i > 25M_\odot$, is much less severe in stars with $Z = 0.002$ than in stars with $Z = 0.014$.*

24.2.2 The Duration of the Core-fusion Phases

Figure 24.3 shows the duration of the phases of core H-fusion, core He-fusion, and core C-fusion in stars of $M_i = 1$ to $120M_\odot$ and $Z = 0.02$ and 0.001 (Schaller et al. 1992). For stars with $M_i \gtrsim 25M_\odot$, H-fusion takes a few million years, He-fusion a few 10^5 yr, and C-fusion about 10^4 yr. Metallicity has very little effect on the duration of the phases.

Q (24.3) *Explain why the slope of these relations gets less steep toward higher mass.*

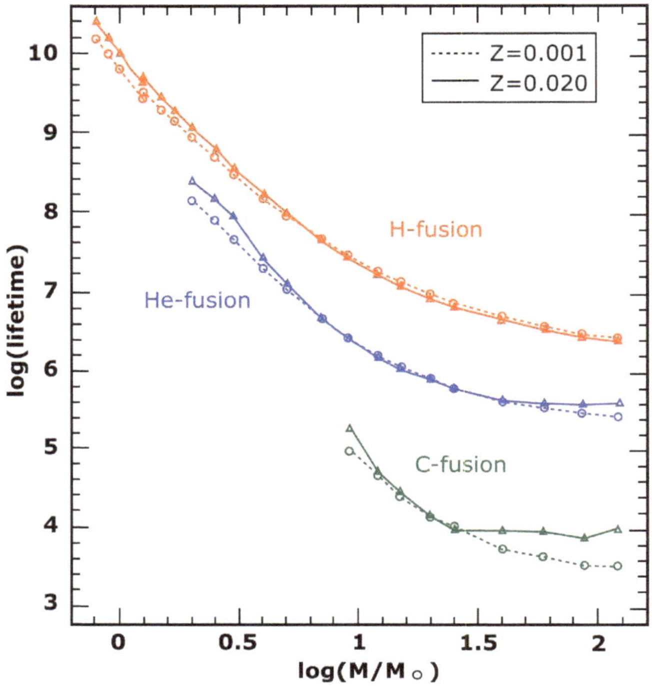

Figure 24.3. Lifetimes of H-fusion, He-fusion, and C-fusion in stars of $M_i = 1$ to $120 M_\odot$ for two metallicities: $Z = 0.02$ and 0.001 (Reproduced from Schaller et al. 1992, with permission. © ESO.)

24.3 The Evolution of a $60 M_\odot$ Star with Mass Loss

Let us consider the evolution of a $60 M_\odot$ star as an example of a massive star. Figure 24.4 shows: the evolutionary track in the HRD (top panel), the changes in internal structure in the Kippenhahn diagram (middle panel), and the changes in the surface composition (lower panel). The letters on the evolutionary track correspond to the phases indicated in the other panels.

This figure shows the major properties of high-mass stellar evolution.

- The mass of the He core after the MS phase is larger than the Schönberg–Chandrasekhar limit of $M/M_i \approx 0.10$ (Section 14.1), so the core immediately contracts after the MS and He fusion starts in the core shortly after the MS (see the KD). This is different from low-mass stars, which have a separate phase of H-shell fusion: the RGB phase.
- In the very short time between the end of the core H-fusion and start of the core He-fusion, shell H-fusion is the main energy source, so the shell H-fusion occurs in a thick layer. As soon as core He-fusion starts, the shell H-fusion retreats to a less massive shell.

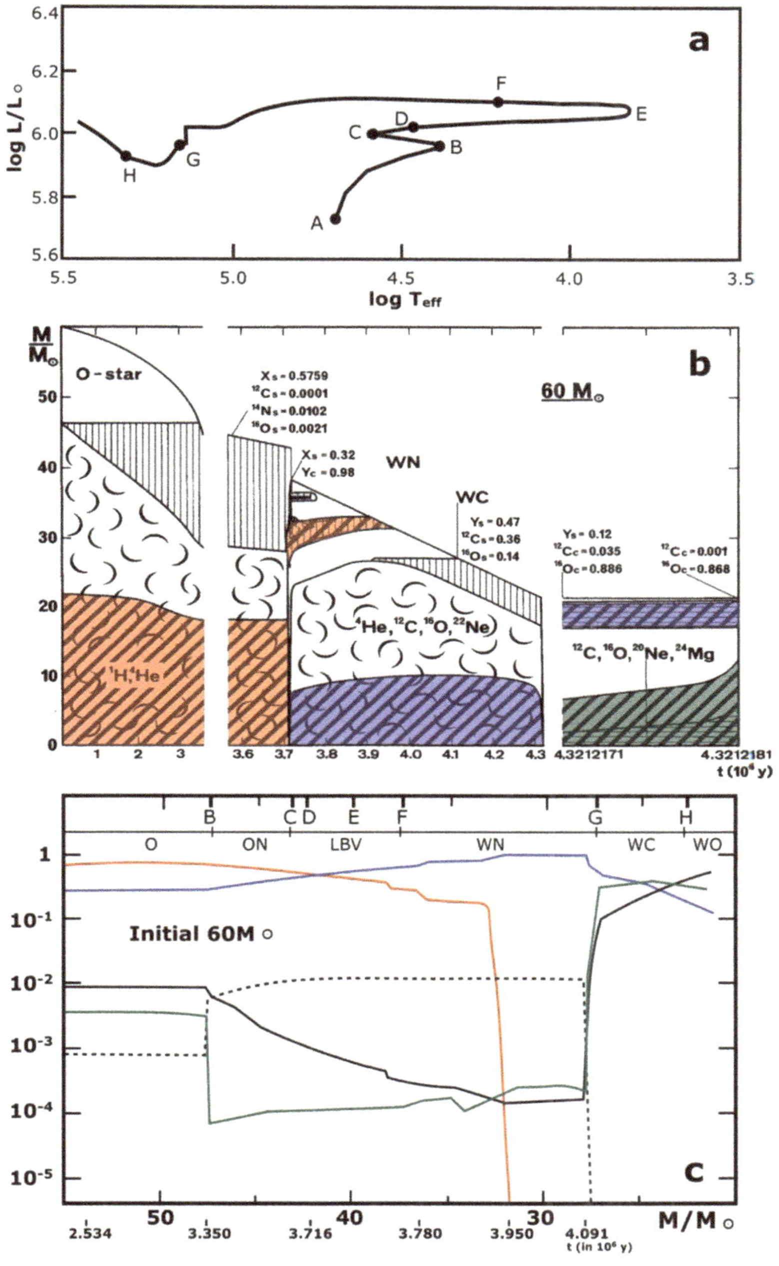

Figure 24.4. (a) Evolutionary track of a solar-metallicity star of $M_{\rm i} = 60\ M_{\odot}$ with mass loss and convective overshooting, with $\ell_{\rm os} = 0.25\ell_P$, and without rotational mixing. (b) Kippenhahn diagram showing the evolution of the internal structure. Hatched areas indicate fusion layers, curls indicate convective zones, and vertical striped regions indicate modified abundances. The letters correspond to the locations in the HRD. The decrease in the upper limit is due to mass loss. Corresponding spectral types are indicated. (c) Changes in the surface abundance of H (red), ^{4}He (blue), ^{12}C (green), ^{16}O (black), and ^{14}N (black dashed), expressed in mass fractions. The abscissa shows both the time (in Myr) and the remaining mass. (Reproduced from Maeder & Meynet 1987, with permission. © ESO.)

- At around 3.7 Myr, the mass of the star decreases steeply in about 10^4 yr. This is due to the very high mass-loss rate of 10^{-4} to 10^{-3} $M_\odot$ yr^{-1} adopted in the calculations to mimic the mass loss by the LBV eruptions.
- The surface composition starts to change during the late MS phase, when mass loss has stripped the star down to $\sim 47 M_\odot$ (see abscissa), the original mass of the convective core. This results in a large increase of the abundance of N at the expense of C and O. These are the observed ON stars.
- Shortly after the LBV phase, when the mass decreases from about 43 to $38 M_\odot$, the star describes a fast loop (D, E, F) in the HRD that takes only about 6×10^4 yr. At phase F, the star has a He-rich and N-rich atmosphere and contracts into a Wolf–Rayet star of type WN.
- At phase G, the atmosphere becomes C-rich. At that time, the star has contracted to a small radius and high T_{eff}, so it is a WC star. At phase H, the atmosphere becomes O-rich and the star becomes a WO star.

Q (24.3) *Correlate the changes in surface composition with the track and with the Kippenhahn diagram and try to explain them.*

24.4 The Conti Scenario

Before 1970, the predicted evolution of massive stars could not explain the range of different types of massive stars with strange properties that were observed: the N-rich ON stars, the Of stars with many emission lines, the N-rich Wolf–Rayet stars (WN), the C-rich Wolf–Rayet stars (WC), the luminous blue variables (LBV), etc. The evolutionary connection of these stars was completely unknown! This situation improved dramatically when the first rocket spectroscopic UV observations of three hot supergiants in the Orion Belt indicated a high mass-loss rate and subsequent satellite UV spectra showed that the phenomena of mass loss is ubiquitous among massive stars. (For a review of the early history of mass loss from massive stars and its effect on stellar evolution, see, e.g., Lamers 2008). Within a decade, evolutionary calculations showed that high mass-loss rates not only changed the evolution of massive stars but could also explain the appearance of the observed nuclear products at the stellar surface.

In 1976, the U.S. astronomer Peter Conti suggested a scheme for the evolution of massive stars, based partly on observations and partly on predictions, that connected the different types of stars in evolutionary sequences. This is called **the Conti scenario** (Conti 1975; Maeder & Conti 1994). This scenario was later slightly updated (Maeder 2009). The present version of this scenario is shown in Table 24.1.

The different types of stars in this scenario are as follows.

O = O stars without emission lines (low $\dot{M} < 10^{-6}$ $M_\odot$ yr^{-1}).
Of = Ostars with emission lines (high $\dot{M} > 10^{-6}$ $M_\odot$ yr^{-1}).
BSG = Blue supergiant (O, B, A).
YSG = Yellow supergiant (F, G).
RSG = Red supergiant (K, M).
LBV = Luminous blue variable (with eruptions).

Table 24.1. The Conti Scenario That Describes the Evolutionary Connections of the Different Types of Stars. (Based on Smith et al. 2014)

Conti Scenario		
$M_i/M_\odot$	Spectral types	Supernova
	Always Blue	
$\gtrsim 90$	: O – Of – WNL – (WNE) – WCL – WCE -	SN (Hypernova?)
60 - 90	: O – Of/WNL – LBV - WNL – WCL - WCE -	SN IIn ?
40 - 60	: O – BSG – LBV - WNL – (WNE) - WCL - WCE -	SN Ib
	WCL – WCE – WO -	SN Ic
	Blue – Red – Blue	
30 - 40	: O – BSG – RSG – WNE – WCE –	SN Ib
	Blue - Red	
25 – 30	: O - BSG – RSG – BLG –blue loop- RSG –	SNIIL
15 – 25	: O – RSG -	SNIIp
10 – 15	: O – RSG – blue loop Cepheid – RSG -	SNIIp

WN = WR star with N-rich wind, He-rich, some H (high $\dot{M} > 10^{-5}\ M_\odot\ yr^{-1}$).
WNL = Late-type WN (spectral type WNL7–WNL8).
WNE = Early-type WN (spectral type WN2–WN6).
WC = WR star with C-rich wind, He-rich, no H (high $\dot{M} > 10^{5}\ M_\odot\ yr^{-1}$).
WCL = Late-type WC (spectral type WC7–WC9).
WCE = Early-type WC (spectral type WC4–WC6).
WO = WR star with O-rich wind.
SN = Supernova (types are discussed in Section 27).

NB 1: The limits of 30, 40, and $60M_\odot$ are uncertain. They depend on composition and rotation because the mass-loss rates are smaller in low-metallicity stars and rotation induces mixing.

NB 2: Recent observations have shown that some Type IIn supernovae are from LBV progenitors. This suggests that massive stars may also end their lives during the LBV phase (Maeder 2009).

NB 3: The evolution described here is for single stars. In narrow binary systems, mass transfer may result in stripping of the most massive component. This may produce Wolf–Rayet stars that originate from stars with an initial mass $M_i < 40M_\odot$. (Crowther, P. A. 2007).

NB 4: We discussed the evolution of stars up to $120M_\odot$. The evidence for more massive stars, possibly up to $300M_\odot$, and their evolutions, has been described in (Vink et al. 2015).

24.5 Summary

1. The evolution of massive stars is strongly affected by their high mass-loss rates. Mass loss during the MS phase leads to lower luminosities and longer MS lifetimes than those seen in stars with conservative evolution. Mass loss also leads to a widening of the MS in the HRD at high luminosity.
2. Mass loss results in stripping of the stars with nuclear products appearing at the surface. This explains the ON stars at the end of the MS phase.
3. LBVs play a crucial role in the evolution of stars of $M_i \gtrsim 40M_\odot$. The stripping of stars during the LBV phase results in Wolf–Rayet stars.
4. Mass loss explains the occurrence of N-rich, C-rich, and O-rich WR stars by stripping massive stars to ever deeper layers.
5. The Conti scenario explains the relation between the different types of massive stars in an evolutionary sequence (the resulting supernovae are discussed in Chapter 27).

Exercises.

24.1 Show that the calculated duration of the MS phase of a star with $M_i = 30M_\odot$ with different mass-loss rates, as shown in Figure 24.1, agrees with expected values.

24.2 Measure the amount of mass that takes part in the H-, He-, and C-fusion phases in a star $M_i = 60M_\odot$ and the mean luminosity at these phases (Figure 24.4). Take into account the contribution by the shell fusion.

Use these data to predict the duration of these phases and compare those with Figure 24.3.

In case of disagreement, comment on the reason.

24.3 What is the expected slope of the log(t) versus log(M) relation in Figure 24.3 for H-fusion in low-mass stars with $L \lesssim 2L_\odot$.

Why does the slope of the log(t) versus log(M) relation in Figure 24.3 for H-fusion flatten toward more massive stars?

24.4
(a) Indicate the changes in the surface composition of the $M_i = 60M_\odot$ star along its evolutionary track.
(b) Indicate the adopted mass-loss rates along the evolutionary track.
(c) Indicate the mass of the star along the evolutionary track.

24.5 How would the surface evolution along the evolutionary track have changed if the adopted mass-loss rates were increased by a factor of 2 or decreased by a factor of 2?

References

Brott, I., Evans, C. J., & Hunter, I., et al. 2011, A&A, **530, 116**

Conti, P. S. 1975, MSRSL, **9, 193**

Crowther, P. A. 2007, ARAA, **45, 177**

de Loore, C., de Greve, J. P., & Lamers, H. J. G. L. M. 1977, A&A, **61, 251**

Ekstrom, S., Georgy, C., & Eggenberger, P., et al. 2012, A&A, **537, 146**

Georgy, C., Ekström, S., & Eggenberger, 2013, A&A, **558, 103**

Lamers, H. J. G. L. M. 2008, in ASP Conf. Ser. 388, Mass Loss from Stars and the Evolution of Star Clusters, ed. A. de Koter, L. J., Smith, & L. B. F. M. Waters, (San Francisco, CA: ASP), 443

Lamers, H. J. G. L. M. & Cassinelli, J. P. 1999, Introduction to Stellar Winds (Cambridge: Cambridge Univ. Press,)

Maeder, A. 2009, Physics, Formation and Evolution of Rotating Stars (Berlin: Springer)

Maeder, A. & Conti, P. S. 1994, ARAA, **32, 277**

Maeder, A. & Meynet, G. 1987, A&A, **182, 243**

Schaller, G., Schaerer, D., Meynet, G. & Maeder, A. 1992, A&AS, **96, 269**

Smith, N., Mauerhan, J. C. & Prieto, J. L. 2014, MNRAS, **438, 1191**

Vink, J.S., Heger, A. & Krumholtz, M.R., et al. 2015, Very Massive Stars in the Local Universe (Cambridge: Cambridge Univ. Press)

Chapter 25

Rotation and Stellar Evolution

Massive stars are born rapidly rotating. This is due to the fact that magnetic braking in their early phases is much less important than in lower-mass stars, which have a convective envelope and strong magnetic fields. The distribution of equatorial rotation velocities of massive main-sequence stars shows a peak near about 200 km s^{-1} and extends to 400 km s^{-1}. This corresponds to rotation periods of only a few days for massive main-sequence stars, compared to the Sun's rotation period of 24.5 days. As a result, the effective gravity at the equator of massive stars is reduced drastically by the combined action of rotation and radiation pressure. The most rapidly rotating stars may even reach the critical limit of zero effective gravity at their equator during their post-MS evolution. This results in very high mass-loss rates that change their evolution. Fast rotation also influences the internal structure of massive stars as it creates meridional circulation. This circulation produces chemical mixing, even in layers that are stable against convection. In this section, we discuss the effects of fast rotation on the shape and the surface temperature distribution of massive stars. Fast rotation produces a nonspherical wind that may result in slowing down or spinning up a star, depending on the ratio between the mass loss from the poles and from the equator. We compare the evolution of fast rotating stars with that of slowly rotating stars.

25.1 The Critical Velocity of Rotating Stars

Surfaces of constant effective gravity in rapidly rotating stars, where $g_{\rm eff} = g_{\rm grav} - g_{\rm centr}$, are not spherical but oblate due to the centrifugal acceleration, $g_{\rm centr} = {\rm v}^2/R$. The **critical rotation velocity, $\mathbf{v_{crit}}$**, is defined as the rotation velocity, where $g_{\rm eff} = 0$ at the equator. It can also be expressed in terms of the **critical angular velocity, $\mathbf{\Omega_{crit}}$**. The values of $\rm v_{crit}$ and $\Omega_{\rm crit}$ follow from the condition $g_{\rm centr} = g_{\rm grav}$:

$$\frac{{\rm v}_{\rm crit}^2}{R_{\rm eq}} = \frac{GM_{\rm eff}}{R_{\rm eq}^2} \rightarrow {\rm v}_{\rm crit} = \sqrt{\frac{GM_{\rm eff}}{R_{\rm eq}}} = \sqrt{\frac{GM_*(1-\Gamma_e)}{R_{\rm eq}}}, \tag{25.1}$$

doi:10.1088/978-0-7503-1278-3ch25

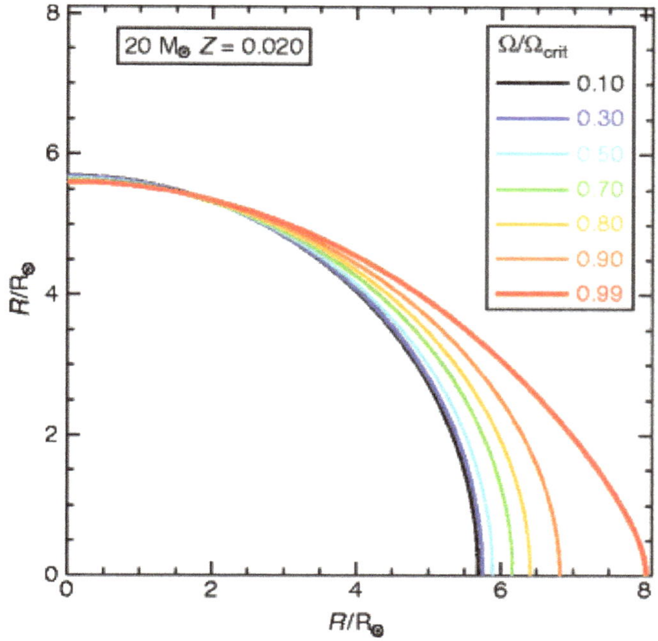

Figure 25.1. Shape of a $20M_\odot$ main-sequence star of solar metallicity for several values of Ω/Ω_{crit}. (Courtesy of Sylvia Ekstrom)

where R_{eq} is the equatorial radius and $M_{eff} = M_*(1 - \Gamma_e)$, where $\Gamma_e = \sigma_e L/4\pi cGM$ is the Eddington factor (Equation (6.14)) to correct for radiation pressure by electron scattering. Using $v_{rot} = \Omega R_{eq}$ we find

$$\boxed{\Omega_{crit} = \sqrt{GM_*(1 - \Gamma_e)/R_{eq}^3}}. \tag{25.2}$$

Figure 25.1 shows a quadrant of the shape of a $20M_\odot$ star at various values of the angular velocities, expressed as Ω/Ω_{crit}. Note that the equatorial radius increases steeply for values of Ω/Ω_{crit} approaching unity, but that the polar radius is only slightly affected by the rotation rate. This implies that the volume of a star increases with rotation rate while the density decreases, especially in the envelope near the equator. The maximum ratio between the equatorial and polar radius of $R_{eq}/R_p = 1.5$ is reached when $\Omega \rightarrow \Omega_{crit}$.

25.2 The Von Zeipel Effect

The surface temperature of rapidly rotating stars is not constant but varies with latitude. This can be understood by considering equipotential surfaces. In rapidly rotating stars, surfaces of constant potential are oblate. The potential at a given radius inside a rotating star is

$$\Phi(r, \theta) = -\frac{G\,M_r}{r} - \frac{1}{2}\Omega^2 r^2 \sin^2\theta. \tag{25.3}$$

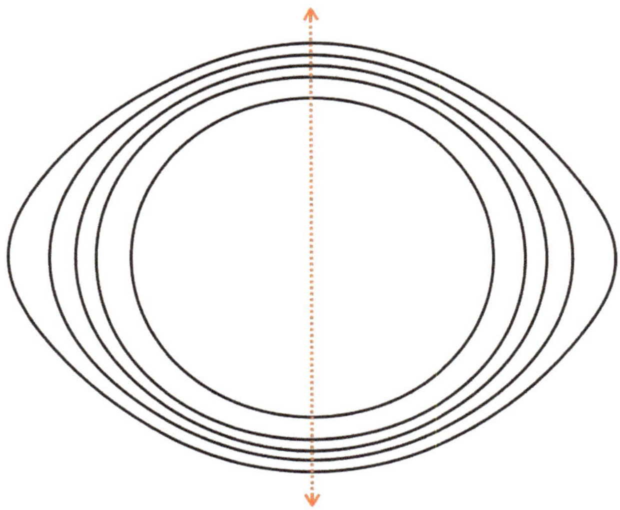

Figure 25.2. Equipotential surfaces for a rigidly rotating star with a rotation of $\Omega\ R_p = 0.54\ (GM/R_p)^{0.5}$ where R_p is the radius at the pole. The spin axis is vertical and shown by the dotted red line. (Reproduced from Lamers & Cassinelli 1999 © Cambridge University Press. Reprinted with permission.)

Figure 25.2 shows equipotential surfaces in a rigidly rotating star. The lines of constant potential are closer together near the poles than near the equator.

Surfaces of constant potential are also surfaces of equal pressure and equal density. Surfaces of constant pressure and density are also surfaces of constant temperature, because $T \sim P/\rho$. So, the isothermal surfaces are closer together at the pole than at the equator. This implies that the T-gradient, and hence the radiative flux, which is proportional to dT^4/dr (Equation (6.3)), is higher at the poles than at the equator! This is expressed in the Von Zeipel theorem, named after the Swedish astronomer Edvard von Zeipel (1873–1959), who proved it in 1924.

The **Von Zeipel theorem** states that in a rotating star the local radiative flux is proportional to the local effective gravity

$$\boxed{\begin{aligned} F_{\text{rad}}(\Omega,\ \Theta) &= \sigma T_{\text{eff}}^4(\Omega,\ \Theta) \\ &= \frac{L}{4\pi R(\Omega,\ \theta)^2} = \frac{L}{4\pi G M_*} g_{\text{eff}}(\Omega,\ \Theta) \\ &\rightarrow T_{\text{eff}}(\Omega,\ \Theta)\ \sim\ g_{\text{eff}}(\Omega,\ \Theta)^{1/4} \end{aligned}}, \tag{25.4}$$

where Ω is the angular velocity and Θ is the stellar latitude ($\Theta = 0$ and π at the poles and $\pi/2$ at the equator). The total luminosity of a rapidly rotating star is

$$L_* = 4\pi \int_0^{\pi/2} R^2(\Theta) F(\Theta)\ \sin\ (\Theta)\ d\Theta. \tag{25.5}$$

As a result, **T_{eff} and the spectral type at the poles are hotter than at the equator.** This implies that the spectral type and the luminosity derived from observations will depend on the orientation of the rotation axis with respect to the line of sight. This may produce an artificial widening of the observed main sequences in clusters if the stars are rapidly rotating and have randomly directed rotation axes.

Q (25.1) *How does this affect the determination of spectroscopic distances (from M_V and spectral type) for rapidly rotating stars?*

25.3 Nonspherical Mass Loss of Rapidly Rotating Stars

The winds from hot stars are driven by radiation pressure. Rapidly rotating stars are more luminous at the pole, where $T_{\rm eff}$ is higher, so they will have a higher mass flux from the pole than from the equator. At the same time, the terminal wind velocity v_∞ will also be higher at the pole than at the equator, because $v_\infty \sim v_{\rm esc}$ (see Section 15.2.2). This produces a **nonspherical wind with a mass flux and wind velocity that depend on stellar latitude**. The surface integrated mass-loss rate of a rotating massive star is larger than that of a nonrotating star in the same evolutionary phase.

The latitude dependence of the mass-loss rate of a rapidly rotating star of $100M_\odot$ is shown in Figure 25.3. The color indicates the distribution of $T_{\rm eff}$ and the size indicates the mass-loss rate as a function of stellar latitude. In general, the mass-loss rate is highest at the poles where $T_{\rm eff}$ is highest. This is shown in the left panel of Figure 25.3; however, at certain polar values of $T_{\rm eff}$ the mass-loss rate is higher at the equator than at the poles. This happens if $T_{\rm eff}$ at the pole is about 25,000 K and if $T_{\rm eff}$ and $v_{\rm esc}$ are both significantly lower at the equator than at the poles due to fast rotation. In that case, the mass-loss rate from the equator will be higher than the rate from the pole. This is shown in the right panel of Figure 25.3. The reason is that the ions that drive the wind are different at $T_{\rm eff} \gtrsim 21{,}000$ than at $T_{\rm eff} \lesssim 21{,}000$. This is

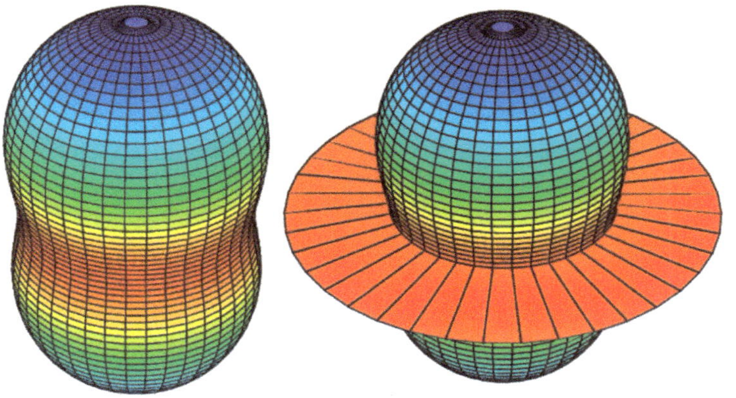

Figure 25.3. Left: mass flux from a rotating star of $100M_\odot$ and $L = 3 \times 10^6\ L_\odot$ with a rotational velocity of 80% of the critical velocity and $T_{\rm eff} = 30{,}000$ K at the pole. The colors indicate the $T_{\rm eff}$ distribution along the stellar surface and the size in any direction indicates the mass-loss rate. The mass-loss rate is higher at the poles than at the equator. Right: the same star with $T_{\rm eff} = 25{,}000$ K at the pole and $T_{\rm eff} < 20{,}000$ K at the equator. This difference in $T_{\rm eff}$ drives a rotation-induced bi-stability disk (Lamers & Pauldrach (1991) and Pelupessy et al. (2000)). (Reproduced from Maeder & Desjacques 2001, with permission. © ESO.)

the effect of the **bi-stability jumps** observed in the ratio v_∞/v_{esc} (Figure 15.2) and discussed in Section 15.2.2.

Maeder (2009) has derived the total mass-loss rates of rapidly rotating stars by integrating the latitude-dependent mass loss over the full stellar surface. This results in

$$\frac{\dot{M}(\Omega)}{\dot{M}(0)} \cong \left[1 - \frac{4}{9}\left(\frac{v}{v_{crit}}\right)^2\right]^{\left(\frac{7}{8}-\frac{1}{\alpha}\right)}, \tag{25.6}$$

where $\dot{M}(0)$ is the mass-loss rate for a nonrotating star (i.e., the mass-loss rate predicted by Vink et al. 2001 described in Section 15.4.1). The empirically derived values of α (from Lamers et al. 1995) depend on T_{eff} as

$\alpha = 0.52$ if $\log T_{eff} > 4.35$,
$\alpha = 0.24$ if $\log T_{eff} \approx 4.30$,
$\alpha = 0.17$ if $\log T_{eff} \approx 4.00$,
$\alpha = 0.15$ if $\log T_{eff} \approx 3.90$.

This expression is valid for stars with an Eddington factor $\Gamma_e < 0.25$. For stars with higher Eddington factors, the mass loss increases much more steeply with v/v_{crit}.

Q (25.2) *Why does the mass-loss rate of massive stars increase much more steeply with v/v_{crit} if Γ_e approaches unity?*

Figure 25.4 shows the increase in mass loss of stars that reach their critical rotation rate, Ω_{crit} (Equation (25.2)), at the end of the main-sequence phase (TAMS). The upper panel shows the values of Γ_e at the TAMS for stars of different initial masses. Stars with initial masses of $M_i > 60 M_\odot$ reach $\Gamma_e > 0.6$ at the TAMS. Even with a relatively small initial rotation velocity, these stars may reach critical rotation during or shortly after the main-sequence phase. The lower panel shows the increase in mass-loss rate compared to that of nonrotating stars at the TAMS. The increase of the mass-loss rate depends on T_{eff} at the TAMS because the value of α depends on the ionization state in the radiation driven wind (Section 15.2.2). The figure shows that the mass-loss rate of rotating massive stars may increase by orders of magnitude when they approach critical velocity.

Stellar winds remove angular momentum from rotating stars. The net effect depends on the latitude dependence of the mass-loss rate. As discussed in previous paragraphs, the radiation driven mass-loss rate from the polar region is higher than from the equatorial regions for most hot stars. Its effect can be described by the loss of specific angular momentum, $j = v_{rot}(\Theta) \times r_{proj}$ per gram of gas, where r_{proj} is the distance from the rotation axis. The specific angular momentum at the polar region is much smaller than at the equatorial region because both $v_{rot}(\Theta)$ and r_{proj} are smallest at the poles. For a solidly rotating star, j at the polar region is smaller than the mean specific angular momentum of the star. High mass loss from the pole therefore results in a decrease of the stellar mass with little loss of angular

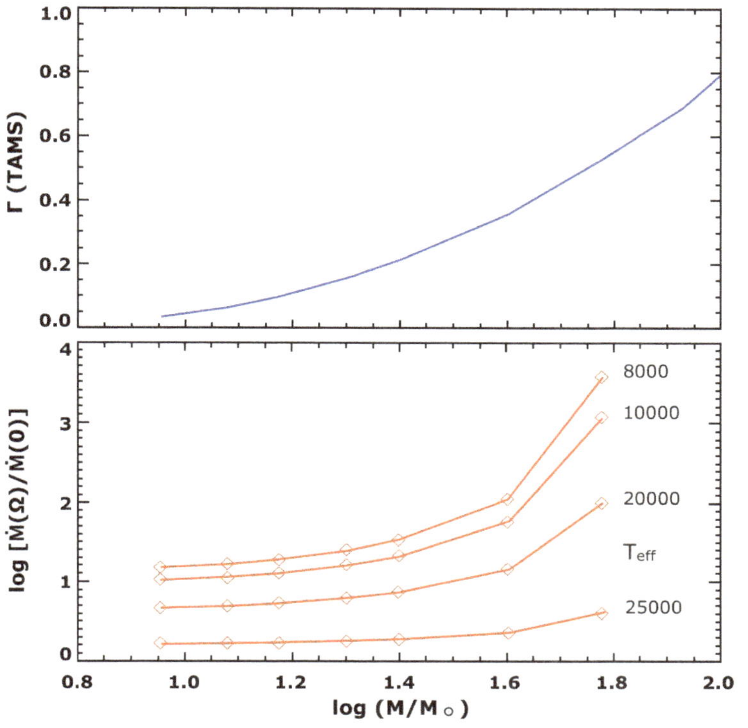

Figure 25.4. Top: the value of Γ_e at the TAMS of stars as a function of the initial stellar mass (blue). Bottom: the increase in mass-loss rate when the star reaches critical rotation at the TAMS for different values of $T_{\rm eff}$ at the TAMS (red). (Reproduced from Maeder 2009. © Springer-Verlag Berlin Heidelberg 2012.)

momentum, so the mean specific angular momentum of the star (i.e., $\bar{j} \equiv J/M$, where J is the total angular momentum of the star) increases during the evolution. This implies that the *rotation velocity will increase during the evolution.* On the other hand, a star with enhanced mass loss from the equatorial regions will lose so much specific angular momentum that $\bar{j}$ decreases: *the rotation rate will decrease during the evolution.* This shows that

- in rapidly rotating stars with enhanced polar mass loss, the ratio $v_{eq}/v_{\rm crit}$ will increase during the evolution, and
- in rapidly rotating stars with enhanced equatorial mass loss, the ratio $v_{eq}/v_{\rm crit}$ will decrease during the evolution. We have seen above that this happens only in a specific $T_{\rm eff}$ range, at about 25,000 K.

A rapidly rotating star with enhanced polar mass loss may reach a critical rotation rate (i.e., $\Omega = \Omega_{\rm crit}$) when it evolves, due to the combined effects of radiation pressure and rotation. This limit is called the **Omega–Gamma limit**, or **ΩΓ limit** (Langer 2012). Rapidly rotating massive stars may reach this limit before they reach the atmospheric Eddington limit.

Q (25.3): *How does the equatorial velocity of a massive star of about 20 $M_{\odot}$ with spherical mass loss vary as the star evolves in the HRD from the ZAMS to the red supergiant phase? Assume for simplicity that the star is rotating as a solid body.*

25.4 Mixing by Meridional Circulation

Rapid rotation induces meridional circulation due to shear forces between layers of different rotation speed. Meridional circulation is a flow pattern toward or away from a pole along meridional lines (i.e., lines of constant longitude). This effect is strongest during the main sequence phase when the rotation speed is still high. At later phases, a star spins down because it expands, or because it loses angular momentum through an equatorial wind (unless the wind is strongly enhanced in the polar direction). Figure 25.5 shows the circulation pattern in an MS star of $20M_{\odot}$.

Meridional circulation produces very efficient mixing. In nonrotating or slowly rotating massive stars, mixing occurs only in the core due to convection and convective overshooting, but these stars do not have convective envelopes that would bring nuclear material to the surface. However, meridional circulation in rapidly rotating stars can bring nuclear products from the top of the convective core all the way up to the surface during the MS phase.

Figure 25.6 shows the predicted surface N abundance of LMC stars in the range of 8 to $60M_{\odot}$ with different initial rotation velocities v sin (i), where v is the equatorial rotation velocity and i is the inclination of the rotation axis ($i = 0$ for pole-on stars). The N abundance is expressed in log(N/H) +12. The initial abundance is log(N/H) = 6.9. The predictions are from a population synthesis model by Brott et al. (2011a), based on stellar evolution calculations with an assumed distribution of

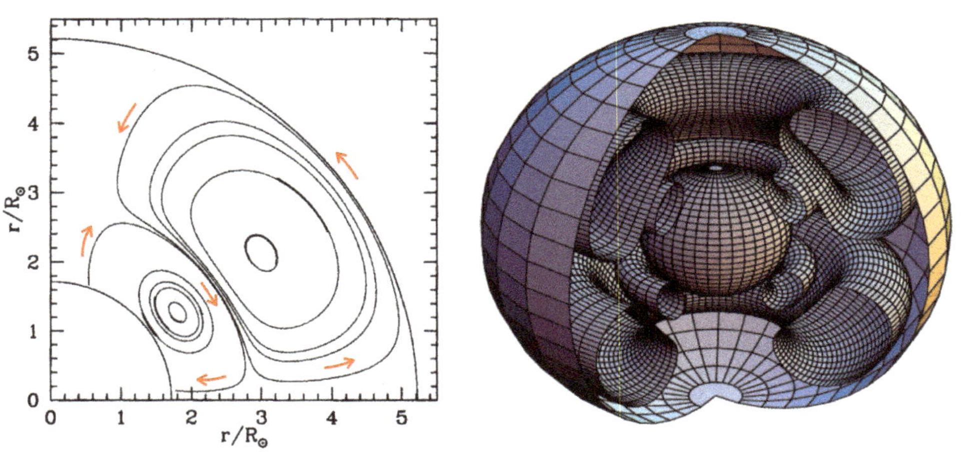

Figure 25.5. Two representations of the meridional circulation currents in a $20M_{\odot}$ MS star, halfway between the ZAMS and TAMS. The initial rotation velocity is 300 km/s. Left: the circulation pattern in 2D. Right: 3D demonstration of the same pattern. The inner sphere is the convective core. The inner circulation cell is rising toward the pole and descending along the equator. The outer circulation is rising along the equator and descending along the poles. (Reproduced from Meynet & Maeder 2002, with permission. © ESO.)

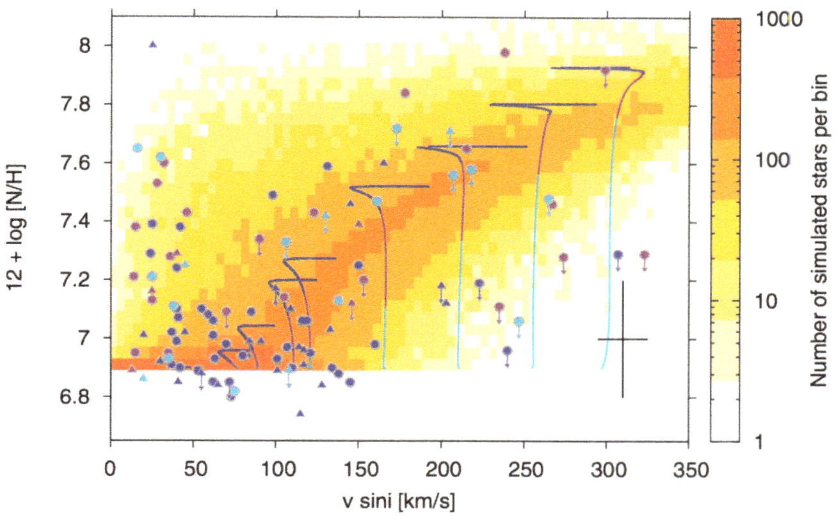

Figure 25.6. Predicted and observed surface N abundance, as a function of the projected rotational velocity *v* sin (*i*). The N abundance is expressed in log(N/H) +12. The colored background shows the predictions. The darker orange region shows the predicted trend for the majority of the stars, with the N abundance increasing with v sin (*i*). The lines show the calculated trends for evolutionary models of a $13M_{\odot}$ star with different initial values of v. The low point of the track is at $t = 0$. The dots are the observed values of stars with different surface gravities. The color coding of the observations and the tracks refer to the surface gravity: $\log g \gtrsim 4.1$ dex (light blue), $3.7 \lesssim \log g < 4.1$ dex (purple), and $3.2 \lesssim \log g < 3.7$ dex (dark blue). (Reproduced from Brott et al. 2011a, with permission. © ESO.)

masses, initial rotation velocities, and random orientation axes. The observations agree roughly with the predicted trend, although there are notable deviations. In particular, the large number of N-enhanced stars at low v sin (*i*) is surprising. Part of the discrepancy may be due to the presence of unresolved binaries in the observed sample.

Q (25.4): *Explain the shapes of the predicted rotation tracks in* Figure 25.6 *of* $13M_{\odot}$ *stars with different initial rotation velocities.*

25.5 The Effect of Rotation on the Evolution of Massive Stars

Rapid rotation affects the evolution of massive stars in two ways.

- Rotation induced mixing will result in a more chemically homogeneous structure than in a nonrotating star. We have seen in Sections 13.2 that a chemically homogeneous star evolves upward and to the left in the HRD during core H-fusion, whereas a non-rotating star moves to the right as the star becomes more chemically stratified.
- Rapid rotation reduces the effective gravity and increases the volume and the mean radius of a star, which results in a lower T_{eff} than nonrotating stars.

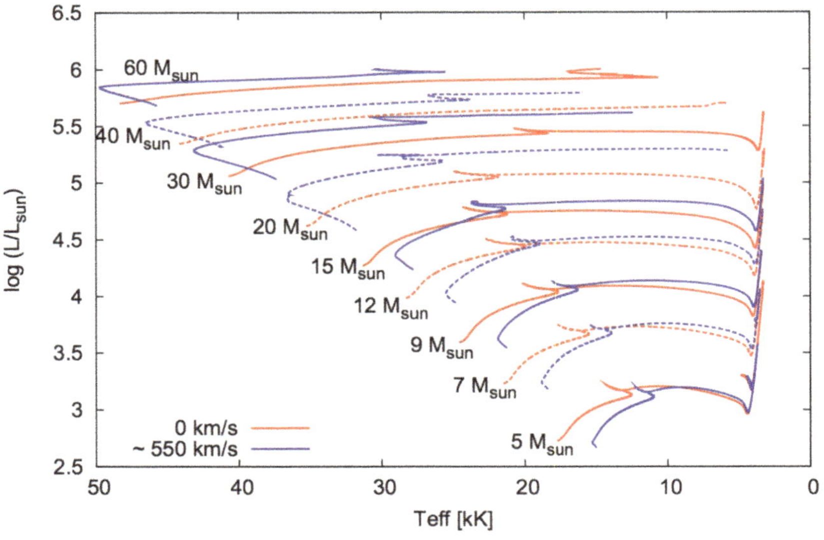

Figure 25.7. Evolutionary tracks of stars of $5 < M/M_{\odot} < 60$ for non-rotating stars (red) and fast-rotating stars with an initial equatorial rotation velocity of 550 km/s (blue).The horizontal axis is the surface-averaged value of $T_{\rm eff}$. For clarity, the tracks are shown in solid and dotted lines that alternate with initial mass. (Reproduced from Brott et al. 2011b, with permission. © ESO.)

- Rapid rotation has no direct effect on the nuclear energy production in the center of the star.

Figure 25.7 shows the predicted evolutionary tracks of stars with initial masses between 5 and $60M_{\odot}$ without rotation (red tracks) and with a high initial equatorial rotation velocity of 550 km s^{-1} (blue tracks). The models have galactic abundances.

Notice the following differences between rotating and nonrotating stars.

- Rapidly rotating stars on the ZAMS have the same luminosity as non-rotating stars, but their $T_{\rm eff}$ are lower. This is because rapidly rotating stars have larger surface areas than their non-rotating counterparts (see Figure 25.1).
- Nonrotating stars move to the right during the MS phase. This is a consequence of chemical stratification. On the other hand, the tracks of rapidly rotating stars first move upward to the left along the MS before bending to the right. This is because rotation induces mixing, which keeps the star chemically homogeneous for some time (see Section 13).
- Near the end of the core H-fusion phase, chemical stratification results in the tracks moving to the right.
- For stars with $5 < M_i/M_{\odot} < 15$, the post-MS evolution of stars with and without rotation is similar.
- For stars more massive than $15M_{\odot}$, the post-MS luminosity of the rapidly rotating star is higher, by about a factor of 2, than that of a nonrotating star. This is because the rotation-induced mixing resulted in a more massive He core.

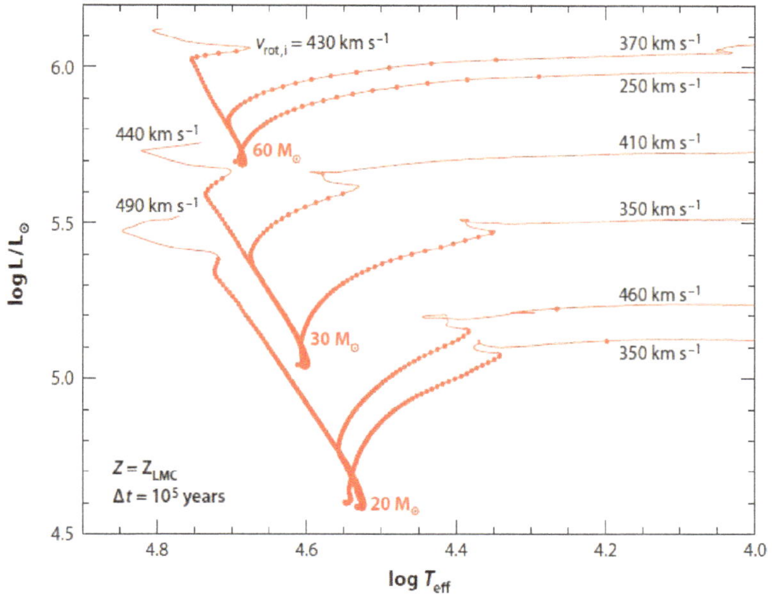

Figure 25.8. Evolutionary tracks of LMC stars with various rotation rates. For initial rotation rates above a critical limit, stars evolve quasi-homogeneously. (Reproduced from Langer 2012, with permission)

- Increased mass loss by rotation implies that tracks for the most massive stars do not reach as far to the right as they do for nonrotating stars.

25.6 Homogeneous Evolution

If the rotation velocity is above a certain limit, mixing can become so strong that the star evolves almost homogeneously, such that at the end of the core H-fusion the star is on the He MS. This limit depends on the metallicity for the following reason. Compared to galactic disk stars, stars of low metallicity have smaller mass-loss rates. This implies that they lose less angular momentum by stellar winds and maintain their fast rotation and mixing longer than Galactic stars. Figure 25.8 shows the evolution tracks of rotating stars of 20, 30, and $60M_\odot$ with LMC metallicity, Z (LMC) = $0.3Z_\odot$, and LMC mass-loss rates. For each of these masses, there is a rotation limit above which the star evolves homogeneously. This critical limit is ~480, ~430, and ~400 km s^{-1} for LMC stars of 20, 30, and $60M_\odot$, respectively.

25.7 Summary

1. A rapidly rotating star is not spherical but oblate, with a higher temperature at the poles than at the equator: the Von Zeipel effect.
2. The winds of rapidly rotating stars are not spherical. In most stars, radiation-driven mass loss is enhanced at the poles due to the higher radiative flux. In some temperature regimes, the bi-stability of radiation driven winds results in enhanced mass loss from the equator. This is the case if T_{eff} is about 25,000 K

at the poles and $< 21{,}000$ K at the equator. In all cases, the wind velocity is lower at the equator because it scales with the local escape speed.

3. Rotating massive stars lose angular momentum to their radiation driven winds. If the equatorial mass loss is higher than the polar mass loss, the specific angular momentum of the star decreases and the rotation rate slows down.
4. If the polar mass loss, with its low specific angular momentum, is higher than the equatorial mass loss, then the specific angular momentum of the star increases. (This might play a role in the formation of gamma-ray bursts, which require rapidly rotating cores at the end of their evolutions.)
5. Rapid rotation in a star produces meridional circulation, which can lead to severe mixing. Stars rotating above a critical initial rotation rate will evolve quasi-homogeneously during their core H-fusion phase. Their tracks move upward and to the left from the H-rich ZAMS to the ZAMS for He stars.
6. Rotation induced mixing may bring nuclear products to the surface during or shortly after the MS phase. The first element to show an enhanced abundance is N, produced by the CNO cycle.
7. Rapidly rotating stars of high luminosity may become unstable by the combination of the large radiation pressure (Γ-effect) and centrifugal force (Ω-effect). This results in an $\Omega\Gamma$-limit in the HRD, which is at lower luminosity than the Eddington limit for nonrotating stars. This may play a role in the instability of LBVs.

Exercises

25.1 The distribution of the mass-loss rates of rapidly rotating stars is shown in Figure 25.3.

(a) Explain the latitude distribution of the mass-loss rate as a function of latitude (θ).

(b) What is the latitude distribution of the terminal velocity?

(c) Explain that the density in a rotation-induced bi-stability disk can be much higher than that in the wind from the poles.

25.2 The angular momentum of a rotating shell with radius (r, $r + dr$), density $\rho(r)$, and angular velocity $\Omega(r)$ is $L = (8\pi/3)\Omega(r)\rho(r)r^4\, dr$. Derive an expression for the total angular momentum of a uniformly rotating star, with angular velocity Ω, that consists of a core with density ρ_c at $0 < r < r_c$, and an envelope with a mean density ρ_{env} at $r_c < r < R$.

25.3 At the end of the MS phase, a star of $M_i = 60M_\odot$ has a mass of $45M_\odot$ a radius of $R = 40R_\odot$ and a He core of $M_c = 30M_\odot$ with $R_c = 0.5R_\odot$.

By what factor will the equatorial rotational velocity decrease if the star expands during the shell-H fusion phase as a solid rotator, i.e., Ω independent of r, by a factor of 10 in radius.

Assume that the star originally rotated as a solid body with an angular velocity Ω.

Adopt mean densities for the core and the envelope.

25.4 Assume that a solidly rotating constant mean density star of $30M_\odot$ and a radius of $50R_\odot$ with an equatorial rotation speed of 200 km/s^{-1} suddenly loses 20% of its mass due to a strong stellar wind, and then readjust its size to its initial radius of $50R_\odot$.

Calculate the equatorial rotation speed after the mass has been lost in three cases.

(a) Suppose that the wind is only from the polar region.
(b) Suppose that the wind is only from the equatorial region.
(c) Suppose that the mass flux of the stellar wind is spherically symmetric. Assume that the radius did not change.

References

Brott, I., de Mink, S. E., & Cantiello, M. et al. 2011b, A&A, **530, 115**

Brott, I., Evans, C. J., & Hunter, I. et al. 2011a, A&A, **530, 116**

Ekstrom, S., Georgy, C. & Eggenberger, P. et al. 2012, A&A, **537, 146**

Lamers, H. J. G. L. M., & Cassinelli, J. P. 1999, Introduction to Stellar Winds (Cambridge: Cambridge Univ. Press),

Lamers, H. J. G. L. M. & Pauldrach, A. W. A. 1991, A&A, **244, 5L**

Lamers, H. J. G. L. M., Snow, T. P. & Lindholm, D. M. 1995, ApJ, **455, 269**

Langer, N. 2012, ARA&A, **50, 107**

Maeder, A. 2009, Physics, Formation and Evolution of Rotating Stars (Berlin: Springer),

Maeder, A. & Desjacques, V. 2001, A&A, **372 L9**

Pelupessy, I., Lamers, H. J. G. L. M. & Vink, J. S. 2000, A&A, **359 695**

Vink, J. S., de Koter, A., & Lamers, H. J. G. L. M. 2001, A&A, **369 574**

Henny J.G.L.M. Lamers and Emily M. Levesque

Chapter 26

Late Evolution Stages of Massive Stars

Massive stars go through all of the fusion phases discussed in Section 8.8, from H-fusion all the way to Si-fusion. Each fusion phase is shorter than the previous one because neutrino emission removes an increasingly large fraction of the energy at high temperatures. Thermal equilibrium of the star requires that this loss is compensated by increasing the reaction rate, thus reducing the duration of each phase. Each fusion phase requires a higher ignition temperature than the previous one. This implies that successive nuclear fusion phases in the core occur in smaller and smaller fractions of the mass of the star. A massive star therefore develops an onion skin model, with each layer having a different chemical composition. The most massive nuclei are in the center and the least massive nuclei are at the surface.

26.1 Late Fusion Phases

The late evolutionary phases of stars more massive than about $M_i \sim 8M_\odot$ proceed at an increasing speed. This is partly due to the fact that the mass defect ε_n of the fusion reactions decreases with increasing atomic mass, but at the latest phases of O-fusion and Si-fusion it is also due to the loss of energy by neutrinos. For a star to remain in equilibrium, its energy production has to be $L_{\rm nucl} = L_{\rm rad} + L_\nu$, where $L_{\rm rad}$ is the energy production required to maintain hydrostatic and thermal equilibrium and L_ν is the neutrino loss. The timescale of each phase is approximately

$$t = M_{\rm nucl}\varepsilon_{\rm n}c^2/(L_{\rm rad} + L_\nu), \tag{26.1}$$

where $M_{\rm nucl}$ is the mass of the nuclear material available for fusion and $\varepsilon_n = \Delta m/m$ is the mass defect of the reaction (Section 8.1). The ratio $L_{\rm rad}/L_\nu \ll 1$ at $T_c \gtrsim 10^9$ K during the C-, Ne-, O-, and Si-fusion phases, so the timescale is

$$t \approx M_{\rm nucl}\varepsilon_{\rm n}c^2/L_\nu. \tag{26.2}$$

doi:10.1088/978-0-7503-1278-3ch26

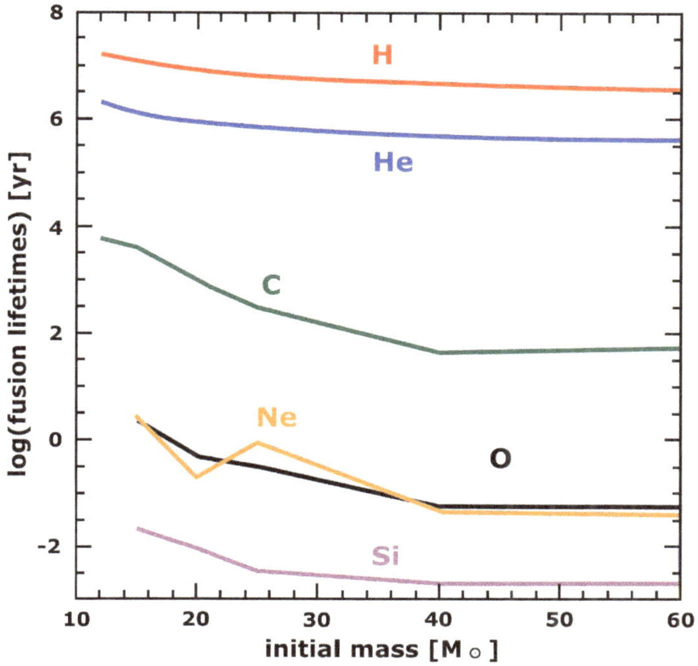

Figure 26.1. Lifetimes of the different fusion phases for nonrotating stars of solar metallicity, with an initial mass in the range of 10 to $60M_\odot$. (Reproduced from Hirschi et al. 2004, with permission. © ESO.)

The lifetime is shortened by a factor of $L_{\rm rad}/L_\nu$ compared to the situation without neutrino losses. This ratio is on the order of 10^{-1}, 10^{-3}, 10^{-4}, and 10^{-6} for the fusions of C, Ne, O, and Si, respectively (see Table 8.4).

Figure 26.1 shows the lifetimes of the different fusion phases for nonrotating stars of solar metallicity in the range of 12 to $60M_\odot$. Core He-fusion is about 10 times shorter than core H-fusion. Core C-fusion lasts about 10^{-4} to 10^{-5} times as long as the core H-fusion, i.e., about 10^2 to 10^3 yr. The O-fusion phase and the Ne photodisintegration phase both last about a month to a year. The last phase, core Si-fusion, lasts only about a day! Stars with initial masses less than $\sim 14M_\odot$ do not reach Ne-, O-, or Si-fusion.

Q (26.1): *Suppose that a star of $M_i = 50M_\odot$ is a Wolf–Rayet star with a radius of about $10R_\odot$ after the C-fusion phase in the core. Does the envelope of the star have time to react to the subsequent internal changes?*

26.2 The Internal Evolution

The core temperature of massive stars during their late fusion phases is set by the equilibrium between the nuclear energy generation rate per gram per second, ϵ_n

(Equation (8.1)), and the sum of the neutrino cooling, ϵ_ν, and the energy generation, $\epsilon_{\rm therm}$, needed to keep the star in hydrostatic and thermal equilibrium.

$$\epsilon_n = \epsilon_\nu + \epsilon_{\rm therm} \approx \epsilon_\nu. \tag{26.3}$$

This equilibrium is shown in Figure 26.2.

The evolution of the core of a star of $60M_\odot$ in the (ρ_c, T_c) diagram is shown in Figure 26.3. The figure shows the expected trend of $T_c \sim \rho_c^{1/3}$ that we derived in Section 8.10 (Equation (8.22)). During the core fusion phases, the star settles into an equilibrium that is determined by the values of ρ_c and T_c needed to produce the net energy, $L_{\rm nucl} - L_\nu$, that is required for the star to remain in hydrostatic and thermal equilibrium.

During the late phases, the conditions in the center of the core are close to the boundary between the equations of state for ideal gas and relativistic electron degeneracy (Figure 4.4).

We have argued before, in Section 17.1, that the ignition of fusion in a degenerate core would lead to an explosion. This is because, contrary to ignition in a nondegenerate gas, igniting a reaction in degenerate gas does not lead to an increase in pressure and expansion, but only in a runaway reaction rate. This is *not* the case for the ignition of nuclear fusion at very high temperatures of $T \gtrsim 10^9$ K because of high neutrino losses. Even a small increase above the ignition temperature results in increased cooling by neutrino losses, so very strong neutrino cooling prevents a runaway of the reaction rate.

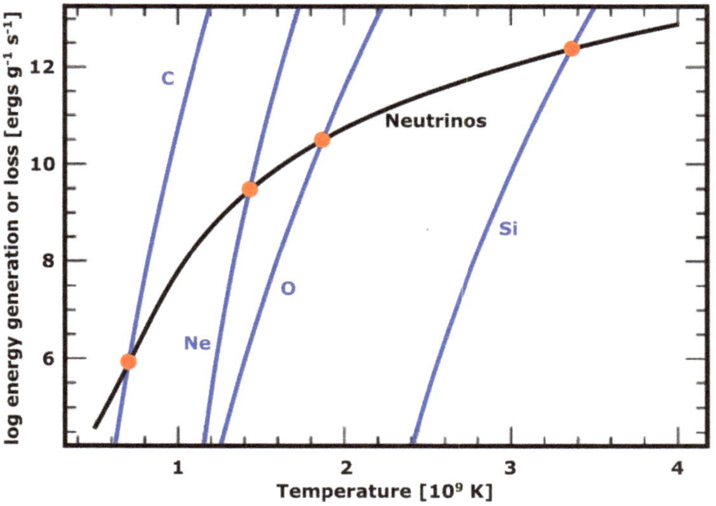

Figure 26.2. Energy generation by several fusion reactions (blue lines) and the energy loss by neutrinos (black line) as a function of temperature in the cores of massive stars. The density is assumed to vary as $\rho \cong 10^6 \times (T/10^9)^3$ g cm^{-3} (Equation (8.22)). Stable fusion, indicated by the red dots, occurs when $\epsilon_n \approx \epsilon_\nu$. (Reproduced from Woosley et al. 2002. Copyright 2002 by the American Physical Society.)

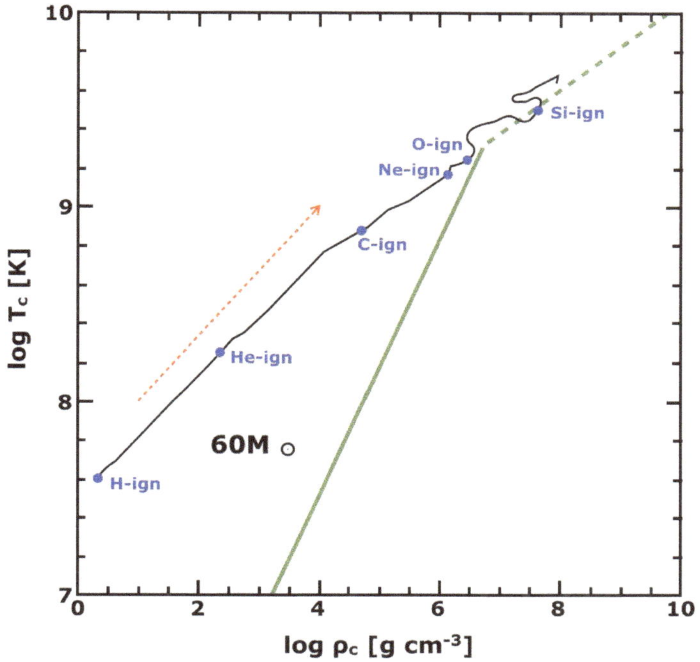

Figure 26.3. Central evolution of a nonrotating star of solar metallicity with an initial mass $60M_\odot$ in the $(\rho_c T_c)$ diagram. The dashed red arrow shows the predicted trend of $T_c \sim (\rho_c)^{1/3}$. The full green line is the boundary between nonrelativistic electron degenerate gas and ideal gas and the dashed green line is the limit between relativistic degenerate gas and ideal gas for $\mu_e = 2$. The conditions at the ignitions of several core fusion phases are indicated in blue. (Reproduced from Hirschi et al. 2004, with permission. © ESO.)

26.3 Pre-supernovae

Near the end of the life of a star, the subsequent fusion phases result in a chemical structure of nested shells. Just before a massive star ends as a supernova, it has an onion-skin chemical structure with successive layers of fusion products.

This is shown in Figure 26.4 for a star of solar composition with an initial mass of $60M_\odot$ and an initial rotational velocity of 300 km s^{-1}. After the Si-fusion phase it has a mass of $14.6M_\odot$. The star has lost $45.4M_\odot$, mainly in the form of H and He. The star consists of four zones, each with its own chemical composition. From outside in, these zones are the result of He-fusion, C-fusion, O-fusion, and Si-fusion.

Figure 26.5 shows the distribution in mass that has taken part in the different fusion phases, as a function of initial mass for stars with solar metallicity, a convective overshooting parameter of 0.1, and an initial rotation velocity of either 0 or 300 km s^{-1}.

Notice the following trends.

- At the end of the Si-fusion phase, rapidly rotating stars have lost considerably more mass than nonrotating stars (see Section 25.3).

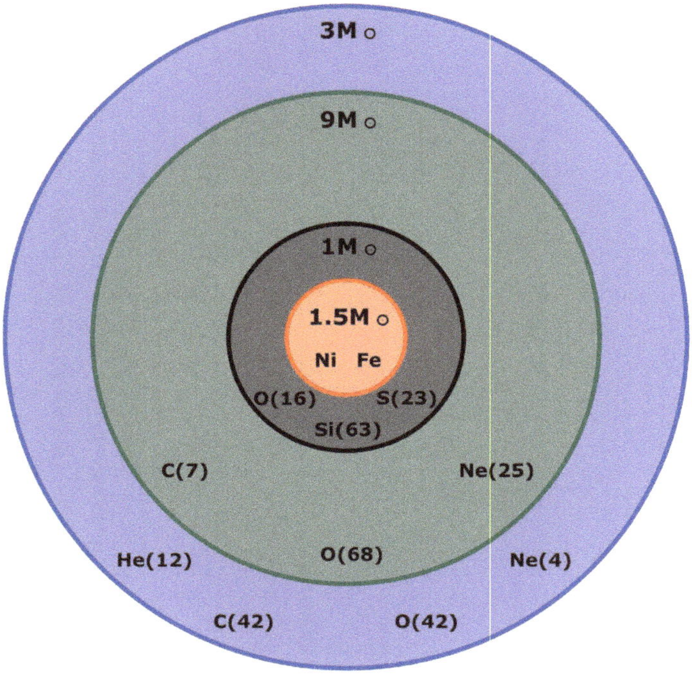

Figure 26.4. Schematic figure of the mass distribution of a rapidly rotating star of solar abundance with an initial mass of $60M_{\odot}$ just before the supernova explosion. The zones are the result of He-fusion, C-fusion, O-fusion, and Si-fusion. The mass of each zone and its chemical composition is indicated, with the values in parentheses indicating the mass percentage of each element. (Reproduced from Hirschi et al. 2004, with permission. © ESO.)

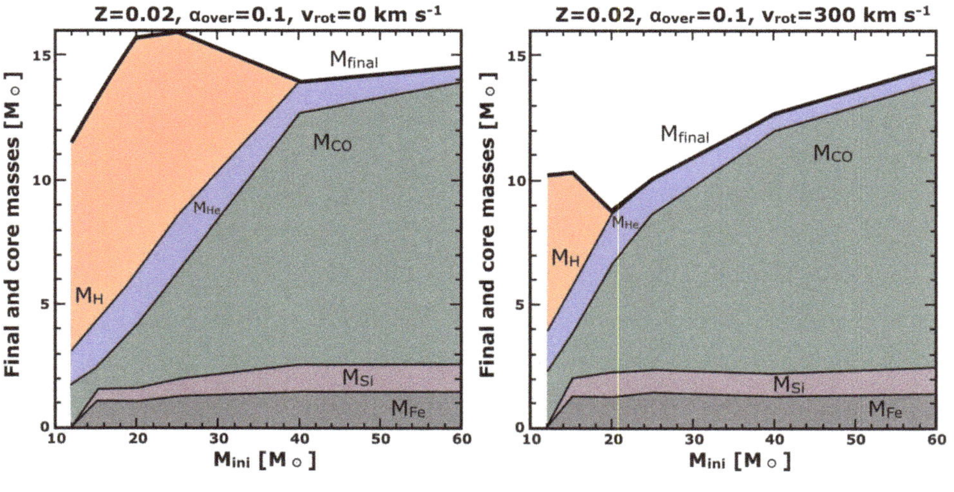

Figure 26.5. Mass of the various regions at the end of the Si-fusion phase as a function of initial stellar mass for stars of solar composition. Left: nonrotating stars. Right: stars with initial rotation velocities of 300 km s^{-1}. Thick line at top: final mass at the end of Si-fusion. Red: H-rich envelope; blue: He-rich layer; green: C-, O-, and Ne-rich layers; purple: Si- and S-rich layer; gray: Ni- and Fe-core. (Figure is from Hirschi et al. 2004.)

- Nonrotating stars with $M_i < 40M_\odot$ still have a substantial H envelope when they explode. They end their lives as red supergiants and produce H-rich SNe. For rapidly rotating stars, this limit is $20M_\odot$. The difference is due to the larger mass-loss rate of the rotating stars.
- All stars have about 1 or $2M_\odot$ of He at the end of their lives. Stars with $M_i > 40M_\odot$ have produced much more He, but a large fraction of it was lost during the WR phase (see Section 24.3).
- Rapidly rotating stars with $12 \lesssim M_\mathrm{i} \lesssim 40M_\odot$, experience rotation-induced mixing and have higher mass fractions of C-, O-, and Ne-fusion products than nonrotating stars. For rapidly rotating stars of $M_\mathrm{i} \gtrsim 40M_\odot$, the very high mass-loss rates experienced during the LBV and WR phases reduce the amount of mass of the final He shell and the C/O layer, as compared to nonrotating stars.
- Most of the mass at the end of the Si-fusion phase is in the form of C and O.
- Stars of $M_\mathrm{i} \lesssim 15M_\odot$ do not reach Ne-, O-, and Si-fusion.
- The Fe core and the Si shell surrounding it make up about $2M_\odot$ in all stars of $M_\mathrm{i} > 15M_\odot$.
- The masses of the Fe core and the Si shell are almost independent of rotation and mass loss. This is because the conditions in the core, i.e., ρ_c and T_c, are hardly affected by the masses of the outer layers.

26.4 Summary

1. Stars with initial masses in excess of $\sim 12M_\odot$ go through all nuclear fusion and photodisintegration phases. The duration of each next phase is shorter than the previous one, with the phases after C-fusion lasting less than about a year. This is due to strong neutrino losses.
2. Because each fusion phase requires a higher central temperature, and takes place in a smaller fraction of the stellar mass, the star develops a chemical onion-skin structure with the most massive element concentrated toward the center. The amount of mass and composition in each shell depends on the internal mixing of the star.
3. Massive stars have lost a substantial fraction of their mass by means of a stellar wind or by eruptions during the luminous blue variable phase. This mass-loss rate is strongly dependent on the rotation of the star. Stars with initial masses less than $40M_\odot$ still have a H envelope, but rapidly rotating stars have lost more mass. For stars with an initial rotation velocity of 300 km/s, this limit goes down to $20M_\odot$. Stars more massive than that have also lost part of their He envelope due to mass loss.

Exercises

26.1 Figure 26.4 shows the composition of a star with $M_i = 60M_\odot$ in the pre-SN phase. Each shell contains a mixture of chemical elements. Explain these mixtures.

26.2 Explain why most of the mass of stars with $M_i > 20 M_\odot$ is in the form of C and O.

26.3 Explain why the difference in structure of pre-SN stars between non-rotating and fast-rotating stars is much larger in the mass range of $M_i <$ 40 $M_\odot$ than for $M_i > 40 M_\odot$.

References

Hirschi, R., Maeder, A., & Meynet, G. 2004 A&A, **425 649**

Woosley, S. E., Heger, A., & Weaver, T. A. 2002, RvMP, **74, 1015**

AAS | IOP Astronomy

Understanding Stellar Evolution

Henny J.G.L.M. Lamers and Emily M. Levesque

Chapter 27

Supernovae

The sudden appearance of bright stars in the sky was attributed in historical times to the visit of a "guest star" or the birth of a new star, hence the name "nova." Later, the intrinsically brightest of these new stars were called "supernovae." Now we know that supernovae are not the beginning but the end of the life of a star.

Massive stars end their lives with a huge explosion. At the end of all possible fusion phases, discussed in the previous chapters, a star has a core consisting of Fe and finds itself without an energy source because Fe-fusion consumes energy instead of generating it. The balance between the inward force of gravity and the outward force from the gradient in gas pressure or radiation pressure is thus finally broken and gravity wins: the core of the star collapses. As the core collapses, the rest of the star explodes as a supernova. In this chapter, we discuss the process of core-collapse and the subsequent explosion. There are different types of supernovae and we discuss their possible origins in terms of stellar evolution. We also consider the possible remnants.

27.1 Light Curves of Supernovae

Supernovae are discovered by their sudden brightening. The light curves of two characteristic types of supernovae (SNe), Type Ia and Type II-P, are shown in Figure 27.1. Notice the steep rise in a few tens of days and the slow decline over about a year. The light curve of a Type II-P supernova has a plateau of several dozen days. Both types reach a peak luminosity of about 10^9 to 10^{10} $L_\odot$, with an absolute magnitude in the B-filter of $M_B \approx -17$ to -20. Compare this with the solar value of $M_B = +5.47$ for the Sun and realize that a star with $M_B = -20.0$ is 1.5 10^{10} times brighter than the Sun and almost as bright as the entire Milky Way at $M_B = -20.4$.

27.2 Core Collapse

Stars with an initial mass $M > 12M_\odot$ go all the way through Si-fusion and develop an Fe core. Because Fe is the last element that creates energy during its formation by

doi:10.1088/978-0-7503-1278-3ch27

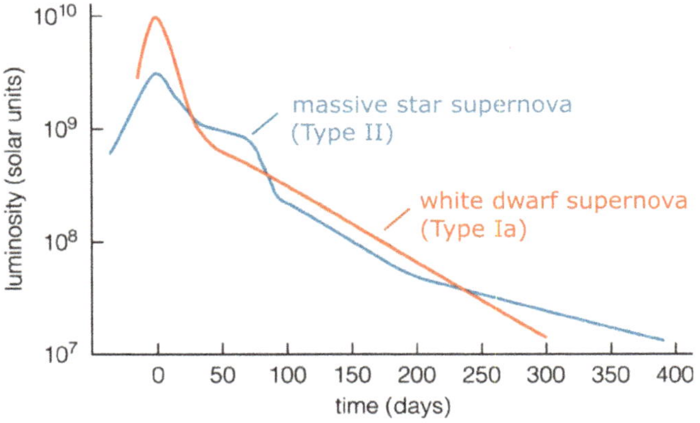

Figure 27.1. Characteristic light curves of two types of SNe: a thermonuclear SN of Type Ia (red) and core collapses of Type II-P (blue) that are discussed in Section 27.3. (Adapted from Weiler & Sramek. Reproduced with permission. © ESO.)

fusion, the core runs out of nuclear energy. At that time, the core has a temperature of $T \sim 4 \times 10^9$ K and a density of $\rho > 10^8$ g cm^{-3} (see Figure 26.3). The core then contracts because it has no energy source and its temperature rises. Soon the core temperature becomes so high, $T > 10^{10}$ K, that the photons are energetic enough to break up heavy nuclei via the reaction

$$^{56}\text{Fe} + \gamma \rightarrow 13\,^{4}\text{He} + 4n. \tag{27.1}$$

Since this is an endothermic reaction, it consumes rather than produces energy. This results in a quick cooling of the core, so the collapse accelerates. Because of the very high density and the resulting high gravity of the core, the collapse occurs on a very short timescale of mere milliseconds.

As the core collapses and the density increases without a significant T increase because of the endothermic nuclear reactions, electrons are captured by protons and form a degenerate neutron gas. Degenerate neutron stars can support a very high pressure as long as the degeneracy is not relativistic. This is similar to the structure of electron degenerate white dwarfs. If the collapsing core has a mass lower than the Oppenheimer–Volkoff limit (OV limit) for neutron stars, which is about $2M_\odot$ (Section 20.5), the pressure of the nonrelativistic degenerate neutron gas can stop the collapse and a neutron star is formed in the center of the collapsing star. If the collapsing core has a mass in excess of about $2M_\odot$, the pressure of the neutron gas cannot stop the collapse, so it continues collapsing until a black hole is formed in the center of the star.

Stars of $8M_\odot \lesssim M_i \lesssim 12M_\odot$ do not reach Si-fusion, so they do not produce an Fe core. They still go into core-collapse at the end of their lives, when at high densities the electrons are captured by heavy nuclei. This reduces the pressure produced by electrons and also results in core-collapse.

27.3 The Core-collapse Supernova Explosion

As the core collapses, the envelope is ejected in the supernova explosion. There are three major effects that are responsible for the ejection of the outer layers of the star.

- **Bouncing shock** at the surface of the neutron star. The matter that falls onto the very compact neutron star experiences a shock. The bounce of this shock is so strong and so energetic that a shock wave runs outward against the infalling material. This results in the ejection of the envelope.
- **Neutrino pressure**. During the collapse, the temperature is so high that photons are energetic enough to create neutrinos via two processes:
  Photoneutrino production $\gamma + e^- \to e^- + \nu_e + \bar{\nu}_e$,
  Pair annihilation $\gamma + \gamma \to \nu_e + \bar{\nu}_e$.
  In the layers just above the neutron star, the density is so high ($\rho \sim 10^{11}$ g cm^{-3}) that the neutrinos can be captured by the infalling gas. This is the case in the layers where the optical depth for neutrinos is $\tau_\nu > 1$. The layer where $\tau_\nu \sim 1$ is called the neutrino photosphere. The capture of the neutrinos by neutrons and heavier particles near the neutrino photosphere suddenly heats up the infalling layers so strongly that the infall stops and is converted into an explosion.
- **Fusion in the infalling shells.** Near the end of its life, the star consists of nested shells (see Figure 26.4), some of which are still fusing. When the core collapses and drags these layers down, the efficiency of the fusion increases dramatically due to the increase in T and ρ. This suddenly creates a large amount of nuclear energy that heats the infalling layers and produces so much gas pressure that the envelope explodes.

These mechanisms work together to eject the layers outside the neutron star. If the neutron star captures more mass than about $2M_\odot$, the collapse continues and a black hole is formed.

27.4 Energetics of Core-collapse Supernovae of Massive Stars

The **potential energy released during the core-collapse**, when its radius decreases from its initial core radius $R_{\rm ci}$ to its final core radius $R_{\rm cf}$, is

$$E_{\rm collapse} \simeq -\frac{GM_c^2}{R_{\rm ci}} + \frac{GM_c^2}{R_{\rm cf}} \simeq \frac{GM_c^2}{R_{\rm cf}} \approx 3 \times 10^{53} \text{ erg}. \quad (27.2)$$

In this estimate, we have assumed that the core with$M_{\rm c} \sim 1.4M_\odot$ (the Chandrasekhar limit) collapses into a neutron star with $R_{\rm cf} \sim 20$ km.

The **potential energy necessary to expel the envelope** with mass $M_{\rm env} = M - M_{\rm c}$, where M is the total mass of the star when the core collapses, is

$$E_{\rm env}^{\rm pot} = \int_{M_c}^{M_*} \frac{Gm}{r} dm \ll \int_{M_c}^{M_*} \frac{Gm}{R_{\rm ci}} dm \simeq \frac{GM^2}{R_{\rm ci}} \approx 3 \times 10^{52} \text{ erg}. \quad (27.3)$$

We have used $M > M_c$ and $M_{env} \sim 10M_\odot$ and $R_{env} >> R_{ci}$. For the initial core radius, we adopted the radius of a WD of $R_{wd} \sim 10^4$ km. This is a severe overestimate because in the middle step we used a radius R_{ci} for the full envelope. A more realistic model, with a mean radius of the envelope of $R_{env} \approx R_\odot$, gives $E_{env}^{pot} \sim 3 \times 10^{50}$ erg.

The envelopes of supernovae are ejected at a velocity on the order of 10^4 km s^{-1}, so the **kinetic energy of the envelope** is

$$E_{env}^{kin} = \frac{1}{2} M_{env} V^2 \approx 6 \times 10^{51} \text{ erg} \quad \text{if } M_{env} \approx 6\, M_\odot \text{ and } V \approx 10^4 \text{ km s}^{-1}. \quad (27.4)$$

The peak luminosity of a Type II SN is typically $\sim 10^9 L_\odot$ and lasts about 60 days, so the **radiative energy of the SN** is about $E_{rad} \approx 10^{49}$ erg. This shows that

$$\boxed{E_{env}^{pot} + E_{env}^{kin} + E_{rad} \ll E_{collapse}}. \quad (27.5)$$

We conclude that only a small fraction of the energy released in the core-collapse is used for ejecting the envelope and emitting light. *Most of the energy comes out in the form of neutrinos!*

Q (27.1) *Could you have guessed that most of the energy is carried away by neutrinos?*

27.5 Observed Types of Supernovae

Originally, supernovae were classified into two types, depending on their emission line spectrum after the explosion.

- **Type I**: no H lines in the spectrum
- **Type II**: H lines in the spectrum

More recent research has resulted in further subdivisions related to the spectral appearance, the evolution of the light curve, and the origin of the supernova (the awkward naming of these types is a reminder that one should never use Roman numbers to indicate types of objects or phenomena because Roman numbers have no decimals).

Here are the current main subdivisions of supernova types:

- **SN Ia: no H lines, Si II lines.**
  These result from the collapse of a white dwarf in a binary system when mass transfer has pushed the mass over the Chandrasekhar limit. The collapse ignites C-fusion in a degenerate core, resulting in a thermonuclear explosion (Section 18.8). The WD explodes and leaves no remnant. Because Type Ia's are all thought to arise from a WD exceeding the Chandrasekhar limit, they form a rather homogeneous group with about the same peak luminosity. They can be used as standard candles for distance determination up to large distances.
- **SN Ib: no H lines, no Si II lines, strong He I lines.**
  The fact that these are found only in star-forming galaxies indicates that they are collapsing massive stars. The lack of H lines and the presence of strong He

lines shows that they are likely explosions of WR stars that have lost their H envelope.

- **SN Ic: no H lines, no Si II lines, absent or weak He lines.**
  This suggests that they are formed from Wolf–Rayet (WR) stars that have lost all their H and even most of their He envelope.
- **SN II: strong H lines.**
  These are the endpoints of massive stars that exploded due to the core collapse of red supergiants. Their initial masses must have been in the range of $8 \lesssim M_i \lesssim 30 M_\odot$.
- **SN II-P: strong H lines and a long brightness plateau after maximum.**
  These result from the core-collapse of red supergiants that still have most of their H envelope. The presence of the thick envelope is the reason for the luminosity plateau (hence "P"), and their initial mass range is therefore estimated to be $8 \lesssim M_i \lesssim 15 M_\odot$. (More massive stars would have lost a larger fraction of their H-rich envelope through stellar winds; this mass range is also supported by observations of Type II-P progenitors in pre-explosion imaging.)
- **SN II-L: H lines and a steep and linear luminosity decline after maximum.**
  These may result from the core-collapse of red supergiants that have lost more H than Type II-P progenitors; their light curves decline linearly with time (hence "L").
- **SN IIb: H lines in early spectra, with later spectra showing He lines and no H.**
  These are thought to result from the core-collapse of red supergiants that have little H left in their outer envelope.
- **SN IIn: H lines, some spectral lines are atypically narrow.**
  The narrow lines are possibly due to interaction by the supernova ejecta with circumstellar material that was produced by strong mass loss shortly before the core collapsed.

A small subset of stripped-envelope supernovae have been observed in conjunction with long duration (>2 seconds) **gamma-ray bursts (GRBs)**. GRBs are extremely energetic SN explosions that produce a very short burst of highly energetic γ-rays, lasting from milliseconds to tens of seconds, followed by an x-ray and optical afterglow lasting days. While the physical mechanism driving this brief flash of high-energy emission is still unclear, it is thought to be a consequence of accretion onto a newly formed rotating black hole produced during the core-collapse of a massive star with a rapidly rotating core.

It should be pointed out that the initial mass ranges that are quoted above depend strongly on the adopted mass-loss rate and core mass of stars. These are uncertain because they may be strongly influenced by rotation (as we have seen in Section 26) and metallicity.

Figure 27.2 shows a schematic division of the supernova types, with an indication of their connection to stellar evolution.

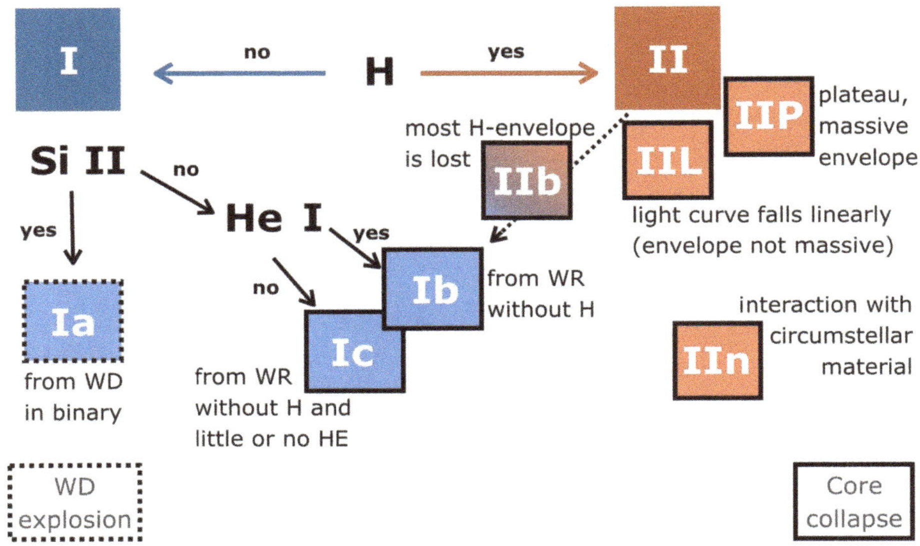

Figure 27.2. Schematic picture of the different types of SNe and their progenitors. Full boxes indicate the core-collapse SNe of massive stars. The dashed box indicates the explosion of a WD in a binary system when it exceeds the Chandrasekhar mass limit. The main characteristic of each type is described. (Reproduced from Turrato 2003. © Springer-Verlag Berlin Heidelberg.)

27.6 The Case of Supernova 1987A

On 1987 February 23, a supernova was detected in the Large Magellanic Cloud (LMC). The blue supergiant Sk −69°202 of spectral type B3 I that was originally of $V = 13.5$ magnitude suddenly increased in visual brightness by a factor of 4×10^3, reaching $V = 4.5$ within a day. This was the first time that the progenitor of an SN was observed and it enabled the first study of a supernova explosion of a star with known properties. The supernova also emitted a large flux of neutrinos that were measured by three neutrino detectors. These detections made it possible to determine the time of explosion accurately and to connect the early photometric and spectroscopic observations to the moment of core-collapse.

We have argued before that massive stars undergo core-collapse when they are either red supergiants with a substantial H-rich convective envelope (e.g., Section 23.2) or as a blue Wolf–Rayet star without a H envelope (e.g., Section 24.4). Surprisingly, the progenitior of SN 1987A is in-between these two possiblities. As a B3 supergiant with $T_{\rm eff} \sim 20{,}000$ K, it was too warm to be a red supergiant and too cool to be a Wolf–Rayet star. Moreover, the spectrum was clearly not that of an RSG nor that of a WR star, but of a normal B3 supergiant with a H envelope. This showed that the progenitor was a massive star and that the explosion was due to core-collapse. The SN spectra showed H lines, defining it as a type II supernova.

The fast initial rise in brightness within a day was due to the shock from below breaking out through the surface of the H-rich envelope. As the shock temperature

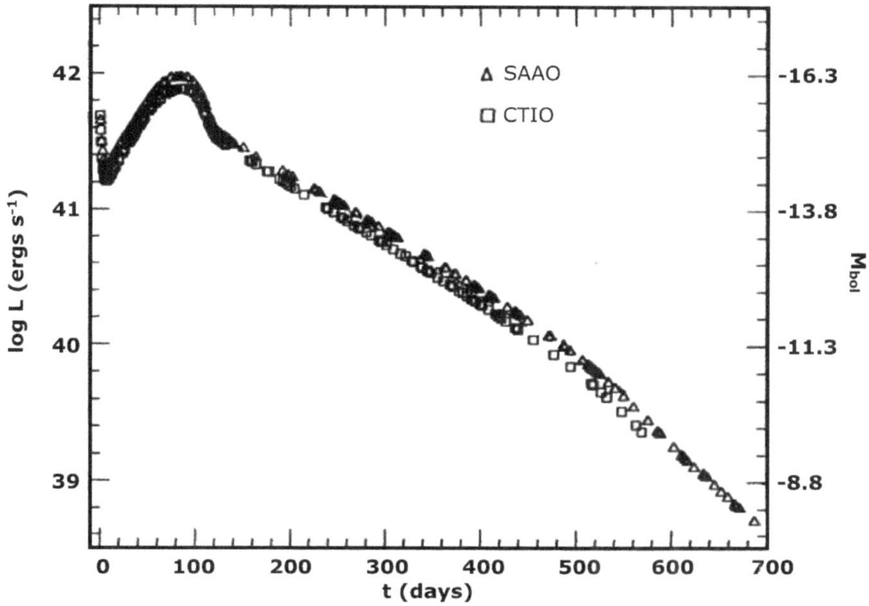

Figure 27.3. Bolometric light curve of SN 1987A expressed in log L, based on observations with the Cerro Tololo Inter-American Observatory (CTIO) and the South African Astronomical Observatory (SAAO). The plateau around day 80 is caused by heating of the expanding envelope by radioactive decay of ^{56}Ni. The gradual decrease later on is due to radioactive decay of ^{56}Co. (Figure adapted from Arnett et al. 1989.)

was high, 10^5 to 10^6 K, a flash of UV radiation was produced that ionized the circumstellar material that was later observed as a three-ring structure. Early spectra, taken within a day of the discovery, revealed strong H lines with velocities exceeding 30,000 km s^{-1}. For summaries, see, e.g., Arnett et al. (1989), Podsialowski (1992), and Woosley et al. (1997).

Figure 27.3 shows the subsequent bolometric light curve of SN1987A during about two years. After the initial rise within a day, the flux drops in about 10 days due to rapid adiabatic cooling. After about 20 days, the luminosity rises again. This is due to heating of the gas by radioactive decay of ^{56}Ni formed in the layers surrounding the core-collapse. The half lifetime of ^{56}Ni is only $\tau_{1/2} = 6$ days, so after about day 80 the luminosity starts to decrease due to the diminishing radioactive decay. After day 100, the luminosity is provided by the radioactive decay of ^{56}Co, which has a halflife time of $\tau_{1/2} = 77$ days. This explains the logarithmic decrease in luminosity by $\Delta \log L \approx -1.6$ between day 140 to day 500.

Studies of the pre-SN spectra of Sk $-69°202$ suggest that it was a star with an initial mass of $16 \lesssim M_i \lesssim 22 M_\odot$ with a He core of about 5 to $7 M_\odot$. The spectrum after outburst shows that the envelope contained about $10 M_\odot$ of H. The ejecta, with a velocity of $\gtrsim$30,000 km s^{-1}, has an unexpected large He/H ratio of 0.2 by number, i.e., $Y = 0.4$. The spectra also show evidence of enrichment by s-process elements such as barium. Additional evidence of its pre-SN evolution comes from the study of

Figure 27.4. *Hubble Space Telescope* image of the three-ringed structure around SN 1987A. (Figure courtesy of ESA.)

the UV spectra of the circumstellar material, which indicates that the star had previously ejected a low-velocity N-rich wind.

Figure 27.4 shows the three-ringed structure around SN 1987A. The inner and outer rings have semi-major axes of 0.21 pc and 0.45 pc, respectively. The rings are due to the light echo of the circumstellar material that was illuminated by the UV flash at the time of the SN explosion. The rings show the structure of material lost by the star in its previous evolutionary phases as red and blue supergiants. The nonspherical structure indicates that the star had a strong aspherical wind, probably due to fast rotation, in the red supergiant phase.

Various sets of evolutionary models have been used to try to explain the pre-SN evolution of Sk $-69°202$. None of these are completely successful unless some ad hoc assumptions are made. For reviews, see, e.g., Podsialowski (1992) and Woosley et al. (1997). The most successful models are those in which a fast-rotating star has gone through an RSG phase, lost a substantial amount of mass, and contracted to a smaller radius (i.e., to the higher $T_{\rm eff}$ of a blue supergiant). The assumption of fast rotation is attractive for several reasons: (a) it brings CNO products to the surface at an early phase, which would explain the N-rich circumstellar material, (b) rotation-induced mixing helps explain the high He/H ratio of the ejecta, and (c) it explains the system of rings that are observed since the late phase around SN 1987A and may have been formed by aspheric mass loss during the RSG phase. However, explaining

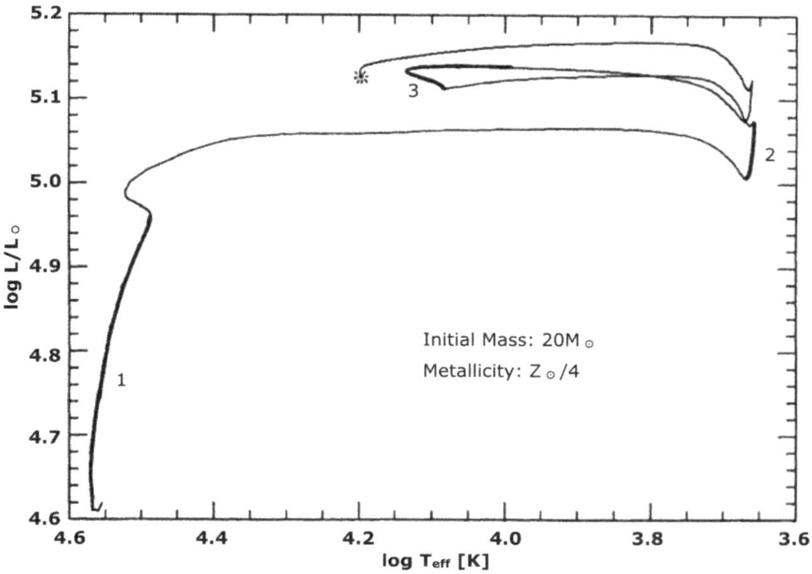

Figure 27.5. Evolutionary track of a $20M_\odot$ star of LMC metallicity that explodes as an SN when it is a blue supergiant. Thin lines indicate fast evolution phases. Thick lines indicate slow evolution phases: 1 = MS phase, core H-fusion (9 Myr); 2 = RSG phase, ignition, and first part of core He-fusion (0.15 Myr); 3 = BSG phase, remaining part of core He-fusion (0.4 Myr). The final blueward loop is due to the contraction of the C-core at the end of the core H-fusion. The star indicates the location in the HRD when core-collapse occurs. (Reproduced from Langer 1991, with permission. © ESO.).

the full set of properties depends crucially on the adopted mass-loss rates in the various phases and on the amount and depth of the mixing.

Figure 27.5 shows a possible evolutionary track of Sk −69°202 before core collapse. The star had an initial mass of $20M_\odot$ and an LMC metallicity of $Z = 0.25\ Z_\odot$. Mass loss, semi-convection, and mixing due to differential rotation were included in the calculation of its evolutionary history. The star evolved from an MS star into an RSG. Core He-fusion was ignited in the RSG. During core He-fusion, the star made a blue loop, as described in Section 23.1. At the end of the core He-fusion phase, the core contracts. At this time, the star had a He- and a H-fusion shell, which both acted as a "mirror." So, as the core contracted, the intershell zone expanded and the envelope contracted. This explains the final blue loop, that started 3×10^4 yr before the explosion. During the subsequent C-fusion and later phases the star was a B-type supergiant until it exploded.

An alternative scenario for the evolution of SN 1987A involves the merging of Sk −69°202 during the RSG phase with a lower-mass companion via a common envelope phase. This scenario has the advantage of explaining the unusual abundance of the circumstellar nebula: He/H = 0.15 by number, corresponding to Y = 0.37, and the observed enhanced N/O ratio. It also explains the rapid rotation of the RSG by the accreted angular momentum of the merged star. During the common envelope phase, the star experiences high aspherical mass loss, which is needed to explain the three-ringed structure of the circumstellar nebula. Dynamical models show that the circumstellar nebula was ejected about 20,000 yr before the explosion (Podsiadlowski 1992 and Menon & Heger 2017).

27.7 The Remnants of Stellar Evolution

We have seen above that stars with different initial masses may end their lives differently. Figure 27.6 shows the final fate of stars with an initial composition of Z = 0.02 as a function ofM_i in a schematic way.

- Stars with initial masses less than about $8M_\odot$ have lost a large fraction of their mass in the AGB wind and end their lives as white dwarfs.
- Stars with $8 \lesssim M_i \lesssim 22M_\odot$ have lost a small fraction of their mass in the RSG wind. They eject a substantial fraction of their mass in the SN explosion and leave a neutron star behind (the apparent sudden disappearance of mass loss by winds in the RSG phase at $M_i = 8M_\odot$ in Figure 27.3 is artificial and a result of the discontinuity in the adopted mass-loss formulae for low- and high-mass stars).
- Stars with $22 \lesssim M_i \lesssim 100M_\odot$ have lost a substantial fraction of their mass in winds during their O star phase, as an LBV, and as a WR star. They eject a small fraction of their mass in the SN explosion and leave a black hole behind.

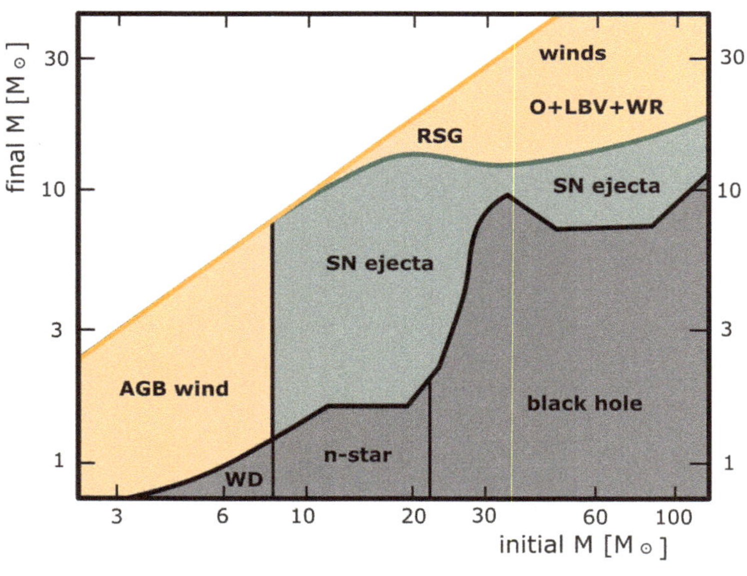

Figure 27.6. Fate of nonrotating stars of solar metallicity as a function of their initial mass. The orange regions indicate mass lost by stellar winds. The green region indicates the mass ejected in the SN explosion. The gray region indicates the mass and type of the remnant. (Reproduced from Podsiadlowski 1992.)

The fate of stars with $M_i > 20M_\odot$ is uncertain. The final fate of stars depends not only on their initial mass but also on metallicity, mass loss, and rotation velocity. For instance, we have shown in Figure 26.5 that the pre-supernova mass of rapidly rotating stars is smaller than that of nonrotating stars. This implies that the amount of mass ejected during the SN explosion is strongly reduced by the fast rotation.

27.8 Summary

1. Stars with an initial mass of $M_i > 8M_\odot$ end their lives with an SN explosion. These SNe are the result of core-collapse when a star has no nuclear fusion source left.
2. If the collapsing core has a mass less than about $2M_\odot$, the collapse stops when a neutron star is formed. If M_c is larger than this limit, the collapse passes through the degenerate neutron phase and the collapse continues to a black hole.
3. The explosion of the outer layers is the combined result of three effects:
   - the bounce of a shock on the hard surface of the neutron star,
   - the pressure due to the capture of neutrinos in the neutrino photosphere, and
   - the sudden increase in the reaction rates of the infalling fusion shells.
4. Most of the potential energy gained in the core-collapse is carried off by neutrinos.
5. Different types of SNe, distinguished by their spectra and light curves, can be related to the nature of their collapsing progenitor stars.
   - Type II supernovae with H lines in their spectra are thought to be collapsing red supergiants or LBVs
   - Type Ib and Ic supernovae without H lines in their spectra are thought to be collapsing Wolf–Rayet stars.
   - Type Ia supernovae are thermonuclear explosions in the degenerate core during the collapse of a WD when its mass goes over the Chandrasekhar limit. This could be caused either by mass transfer from a binary companion, or by merging of two WDs.
6. Supernova 1987A was a core-collapse type II SN that exploded when it was neither a Wolf–Rayet star nor a red supergiant, but a blue B3 supergiant with unusual abundances. The calculated evolution of a single pre-SN star involved mass loss, semi-convection, rotation induced mixing and aspherical mass loss with fine-tuned parameters. Alternatively, a model consisting of a binary of 16 and $7M_\odot$ that merged during the red supergiant phase of the primary is more successful in explaining most of the observed characteristics.

Exercises

27.1 Stellar evolution calculations suggest that solar metallicity stars with $M_i \lesssim 22M_\odot$ will become neutron stars and stars above this limit will end their lives as black holes. In which direction would this limit change if all massive stars were rapidly rotating?

27.2 Massive stars end their lives in a core-collapse SN explosion of type I or II. Explain how the number ratio between types I and types II depends on stellar rotation and metallicity.

27.3 Type Ia SNe are assumed to be the result of mass accretion onto a WD in a binary system when the mass of the WD exceeds the Chandrasekhar limit. There is an alternative suggestion that Ia's may be the result of the merging of two WDs. Can you think of ways to discriminate between these scenarios after the SN has exploded?

References

Arnett, W. A., Bahcall, J. N., Kirshner, R. P., & Woosley, S. E. 1989, ARA&A, **27, 629**

Langer, N. 1991, A&A, 243, 155

Maeder, A. 2009, Physics, Formation and Evolution of Rotating Stars (Berlin: Springer)

Menon, A. & Heger, A. 2017, arXiv: 1703.04918

Podsiadlowski, P. 1992, PASP, **104, 717**

Turrato, M. 2003, LNP, **598, 21**

Weiler, K. W. & Sramek, R. A. 1988, ARA&A, **26, 295**

Woosley, S. E., Heger, A., Weaver, T. A., & Langer, N. 1997, arXiv: astro-ph/9705146

IOP Astronomy

Understanding Stellar Evolution

Henny J.G.L.M. Lamers and Emily M. Levesque

Chapter 28

Principles of Close Binary Evolution

A considerable fraction of all stars, possibly as many as half, are born in a binary system. For stars with $M \gtrsim 30\ M_\odot$, the fraction might be even higher. The evolution of a star in a binary system will be affected by the presence of the companion if the two stars are within a distance of roughly a few stellar radii from each other. This might happen early in the evolution if the stars are born in a tight system, or later on when one or both stars expand. This interaction may lead to mass transfer from a star to its companion. Mass transfer not only changes the evolution of the two stars but also leads to changes in the orbits of the two stars. This chapter deals with the principal effects of binary evolution: periods and angular momentum, stable or run-away mass transfer, orbital changes, and contact phases. In the next chapter, these principles will be applied to understand the formation of various types of observed interacting binaries.

28.1 Periods and Angular Momentum

The **period of a binary** in a circular orbit is described by Kepler's third law

$$\left(\frac{2\pi}{P}\right)^2 = \omega^2 = \frac{G(M_1 + M_2)}{a^3}, \tag{28.1}$$

where P is the orbital period, ω is the angular velocity, M_1 and M_2 are the masses of the two components and a is the separation between their centers of gravity.

The **angular momentum** of the system around the center of gravity is

$$J = M_1 a_1 v_1 + M_2 a_2 v_2, \tag{28.2}$$

where v_1 and v_2 are the orbital velocities of the two stars around the center of gravity and a_1 and a_2 are the distances from the center of gravity with $a_1 + a_2 = a$. The center of gravity is defined by

doi:10.1088/978-0-7503-1278-3ch28

$$M_1 a_1 = M_2 a_2 = \frac{M_1 M_2}{(M_1 + M_2)} a. \tag{28.3}$$

The velocities of the system are

$$v = v_1 + v_2 = \omega a_1 + \omega a_2 = \omega a. \tag{28.4}$$

Substitution of Equations (28.1), (28.3), and (28.4) into (28.2) results in an expression for the angular momentum of a system in circular orbit

$$J^2 = Ga \frac{M_1^2 M_2^2}{(M_1 + M_2)}. \tag{28.5}$$

We have ignored the rotation of the stars and the possible ellipticity of the orbit.

28.2 Equipotential Surfaces of Binaries

Figure 28.1 shows the definition of the parameters and the geometry of a binary system in circular orbits in the equatorial plane that co-rotates with the orbit.

A particle located at r in the co-rotating system has an effective potential energy of

$$\Phi = -\frac{GM_1}{r_1} - \frac{GM_2}{r_2} - \Phi_c, \tag{28.6}$$

where $-\Phi_c = \frac{1}{2}\omega^2 r^2$ is the potential of the centrifugal acceleration at r, with ω given by Equation (28.1). This results in

$$\frac{\Phi}{G} = -\left(\frac{M_1}{r_1} + \frac{M_2}{r_2}\right) - \frac{1}{2}\frac{M_1 + M_2}{a^3} r^2. \tag{28.7}$$

An equipotential surface is the locus of points that have the same value of Φ. A particle can move along an equipotential surface without loss or gain of energy.

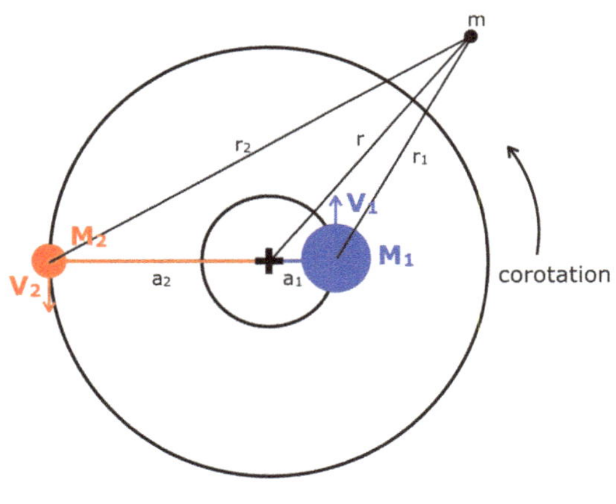

Figure 28.1. Location of points in the equatorial plane of a binary system of masses M_1 and M_2 in a circular orbit. The center of gravity is indicated by a cross.

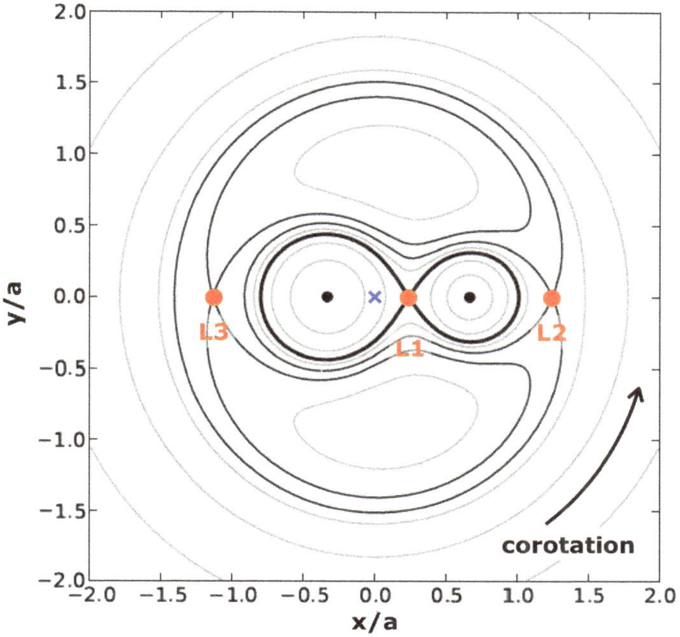

Figure 28.2. Equipotential surfaces in the corotating orbital plane of a binary system with $M_1/M_2 = 2$ as seen from the orbital pole. The coordinates are normalized to the distance between the two stars, a. A blue cross marks the center of gravity. The thick inner 8-shaped figure is the Roche lobe. The location of the Lagrangian points L_1, L_2, and L_3 are indicated.

Figure 28.2 shows the equipotential surfaces in the orbital co-rotating plane of a binary system with a mass ratio of $M_1/M_2 = 2$. The shape of the equipotential surfaces, scaled to a, depends only on the mass ratio of the components. The deepest potentials are those closest to the stars. Moving to a surface with less negative potential will require energy.

One particular equipotential surface is shown by a thick line. This is the **Roche lobe**, which is the tightest equipotential surface that includes both stars. It is named after the French mathematician Edouard Roche (1820–1883). The place where the Roche lobe crosses the connecting line between the two stars is called the **inner Lagrangian point L_1**, named after the Italian–French mathematician Joseph-Louis Lagrange (1736–1813), who contributed to many fields in mathematics, in particular, the theory of mechanics. The Roche lobe is the tightest surface where gas can flow freely from star 1 to star 2 without gaining or losing energy. Each of the components has its own Roche lobe, with the two lobes meeting at L_1. Other interesting points are the outer Lagrange points L_2, close to M_2, and L_3, close to M_1. Matter outside L_2 and L_3 is still bound to the binary, but it cannot maintain corotation.

Q 28.1 *Follow the equipotential surface that goes through L_2 and L_3 and try to understand its shape. Argue that matter within this surface must co-rotate with the system. Hint: consider the energy.*

We have described the shape of the equipotential surfaces in the orbital plane. In reality, they are 3D surfaces whose shape can be derived by means of the 3D version of Equation (28.7) by adding the height, z, above or below the orbital plane to Equation (28.7). This yields

$$\frac{\Phi}{G} = -\left(\frac{M_1}{r_1^z} + \frac{M_2}{r_2^z}\right) - \frac{1}{2}\frac{M_1 + M_2}{a^3}r^2, \tag{28.8}$$

where $(r_1^z)^2 = r_1^2 + z^2$ and $(r_2^z)^2 = r_2^2 + z^2$ are the distances in the 3D corotating Cartesian coordinate system and r is the distance from the axis of rotation.

Close to the orbital plane, the 3D Roche lobe has a shape that closely resembles that of the 2D shape, rotated around the axis connecting the stars. Further away from the plane, it has an open structure, resembling a hollow tube that is perpendicular to the orbital plane.

We will show below that the effect of binary evolution depends on the volume of the Roche lobe of each star. The volume within the 3D Roche lobe of a component of a binary can be expressed in terms of an **effective Roche lobe radius, R_L**, that is defined by the volume

$$\text{Volume (Roche lobe)} = \frac{4\pi}{3}R_L^3, \tag{28.10}$$

where R_L will be different for both components. For mass ratios in the range of $0.1 < M_1/M_2 < 10$, the result can be approximated by

$$\frac{R_{L,1}}{a} \cong \frac{0.49x^{2/3}}{0.6x^{2/3} + \ln(1 + x^{1/3})} \quad \text{with } x = M_1/M_2 \tag{28.11}$$

and

$$\frac{R_{L,2}}{a} \cong \frac{0.49y^{2/3}}{0.6y^{2/3} + \ln(1 + y^{1/3})} \quad \text{with } y = M_2/M_1 \tag{28.12}$$

proposed by Eggleton (1983).

28.3 Contact Phases

Binaries interact when the size of one of the two components reaches or overflows its Roche lobe. When that happens, matter can flow freely to the other star. This depends on the separation and the radius of the two components.

The left panel of Figure 28.3 shows the potential wells of a binary system with a mass ratio of $M_1/M_2 = 2$ along the line connecting the two stars. M_1 is the more massive star, so its potential well is deeper and wider. The location of the Lagrangian points L_1, L_2, and L_3 is indicated. Colors indicate the situation for different diameters of the two components. Green refers to stars that safely fit inside their Roche lobe: matter cannot flow freely from one star to the other. This is called a **detached system**. Blue refers to the case when star 1 (the more massive one) fills its Roche lobe: gas can freely flow from star 1 and fall onto star 2. This is called a

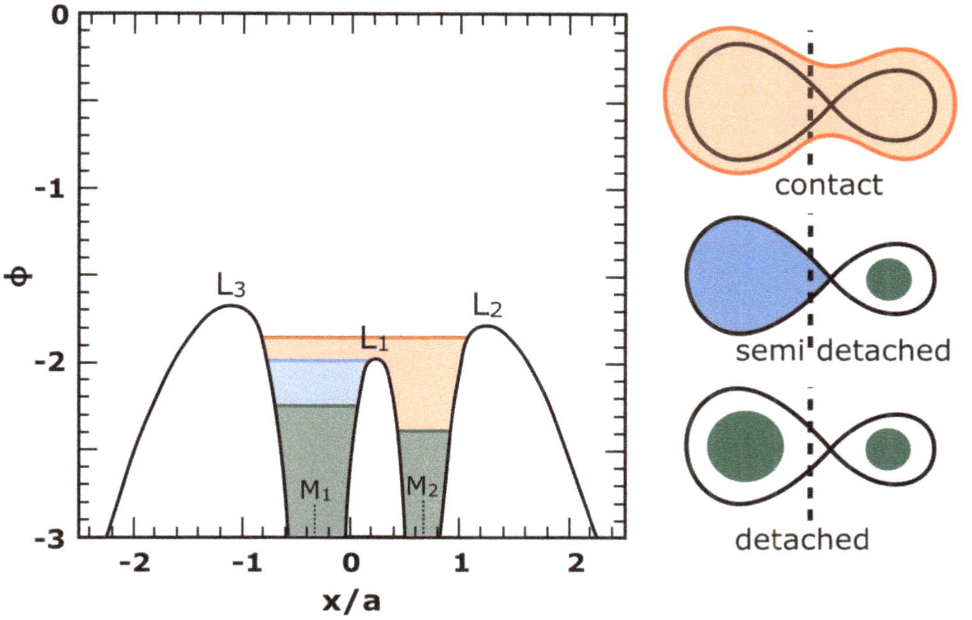

Figure 28.3. Left: the depth of the potential wells of a binary system with $M_1/M_2 = 2$ along a line connecting the centers of the two stars. Vertical axis: the potential (Equation (28.7)) in units of $G(M_1 + M_2)/a$. Horizontal axis: distance from the center of gravity in units of a. The locations of the Lagrange points L_1, L_2, and L_3 are indicated. Green, blue, and red indicate detached, semi-detached, and contact systems, respectively. Right: sketch of the corresponding configurations. The vertical dashed line indicates the rotation axis of the system. (Adapted from Pols 2011.)

semi-detached system. Red refers to stars that both fill their Roche lobe: they are in contact with one another. This is called a **contact system.** The right panel of Figure 28.3 shows the location of the stars within their Roche lobes for these three systems.

We have seen so far that the radius of a star can increase during several evolutionary phases:

- during the main-sequence phase. This radius increase is small.
- during the H-shell fusion phase when the star expands and develops a convective envelope. This happens for low-mass stars during evolution toward the RGB and for massive stars when they expand to become red supergiants.
- during the rise along the Hayashi track with a fully convective envelope (i.e., during the AGB phase of low-mass stars and the increasing luminosity phase for red supergiants).

The evolution of interacting binaries is classified into three cases, according to the evolutionary phase when the first mass transfer occurs.

Case A: when the first contact occurs during the MS phase.
Case B: when the first contact occurs during H-shell fusion.
Case C: when the first contact occurs while the star is on the Hayashi track.

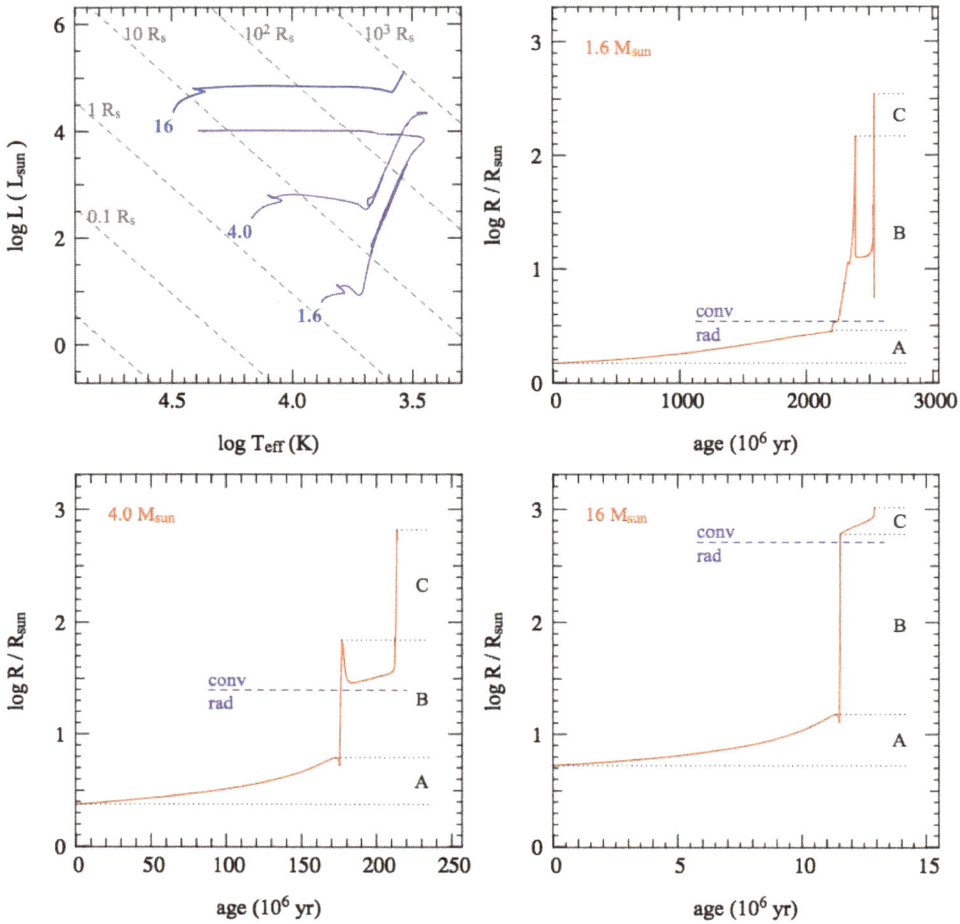

Figure 28.4. Upper left: the HRD with the evolutionary tracks of stars of 1.6, 4, and $16 M_{\odot}$ and lines of constant radius. Other panels: the radius evolution of the three stars. The dotted horizontal lines indicate the range in radius when case A, B, and C contacts occur. The blue dashed line indicates the radius at the time when the outer envelope goes from radiative to convective equilibrium. (© Pols 2011.)

Figure 28.4 (upper left panel) shows the evolutionary tracks of single stars of 1.6, 4, and $16 M_{\odot}$. The other panels show the radius evolution of these stars. Each of these panels shows horizontal lines that separate the evolutionary phases when the star might be in contact with a close companion: case A (during MS), case B (during H-shell fusion), and case C (when the star climbs the Hayashi track). The blue dashed lines separate the early phases when the outer envelope is in radiative equilibrium from the later phases when it is convective. The larger the increase in radius at some phase, the more likely it is that the binary reaches contact at that phase.

Q 28.2 *Describe the radius evolution of the star of* $4 M_{\odot}$ *in terms of an evolutionary track and identify the different evolution phases.*

28.4 Changes in Period and Separation during Mass Transfer

When one of the binary stars reaches its Roche lobe, matter may overflow to its companion. Let us call the *initially* most massive star the **primary**, and the initially least massive star the **secondary**. In the early lifetime of binaries, the primary will reach Roche lobe overflow first, because it evolves on a shorter timescale. In that case, the primary is the **donor (d)** and the secondary is the **accretor (a)**. In later phases, however, the secondary might be the donor that transfers mass to the primary. An example is the case of an expanding low-mass red giant transferring mass to a white dwarf that originates from an initially more massive star.

If all gas lost by the donor is accreted by the receiving star, the mass transfer is called **conservative,** indicating that all mass is conserved in the system. If the receiver is unable to accept all gas from the donor, the mass transfer is **nonconservative** because mass is lost from the system. This could be the case, for instance, when the transfer rate of the donor is so high that the accretor cannot adjust fast enough to accommodate the accretion rate.

Mass transfer changes the period and the separation of a binary system. Consider the time derivatives of the stellar masses, indicated by $\dot{M}_1$, $\dot{M}_2$, etc. In case of conservative mass transfer

$$-\dot{M}_1 = \dot{M}_2 \text{ and } \dot{J} = 0. \tag{28.13}$$

Differentiating Equation (28.5) for the angular momentum results in

$$2\frac{\dot{J}}{J} = \frac{\dot{a}}{a} + 2\frac{\dot{M}_1}{M_1} + 2\frac{\dot{M}_2}{M_2} - \frac{\dot{M}_1 + \dot{M}_2}{M_1 + M_2}. \tag{28.14}$$

The easiest way to derive this equation is by taking the time derivative of the natural logarithm of Equation (28.5) and using $d\ln(x)/dt = \dot{x}/x$. For conservative mass transfer, specified by conditions (28.13), this gives

$$\frac{\dot{a}}{a} = 2\frac{\dot{M}_1}{M_1}\left(\frac{M_1}{M_2} - 1\right). \tag{28.15}$$

If M_1 is the mass of the donor star (so $\dot{M}_1 < 0$), then the orbit will shrink if $M_1 > M_2$ and expand if $M_1 < M_2$. If the donor is more massive than the accretor, the orbit will shrink. The minimum orbit is reached when the stars have equal mass.

The change in a can be derived directly from the conditions that J and $M_1 + M_2$ are both constant. Equation (28.5) for J shows that

$$M_1^2 M_2^2 a = J^2(M_1 + M_2)/G = \text{constant}. \tag{28.16}$$

This implies that after mass transfer, we can express the ratio between the final separation a and its initial value a_i, and between the period P and its initial value P_i, as

$$\frac{a}{a_i} = \left(\frac{M_{1i}}{M_1}\frac{M_{2i}}{M_2}\right)^2 \text{ and } \frac{P}{P_i} = \left(\frac{M_{1i}}{M_1}\frac{M_{2i}}{M_2}\right)^3, \tag{28.17}$$

where the subscript i indicates the initial value and the last equation follows from Kepler's third law.

28.5 Stable and Runaway Mass Transfer

Mass transfer occurs when one of the two components fills its Roche lobe. When mass is transferred from a donor to an accretor, three properties must be considered: the change in radius of the donor, the change in radius of the accretor, and the change in separation. They all contribute to a change in the size of the Roche lobes. Depending on these changes, the mass transfer may be stable, unstable, or even a runaway process, as discussed below.

(1) **Stable mass transfer on the evolution timescale of the donor.**
This occurs when the radius of the donor decreases, due to mass transfer, faster than the size of the Roche lobe. In this case, the transfer of an amount of mass leads to shrinking of the donor radius, moving it back within its Roche lobe. The stellar evolution of the donor will then make the radius expand again until it fills its Roche lobe and again transfers mass to the accretor. This results in stable mass transfer on the timescale of the evolution of the donor. This happens in **case A transfer** when the donor is still on the main sequence with a radiative envelope.

(2) **Runaway mass transfer: dynamically unstable mass transfer**.
This occurs when the transfer of mass results in a shrinking of the Roche lobe, while the donor radius does not shrink fast enough or even keeps expanding. In that case, the mass transfer is so fast that the donor is out of hydrostatic equilibrium. Runaway mass loss occurs in stars with deep convection zones (i.e., if the donor is on the Hayashi track). This happens in **case C mass transfer.**

The reason that interacting binaries on the Hayashi track will suffer dynamically unstable mass loss is because their luminosity is set by the core mass and their T_{eff} is almost constant. This means that their radius is independent of the mass of the envelope. As the AGB star transfers mass to a lower mass companion, their separation decreases, which decreases the Roche lobe volume, so more of the envelope mass of the AGB star will be transferred. This results in a runaway process. It will end when the donor has lost almost all of its envelope and contracts on its way to becoming a WD in the case of the low-mass donor, or a WR star if the donor is massive.

(3) **Unstable mass transfer on the thermal timescale of the donor.**
This situation is in between the two extremes discussed above. In this case, the donor is out of thermal equilibrium (i.e., out of energy balance), but the mass transfer is slow enough for the donor to remain in hydrostatic equilibrium because the timescale for mass transfer is longer than the dynamical timescale. Readjustment to thermal equilibrium occurs on a Kelvin–Helmholtz timescale, so in this case the mass transfer is self-regulating and the timescale is the Kelvin–Helmholtz timescale. This happens in stars with radiative envelopes (i.e., in **case B transfer of massive stars**) that are expanding after the MS but have not yet reached the Hayashi track.

28.6 Summary

1. The Roche lobe is the smallest equipotential surface that includes both stars. The volume within this 3D surface is the Roche volume.
2. When evolution forces a star to increase its size beyond the Roche volume, mass can be transferred to its companion. If the companion can accommodate this mass (i.e., in the case of conservative evolution), the orbit will shrink if the donor is more massive than the accretor and expand if the donor is less massive than the accretor. The minimum separation is reached when the two masses are equal.
3. The transfer may occur in three evolutionary phases of expansion of the donor star:
   - when the donor expands during the main-sequence phase: case A,
   - when the donor expands to become a red giant after the main-sequence phase: case B,
   - when the donor expands while it is on the Hayashi track with a convective envelope: case C,
4. Mass transfer can be stable or unstable:
   - If the donor is a main-sequence star, the mass transfer is stable. This happens in case A transfer.
   - If the donor is an AGB star or a red supergiant with a convective envelope, runaway mass transfer occurs that ends when the donor has lost its convective envelope. This happens in case C transfer.
   - If the donor star is expanding during the H-shell fusion phase, the mass transfer is unstable on a Kelvin–Helmholtz timescale. This happens in case B transfer.

Exercises

28.1 Calculate the change in period and separation of a close binary with conservative mass transfer. Star 1 has a mass of $25M_{\odot}$ and a He core of $10M_{\odot}$ at the end of the MS phase. Star 2 has a mass of $10M_{\odot}$ and is still on the MS. The orbits are circular and the initial period is five days.

   a. Calculate the orbital separation (in $R_{\odot}$).
   b. Show that star 1 filled its Roche lobe shortly after the TAMS (use Appendix D).

   At the end of the MS phase, star 1 expands and mass transfer starts. The mass transfer continues until star 1 has lost its envelope and only the He star remains.

   b. Calculate the minimum period and the minimum separation.
   c. Calculate the final period and separation.

28.2 Suppose that star 1 of Exercise 28.1 loses $10M_{\odot}$ by a stellar wind that is not captured by the secondary. Suppose that this mass loss happens instantaneously.

   a. Calculate the initial angular momentum.
   b. Calculate the angular momentum that is removed from the system.

c. Calculate the new orbital separation and period.
d. Will the system go through case A, case B, or case C contact? Or no contact phase at all?

References

Eggleton, P. P. 1983, ApJ, **268, 368**

Pols, O. 2011, in Binary Stars, part 2,Utrecht University Lecture Notes, **6, https://www.astro.ru.nl/~onnop/education/binaries_utrecht_notes/**

Chapter 29

Close Binaries: Examples of Evolution with Mass Transfer

In this section, we discuss a few examples of close binary evolution with mass transfer. Case A transfer leads the Algol systems and blue stragglers. Case B transfer may lead to a system containing a low-mass Wolf–Rayet star and a more massive O-star. This happens in the case of conservative mass transfer (i.e., all mass lost by the donor is accepted by the receiver). Case C leads to binaries with a common envelope. In this case, the mass transfer is nonconservative (i.e., the receiver cannot accrete all mass transferred from the donor), so part of this mass is lost from the system. Friction on the common envelope causes the binary stars to spiral in. This may result in either the merging of the two stars or the formation of a narrow binary system with two stripped stars. We will also discuss the scenarios for the formation of low-mass X-ray binaries and high-mass X-ray binaries. Novae are erupting variables, consisting of a white dwarf that accretes matter from a companion. The large outbursts of classical novae are due to the sudden ignition of H-fusion at the bottom of the accumulated layer on top of the white dwarf. Dwarf novae show recurrent outbursts of lower luminosity, which are due to variations in the accretion rate.

29.1 Algol Systems: Conservative Case A Mass Transfer

An Algol system consist of a (sub)giant that fills its Roche lobe and a more massive main-sequence star. The existence of these systems was a puzzle for a long time, since the giant was expected to be more massive because the evolution time is shorter for more massive stars. This was called the Algol paradox. It was solved in 1967 by Kippenhahn and Weigert, who showed that Algol systems are the natural result of close binary evolution with mass transfer (Kippenhahn & Weigert 1967).

Figure 29.1 shows the evolution in the HRD and the mass transfer rate of a binary system with a metallicity of $Z = 0.004$ and with initial masses of $M_1 = 10.0$ for the primary and $M_2 = 8.9 M_\odot$ for the secondary in a circular orbit with an initial

doi:10.1088/978-0-7503-1278-3ch29

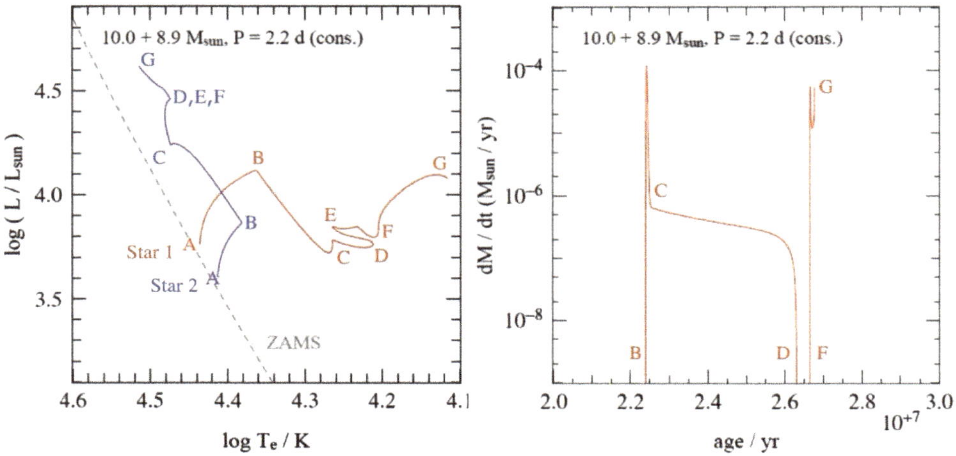

Figure 29.1. Left: the evolutionary tracks of the two components of a binary of $M_1 = 10M_\odot$ (red) and $M_2 = 8.9M_\odot$ (blue) with an initial period of $P = 2.2$ day and a separation of $20R_\odot$. Right: the mass-loss rate of star 1 as a function of time. Letters indicate specific evolutionary phases that are discussed in the text. The stars have a metallicity of $Z = 0.004$. (© Pols 2011.)

separation of $20R_\odot$ and a period of 2.2 days. The stars, with radii of $R_1 = 3.2$ and $R_2 = 3.0R_\odot$, are initially detached but at **point B** star 1 has expanded to a radius of $7.2R_\odot$ and fills its Roche lobe. At that point, the system becomes semi-detached. **Between B and C**, star 1 transfers mass to star 2. The donor, which is still fusing H on the MS, loses mass so its luminosity decreases, whereas star 2 gains mass and becomes more luminous. The transfer is slow so that the stars have time to adjust and stay in hydrostatic and thermal equilibrium. As mass is transferred and the two stars become more equal in mass, the separation shrinks (Section 28.4). The Roche lobes of both stars also shrink but their radii do not. Star 2 remains on the MS and does not fill its Roche lobe. On the other hand, the shrinking Roche lobe of star 1 implies that it must lose mass at a higher rate of $\sim 10^{-5}\ M_\odot\,\mathrm{yr}^{-1}$. At this high transfer rate, star 1 is out of thermal equilibrium. This stops when star 1 becomes the less massive of the two. Continuing mass transfer from now on will increase the separation. Star 1, which remains on the MS, is still filling its Roche lobe so it keeps transferring mass to star 2, albeit at a lower rate. At **point C**, star 1 has lost $4M_\odot$ so $M_1 = 6M_\odot$ and $M_2 = 12.9M_\odot$. **Between C and D**, the mass transfer has settled to the nuclear timescale of star 1 (i.e., the core H-fusion timescale). **Point D** marks the end of the MS phase of star 1: the star shrinks briefly, so the mass transfer stops (**D-E-F**). During the subsequent H-shell fusion phase, **F-G,** the star expands and the resulting high-mass transfer rate occurs on the Kelvin–Helmholtz timescale.

During this evolution, the giant is less massive than its MS companion after phase C, as is observed in **Algol systems**. If this evolution occurs in a cluster and star 2 reaches a mass that is higher than the turn-off mass of that cluster, it will be a **blue straggler** (see Figure 2.7). Blue stragglers are MS stars in clusters that are more massive that the turn-off mass and can only be formed by mass transfer in a close binary system.

29.2 Massive Interacting Binaries: Conservative Case B Mass Transfer

The fraction of main-sequence O stars (luminosity class V) that have one or more bound companions is very high, suggesting that massive stars are often formed in multiple systems (Sana, et al. 2014). This implies that binary interactions play a critical role in massive star evolution from the main sequence straight through core collapse (Sana et al. 2012; De Mink et al. 2014).

Short period binaries with $P \lesssim 20$ days consisting of a massive O star and a less massive Wolf–Rayet star are typical examples of the results of such interactions. The Wolf–Rayet star was originally the more massive one, but it has been stripped by case B mass transfer.

Figure 29.2 shows the evolution of a massive binary with an initial mass of $M_1 = 10$ and $M_2 = 8M_\odot$ and a period of $P = 12$ days. This corresponds to an initial separation of $a = 60R_\odot$. This will result in case B mass transfer because star 1 will fill its Roche lobe in the H-shell fusion phase.

Star 1 fills its Roche lobe at **point B**, when it is crossing the Hertzsprung gap during the H-shell fusion phase, and starts transferring mass to the companion. The expansion of the donor happens on the Kelvin–Helmholtz timescale. As the orbit shrinks, this results in a high mass transfer rate and a steep drop in luminosity. This is because the mass transfer is so fast that the star is out of thermal equilibrium: the core cannot produce sufficient nuclear energy to keep up with the expansion of the envelope, so the luminosity of the star drops dramatically between **points B and C.** At the same time, star 2 is gaining mass and moves up along the MS. At **point C**, the two stars have reached about equal mass, $M_1 = M_2 = 9M_\odot$, and the separation is at

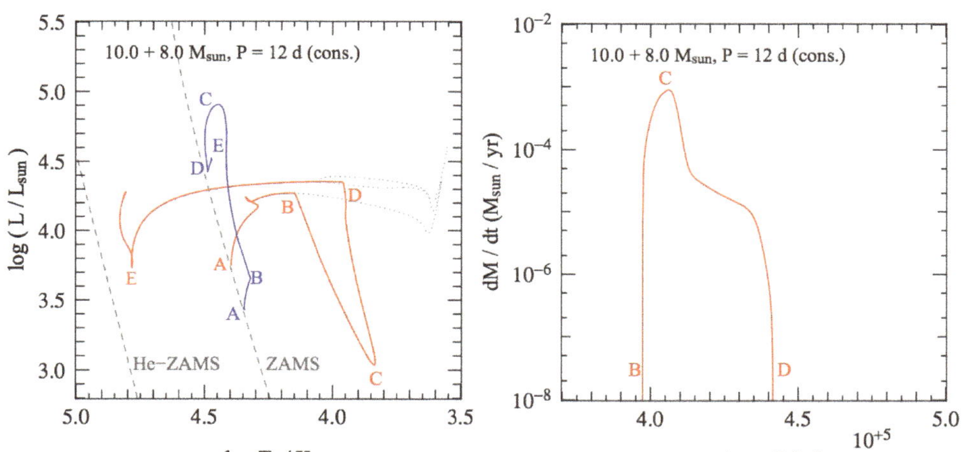

Figure 29.2. Left: the evolution of a binary with an initial mass of $M_1 = 10$ (red) and $M_2 = 8M_\odot$ (blue), an initial period of $P = 12$ days, and a separation of $a = 60R_\odot$. The thin gray line indicates the track of a single star of $10M_\odot$. The locations of the ZAMS and the He main sequences are shown as dashed lines. Right: the mass transfer rate as a function of time. Notice that most of the mass transfer occurs during a short time of only 5×10^3 yr. The letters refer to phases that are discussed in the text. (© Pols 2011.)

its minimum. From then on, the separation increases when mass is transferred to star 2, so the mass transfer rate drops. Star 1 regains thermal equilibrium again and its luminosity increases to the value that agrees with its core mass. At **point D**, when core He-fusion is ignited in star 1, it has lost so much of its envelope that it shrinks and the mass transfer stops. At **point E**, the star has evolved toward the He main sequence as a Wolf–Rayet star. Meanwhile, the companion has reached its final mass and continues its evolution as a massive star, so in the end the system has evolved into a close binary system in which the initially more massive star is now a low-mass WR star and the initial secondary is a more massive O star.

29.3 Common Envelope Stars: Case C Mass Transfer

Figure 29.3 shows a sequence of events that leads to a common envelope system. The initial configuration is shown in A. If mass transfer occurs when the donor is on the Hayashi line, the mass transfer will be unstable (B). This will lead to rapid shrinking of both the orbital separation and the size of the Roche lobes, while the donor star keeps expanding. This results in such a high mass-loss rate that the receiver is unable to adjust. The accreted mass is piled on top of the receiver, which may fill its Roche lobe. As the donor keeps expanding and transferring mass, the orbit keeps shrinking and the Roche lobes of both stars may be overfilled: the stars develop a common envelope (C). As the two stars orbit inside the common envelope, friction heats the envelope and matter is lost from the system (D).

At the same time, friction will also lead to the spiral-in of the two stars (E).

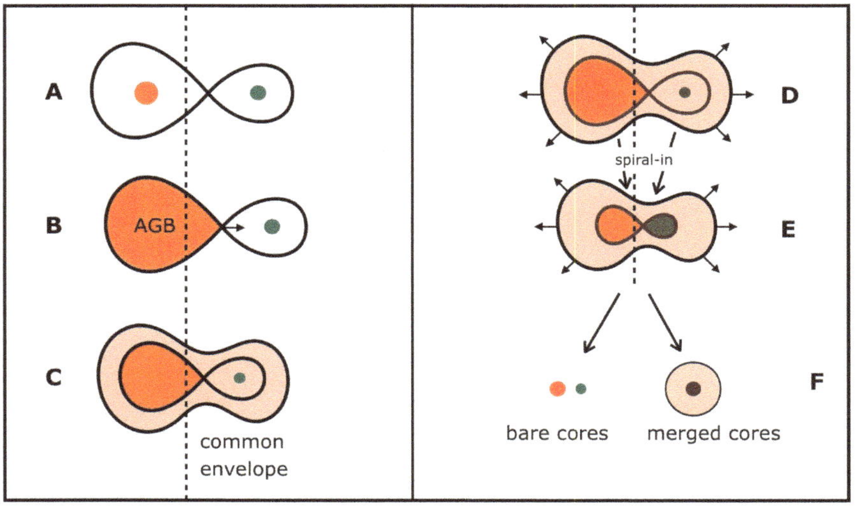

Figure 29.3. Schematic picture of the evolution of a binary that leads to a common envelope phase. The vertical dashed line indicates the location of the center of gravity. A: initial configuration; B: the AGB star fills its Roche lobe; C: formation of a common envelope (light red) due to runaway mass transfer; D: dynamical friction leads to heating of the envelope and mass loss; E: spiral-in due to dynamical friction and ejection of the common envelope; and F: two possible results: bare cores in a very close orbit or the merging of the two stars.

Once the orbital separation is so small that the system consists of the two cores rotating inside a common envelope, friction will slow down the orbital motion of the cores and cause the orbits to shrink even more. This may result in two possible outcomes (F).

- The energy released by the friction and the shrinking of the orbit heats the common envelope to such high temperatures that the envelope escapes, leaving two bare cores in a very close orbit.
- The slow-down of the orbital motion by friction leads to a spiral-in and finally the merging of the two cores surrounded by one envelope (i.e., into one single star).

Which one of these two scenarios is realized depends on the amount of energy that is produced by the friction and the shrinking of the system, $E_{\rm fric}$, minus the radiative losses, $E_{\rm rad}$. The net energy is $E_{\rm net} = E_{\rm fric} - E_{\rm rad}$. If $E_{\rm net}$ is more than the energy needed to unbind the common envelope, $E_{\rm env}$, this common envelope will be expelled and the spiral-in will stop, leaving two separate bare cores. If $E_{\rm net} < E_{\rm env}$, the friction and the spiral-in of the two cores within the common envelope will continue until the two cores merge completely within their common envelope.

29.4 The Formation of High-mass X-ray Binaries

Massive X-ray binaries are the result of merging and spiral-in (Van den Heuvel 1983). As an example of a high-mass X-binary system, we adopt the system LMC X-3. It consists of a B star of $8M_\odot$ and an accreting black hole of $10M_\odot$ in an orbit with a period of 1.7 days. A possible evolutionary scenario that would lead to such a system is shown in Figure 29.4. The letters refer to the phases described below.

a. The initial configuration is a system of $M_1 = 71.9M_\odot$ and $M_2 = 8.6M_\odot$ in an orbit with an initial period of 132.9 days. Star 1 is so massive that it loses about 10% of its mass via a stellar wind.
b. The primary fills its Roche lobe when it leaves the main sequence and its radius expands, which results in runaway mass loss as discussed in Section 28.5. The star goes through a common envelope phase with spiral-in due to dynamical friction. The energy gained by the spiral-in and by friction heats the envelope of star 1 to such a degree that it is expelled. This process happens so fast that star 2 accretes only a small fraction of the mass. Star 2 is spun up by the angular momentum of the accreted mass and develops a rotation induced disk.
c. At the end of the common envelope phase, the primary has lost about half of its mass, from 63.8 to $31.6M_\odot$, leaving only the He core. The mass of star 2 is hardly increased. The orbit has been shrinking from about $500R_\odot$ ($P = 164.3$ days) to less than $10R_\odot$ (P = 0.5 days). The change in orbital radius is so large that the scale in the right side of Figure 29.4 is enlarged drastically.
d. Star 1 is a Wolf–Rayet star and has a high radiation-driven mass-loss rate: its mass decreases from 31.6 to $23.6M_\odot$.

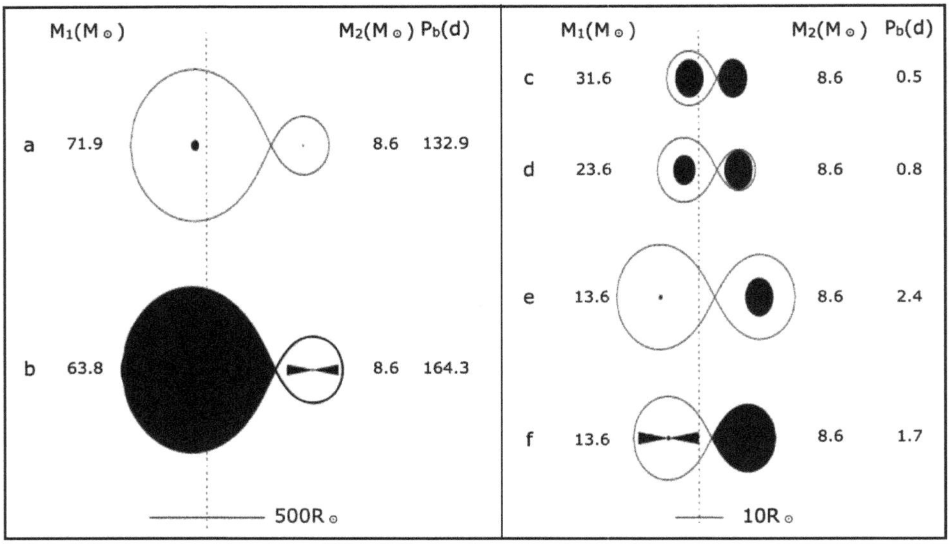

Figure 29.4. Evolution of a close binary system of 71.9 and $8.6M_{\odot}$ with a period of 132.9 days that leads to the formation of a high-mass X-ray binary similar to LMC X-3. Notice the different scales of the initial (left) and later (right) phases. (© Pols and Verbunt 2011.)

e. The Wolf–Rayet star explodes as a Type Ib or Ic SN, leaving behind a black hole of $13.6M_{\odot}$.
f. Star 2 fills its Roche lobe and transfers mass to the black hole. This creates an accretion disk that produces X-rays. The system has thus evolved into a high-mass X-ray binary.

29.5 The Formation of Low-mass X-ray Binaries

Low-mass X-ray binaries are also the result of binary evolution with spiral-in (Van den Heuvel 1983). As an example of a low-mass X-ray binary, we adopt the system Sco X-1, which consists of a neutron star of $1.4M_{\odot}$ and a low-mass $0.42M_{\odot}$ companion with an orbital period of 0.88 days. A possible evolutionary scenario that would lead to such a system is shown in Figure 29.5. The letters refer to the phases that are described below.

a. The system consists of star 1 with $M_1 = 11M_{\odot}$ and star 2 with $M_2 = 1M_{\odot}$ in an orbit with an initial period of 300 days and an orbital separation of $a = 430R_{\odot}$.
b. The primary fills its Roche lobe when it becomes a red giant with a He core of $2.5M_{\odot}$. Because the star has a convective envelope, this leads to runaway mass transfer (Section 28.5) with a common envelope phase and a rapid spiral-in by dynamical friction. This phase is so fast that star 2 accretes very little mass; however, it is spun up by the angular momentum of the accreted mass (this is similar to phase b of the high-mass X-ray binary evolution in Figure 29.4).

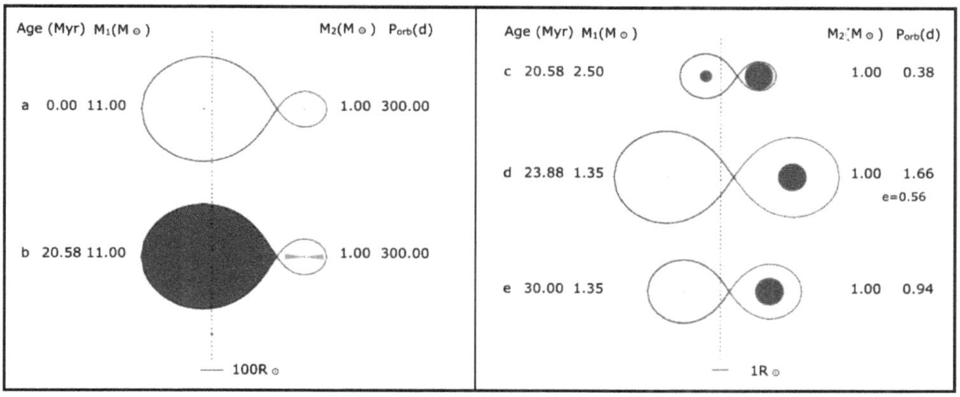

Figure 29.5. Evolution of a binary system of stars with initial masses of $11 M_\odot$ and $1 M_\odot$, in an orbital period of 300 days and a separation of $430 R_\odot$. The evolution of such a system results in a low-mass X-ray binary system quite similar to Sco X-1. Notice the different scales of the initial (left) and later (right) phases. (© Pols and Verbunt 2011.)

c. The energy released by the spiral-in and the friction has expelled the envelope of star 1, leaving behind a He star with a mass of $2.5 M_\odot$. The orbital radius decreased from 430 to $3.4 R_\odot$ and the period decreased from 300 days to 0.38 days.
d. Star 1 explodes as an SN, leaving behind a neutron star of $1.35 M_\odot$. Star 2 is still a main-sequence star. The sudden loss of mass from the system results in a widening of the orbit with an increased period of 1.66 days. Because the explosion of the supernova was asymmetric, the neutron star received a kick velocity. This results in an elliptical orbit with an eccentricity of $e = 0.56$.
e. Due to tidal interaction, the orbit circularizes, resulting in the final configuration of a neutron star with a main-sequence companion in a circular orbit with a period of 0.94 days and a separation of $5.5 R_\odot$. When star 2 expands, the accretion onto the neutron star produces the X-ray luminosity.

29.6 Novae: WDs in Semi-detached Systems

Novae are stars that show recurrent outbursts. They are close binary systems with orbital periods of about 1 to 10 hours, consisting of a WD and a star of spectral type G or later with a mass of about $1 M_\odot$ or less (Gallagher & Starrfield 1978). There are two types of novae: **classical novae** and **dwarf novae**.

In a **classical nova** outburst, the brightness increases within a few hours from that of a typical red dwarf ($M_v \sim +5$, $L \sim L_\odot$) to as high as $M_v = -6$ to -9, which corresponds to a luminosity of about $L \approx 2 \times 10^4 L_\odot$ to $3 \times 10^5 L_\odot$. After maximum, which lasts a few days, the brightness decreases over a period of tens to hundreds of days to its normal level.

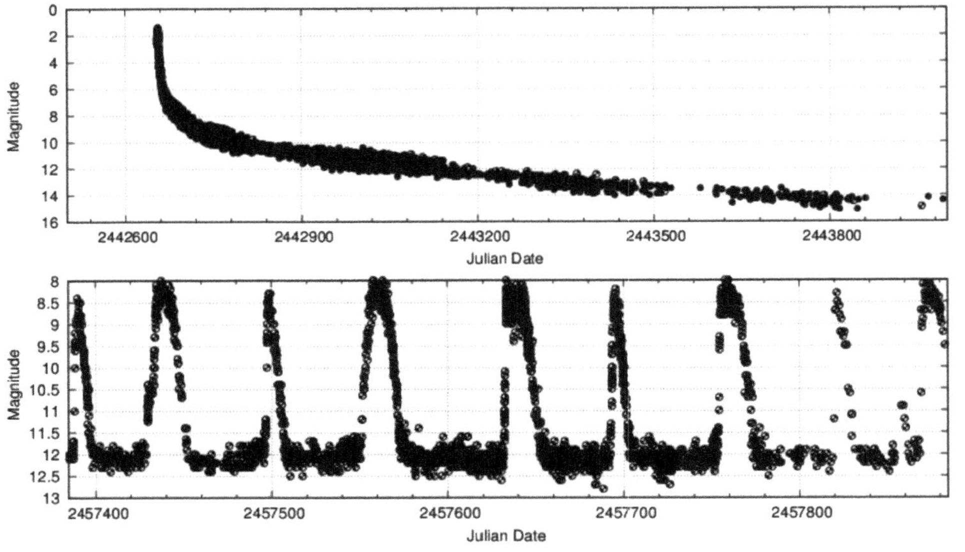

Figure 29.6. Typical light curves of two novae. The top panel is for nova V1500 Cyg that had an outburst on 1975 August 30, when its visual brightness increased by 13 magnitudes. Notice the slow decrease in visual magnitude over a period of about three years. The lower panel is for the recurrent dwarf nova SS Cyg. The recurrence time of the outbursts, when the visual brightness increases by 4 mag, is about 60 days. (Data are from the AAVSO https://www.aavso.org/lcg.)

In a **dwarf nova**, the brightness increases by only a factor of 10 to 100. Such eruptions may occur at intervals of 30 to 300 days. Typical light curves for a classical nova and a dwarf nova are shown in Figure 29.6.

The spectrum of a nova before outburst shows the presence of an accretion disk in the system, indicating that mass transfer occurs from the companion that fills it Roche lobe to the WD. The accumulation of matter onto the WD triggers a thermonuclear outburst on the surface of the WD. The outburst and post-outburst spectrum show wide blueshifted absorption lines: evidence of mass loss from the system with a velocity on the order of 10^3 km s^{-1}.

Let us try to explain the physical effects and energies involved in the outburst of a **classical nova**. For a WD with a mass $M_{\rm WD}$ and radius $R_{\rm WD}$ accreting matter at a rate of $\dot{M}_a$, the accretion energy rate is

$$L_a = GM_{\rm WD}\dot{M}_a/R_{\rm WD}. \tag{29.1}$$

Assume that the nondegenerate photosphere of the WD has a thickness h, a mean particle density n, a mean mass per particle $\mu m_{\rm H}$, and an envelope mass

$$M_{\rm env} = 4\,\pi R_{\rm WD}^2 hn\mu m_{\rm H}. \tag{29.2}$$

The total thermal energy of the envelope is

$$E_{\rm th} = 4\pi R_{\rm WD}^2 \times (3/2\ hnkT) = 3/2 M_{\rm env}(kT/\mu\ m_{\rm H}). \tag{29.3}$$

If the accretion energy is completely converted into thermal energy, the temperature of the envelope will increase at a rate of

$$\left(\frac{GM_{\mathrm{WD}}}{R_{\mathrm{WD}}}\right) \times \dot{M}_{\mathrm{a}} = (3/2\ \mathrm{k}/\mu m_{\mathrm{H}}) \frac{dT}{dr} \times M_{\mathrm{env}} \rightarrow \frac{dT}{dt} = 2/3 \frac{GM_{\mathrm{WD}}}{R_{\mathrm{WD}} M_{\mathrm{env}}} \frac{\mu m_{\mathrm{H}}}{k} \dot{M}_{\mathrm{a}}. \quad (29.4)$$

The temperature increase of a nondegenerate H-rich envelope of $M_{\mathrm{env}} = 10^{-5} M_{\odot}$ and $\mu = ½$ of a CO white dwarf with a mass of 1 $M_{\odot}$ and a radius of $0.011 R_{\odot}$ (Equation (20.2)) is

$$\frac{dT}{dt} \approx 1 \times 10^{13}\, \dot{M}_{\mathrm{a}}\, \frac{\mathrm{K}}{\mathrm{yr}}. \quad (29.5)$$

The observed accretion rates of novae are of the order of 10^{-9} to 10^{-8} $M_{\odot}\,\mathrm{yr}^{-1}$. At such an accretion rate, the envelope will be heated to the ignition temperature for H-fusion via the CNO cycle of 10^7 to 10^8 K in about 10^2 to 10^3 years. At this time, about 10^{-6} to $10^{-5} M_{\odot}$ has been accumulated. For lower accretion rates, the accumulation time increases inversely.

Models show that the H-fusion is ignited at the bottom of the envelope if the envelope has a mass of $\sim 10^{-5}$ to $10^{-4} M_{\odot}$ at a density of about 2×10^2 g/cm^3. This fusion via the CNO cycle produces a luminosity of the order of 10^5 $L_{\odot}$ in a layer with a mass of about $10^{-6} M_{\odot}$ (Starrfield et al. 2000). The amount of energy generated by the fusion of $10^{-6} M_{\odot}$ of H is $\sim 10^{46}$ erg. This energy can sustain a luminosity of 10^5 $L_{\odot}$ for about a year. However, the peak luminosity of a nova only lasts for a few days (in the case of a fast nova) to 100 days (for a slow nova). This implies that a large part of the generated energy is lost in the ejection of a shell. The potential energy of one gram at the surface of the WD is 2×10^{17} erg and the kinetic energy of gas ejected at a speed of a few times 10^3 km/s is 5×10^{17} erg, so the energy created by the fusion is sufficient to eject a shell of about 10^{-6} to $10^{-5} M_{\odot}$, roughly in agreement with the estimates of the observed mass.

The eruptions in **dwarf novae** are not due to the ignition of thermonuclear fusion but to large variations in the accretion rate. The spectra of novae during quiescence show that most of the luminosity comes from the accretion disk. The luminosity of an accretion disk is directly coupled to the accretion rate. At quiescence, the accretion rate is small and on the order of 10^{-11} to 10^{-10} $M_{\odot}\,\mathrm{yr}^{-1}$. During outburst, the rate increases to 10^{-9} to 10^{-8} $M_{\odot}\,\mathrm{yr}^{-1}$. The variation in the accretion rate is probably due to an instability in the accretion disk around the WD, although a variation in the mass transfer rate from the donor star has also been suggested (Sparks et al. 1977).

29.7 Summary

1. Algol systems, consisting of a giant and a more massive main-sequence star, are the result of close binary evolution with case A evolution and conservative mass transfer. The primary fills its Roche lobe when it is still in the core H-fusion phase. The receiver is also an MS star and gets more massive than the donor.

2. This evolution also explains the existence of blue stragglers in clusters with a turn-off mass that is lower than the final mass of the receiver.
3. If mass transfer occurs when the primary is an AGB star or a red supergiant, the shrinking of the orbit combined with the case C very high mass transfer rate results in a common envelope surrounding both stars. Dynamical friction leads to rapid shrinking of the orbit. If the released energy is insufficient to remove the common envelope, the stars will merge into a single star. If the released energy is sufficient to remove the common envelope, the system ends as a very close binary consisting of the two bare cores.
4. Massive binaries consisting of a Wolf–Rayet star and a more massive O star are the result of case B close binary evolution with conservative mass transfer. The primary fills its Roche lobe when it is expanding during H-shell fusion. The shrinking of the orbit combined with the rapid expansion of the star results in a short phase of very high mass transfer that strips the primary down to its He core. The primary becomes a Wolf–Rayet star with a more massive O star companion.
5. High-mass X-ray binaries are the result of the nonconservative evolution of a close binary system consisting of two massive stars that pass through a common envelope phase. The primary fills its Roche lobe and the resulting unstable mass transfer rate is so high that the secondary cannot accept it. The primary loses so much mass that it becomes a Wolf–Rayet star that later explodes as an SN, leaving behind a black hole. When the secondary, whose mass has hardly increased, later fills its Roche lobe and transfers mass to the black hole, the system becomes a high-mass X-ray binary.
6. Low-mass X-binaries form basically in the same way as high-mass X-ray binaries, but originate from a lower-mass system. In this case, the primary leaves a neutron star behind when it explodes. The mass of the secondary remains small because it cannot accrete the mass that is lost in the common envelope phase.
7. Classical novae are close binaries consisting of a WD receiver with a G-type or later donor. When the accumulated H on top of the WD reaches a mass of about $10^{-5} M_{\odot}$ and a temperature of $\sim 10^8$ K, H-fusion is ignited. This results in a sudden increase in brightness up to about $10^3 - 10^5 L_{\odot}$ and the ejection of part of the envelope. The recurrence time of the outbursts is of the order of 10^2 to 10^3 yr.
8. Dwarf novae are also close binaries with an accreting WD and a cool companion. The recurrent outbursts of dwarf novae are due to variations in the accretion rate, triggered by an instability in the accretion disk.

For a more extensive discussion of binary evolution, the reader is referred to the excellent lecture notes of the course "Binary Stars" by our Utrecht colleagues Onno Pols and Frank Verbunt at https://www.astro.ru.nl/~onnop/education/binaries_utrecht_notes/.

Exercises

29.1 Estimate the mass of the two stars at the end of the case A mass transfer phase of the Algol system shown in Figure 29.1 and at the end of the case B mass transfer of a massive binary shown in Figure 29.2.

29.2 Consider a binary system consisting of two stars with 10 and $1M_\odot$ in circular orbits. Calculate the initial period and initial separation for mass transfer to start during core H-fusion, during core He-fusion, and during double-shell fusion of the most massive component.

29.3 The rapid increase in brightness at the beginning of a nova outburst is called the "fireball phase." In this exercise, you are going to calculate the radius and duration of the fireball phase. Assume that the mass ejection occurs at a constant rate, $\dot{M}_{ej} = 3 \times 10^{-6}\ M_\odot\,\mathrm{yr}^{-1}$, at a constant velocity, $V = 1000\ \mathrm{km\,s}^{-1}$, and lasts more than 10 days.

a. Derive an expression for the density at distance r in the shell.

b. Show that the outer radius of the shell $R_{out}(t)$ is much larger than R_{WD} after about 10 seconds. Thus, we can safely adopt $R_{out}(t) = V \times t$.

c. Derive an expression for the radius of the photosphere, $R_{phot}(t)$, defined by $\tau = \frac{2}{3} = \int_{R_{phot}}^{R_{out}} \kappa\rho\, dr$, with $\kappa = \sigma_e = 0.3\ \mathrm{cm}^2\ \mathrm{g}^{-1}$. Show that the photospheric radius varies with time as $\frac{1}{R_{phot}(t)} = \frac{1}{Vt} + \frac{1}{X}$ and derive an expression for X.

d. Plot the value of $R_{phot}(t)$. What is the physical meaning of X?

e. Explain in physical terms why $R_{phot}(t)$ approaches a constant value, whereas R_{out} keeps increasing.

f. Calculate the time t_1 when $R_{phot}(t_1) \approx X$. How large is R_{out} at that time? What is the mass in the envelope at that time?

g. Assume that the temperature of the outflow during the fireball phase remains constant at $T \approx 10^5$ K. How large is the peak luminosity?

References

De Mink, S. E., Sana, H., Langer, N., Izzard, R. G., & Schneider, F. R. N. 2014, ApJ, **782, 7**

Gallagher, J. S., & Starrfield, S. 1978, ARA&A, **16, 171**

Kippenhahn, R., & Weigert, A. 1967, Zeitschrift fuer Astrophysik, 65, 251 (see review by Plavec, M. 1970, PASP, 82, 95)

Pols, O. & Verbunt, F. 2011, i Binary Stars, part 2, Utrecht University Lecture Notes, 228–229 and 256–259

Sana, H., de Mink, S. E., de Koter, A. et al. 2012, Sci, **337,** 444

Sana, H., Le Bouquin, J.-B., Lacour, S. et al. 2014, ApJS, **215, 15**

Sparks, W. M., Starrfield, S. & Truran, J. M. 1977, ASSL, **65, 189**

Starrfield, S., Truran, J. W., & Sparks, W. M. 2000, NewAR, **44, 81**
Van den Heuvel, E. P. J. 1983, in Accretion Driven X-ray Sources, ed. W. H. Lewin, E. P. J. van den Heuvel, (Cambridge: Cambridge Univ. Press), 316

AAS | IOP Astronomy

Understanding Stellar Evolution

Henny J.G.L.M. Lamers and Emily M. Levesque

Chapter 30

Chemical Yields: Products of Stellar Evolution

Stars are formed from interstellar material. During their life cycles, they produce a wide range of chemical elements by nuclear fusion or neutron capture. A considerable fraction of their initial mass is returned to the interstellar medium by means of stellar winds or supernova explosions. The ejected material is partly chemically enriched. This results in a steady chemical enrichment of the interstellar matter. The amount of mass of a given chemical element that is produced in a star and returned to the interstellar medium is called the stellar yield. As H is destroyed in stars, the stellar yield of H is negative. In this section, we will summarize the evolution of single stars and describe the resulting chemical yields. The very presence of life on Earth owes its thanks to the chemical yields of previous generations of stars, before the Sun and the solar system were even born.

30.1 A Summary of the Evolution of Single Stars

The yields of a star depend on its nuclear evolution and on mass loss by winds or SNe. We therefore briefly summarize the evolution of single stars that were discussed in the previous sections.

$0.01 \lesssim M_i \lesssim 0.8 M_\odot$: the lifetime of these stars is larger than the age of the Universe, so they are still in their MS phase. They have not yet contributed to the enrichment of the ISM (Section 13.5)

$0.8 \lesssim M_i \lesssim 8 M_\odot$: after core H-fusion as an MS star and shell H-fusion as a red giant, these stars go through core He-fusion with H-shell fusion as an HB star. They lose only a small fraction of their mass by a stellar wind in the red giant phase. During the subsequent AGB phase, they go through dredge-ups and thermal pulses that bring the products of the CNO cycle, and later the He-fusion products and s-process elements, to the surface (Section 18). AGB stars have a high mass-loss rate produced by radiation pressure on dust and pulsation (Section 15.3). The combination of severe mixing with high mass-loss rates results in large chemical yields.

doi:10.1088/978-0-7503-1278-3ch30

$8 \lesssim M_i \lesssim 25M_\odot$: these stars go through all fusion phases, even through Si-fusion, before exploding as SNe in the red supergiant (RSG) phase. RSGs have deep convective envelopes and severe mass loss, which, together with the SN explosion, results in large chemical yields (Section 23).

$25 \lesssim M_i \lesssim 50M_\odot$: these stars go through all fusion phases and explode as SNe during the Wolf–Rayet phase. During their evolution, they experience high mass-loss rates as RSGs and WN stars, when most of the products of the CNO cycle are ejected, and as WC stars when the products of He-fusion are ejected (Section 24).

$50 \lesssim M_i \lesssim 120M_\odot$: these stars go through all fusion phases and explode as SNe during the Wolf–Rayet phase. Before doing so, they have gone through high mass-loss phases as LBVs with multiple eruptions, where CNO-cycle products are lost, and as WN and WC stars.

30.2 Chemical Yields of Single Stars

Stars lose mass during their evolution due to stellar winds and supernova explosions. The **total amount of mass of a given chemical element** E that is lost by a star during its evolution can be written as

$$M_E = \int X_E^S(t) \times \frac{dM}{dt} dt, \tag{30.1}$$

where $X_E^S(t)$ is the mass fraction of element E at the surface of the star (because mass is always ejected from the stellar surface) and dM/dt is the mass-loss rate at time t. The integral is over the full lifetime of the star.

The **stellar chemical yield** Y_E of element E is defined as the amount of mass of element E that is lost by the star, minus the initial amount of that element in the ejected mass. The star had an initial composition described by the mass fraction X_E^i at the time of formation of the star, so the yield can be written as

$$Y_E = \int \left\{ X_E^S(t) - X_E^i \right\} \times \frac{dM}{dt} dt. \tag{30.2}$$

If the chemical composition of the ejected mass is higher than its initial composition, the yield is positive. This is, for instance, the case for He that is formed by fusion in the star and ejected in the wind. The yield of H is negative, because H is destroyed by nuclear fusion.

The **total chemical yield,** $Y_E^{\rm tot}$, is the chemical yield per star, multiplied by the relative number of the stars (i.e., by the stellar initial mass function). The total chemical yield, in $M_\odot$, can be calculated per unit volume of the galaxy. It can also be calculated as a fraction of a given total stellar mass. In the latter case, $Y_E^{\rm tot}$ is dimensionless.

30.2.1 Chemical Yields of Low-mass Stars

Let us consider the chemical yields of low-mass stars of $1 \lesssim M_{\rm i} \lesssim 6M_\odot$, with initial composition $X = 0.687$, $Y = 0.293$, and $Z = 0.020$, derived from the stellar evolutionary calculations by Karakas (2010). These calculations include the evolution of thermal pulsing AGB stars. Figure 30.1 shows the amount of mass of several chemical elements ejected during the evolution, expressed as a fraction of the initial stellar mass, M_i. Most of the ejected matter is in the form of H and He, with a small contribution of C, N, and O.

Figure 30.2 shows the amount of C, N, and O ejected by the stars as a fraction of the initial mass. The amount of C includes ^{12}C and ^{13}C, N is mainly in the form of ^{14}N, and O is the combination of ^{16}O, ^{17}O, and ^{18}O (the values for the individual isotopes are given in Karakas 2010).

The adopted initial mass fractions are, respectively, 0.293 for He, 0.0034 for C, 0.0011 for N, and 0.0096 for O. The peak in the yield of ^{12}C at $\sim 3M_{\odot}$ is due to the large number of dredge-ups during the thermal pulses of these stars. The number of thermal pulses decreases toward higher mass. At $M > 4M_{\odot}$, the temperature of the H-fusion shell in-between the thermal pulses is higher than about 3×10^7 K, so the H-fusion occurs via the CNO cycle. This increases the abundance of ^{14}N at the expense of ^{12}C and ^{16}O (Section 8.4.3). In AGB stars, this process is called **Hot-Bottom Burning** (see Section 18.5).

Figure 30.3 shows the amount of ejected mass and the yields of He, C, N, and O in $M_{\odot}/M_i$ for low-mass stars. Notice that the yield of He is small, although about 25% of the stellar mass was ejected in He; however, the initial composition was $X_{\text{He}} = 0.293$, so most of the lost He was not the result of H-fusion in the star. A similar

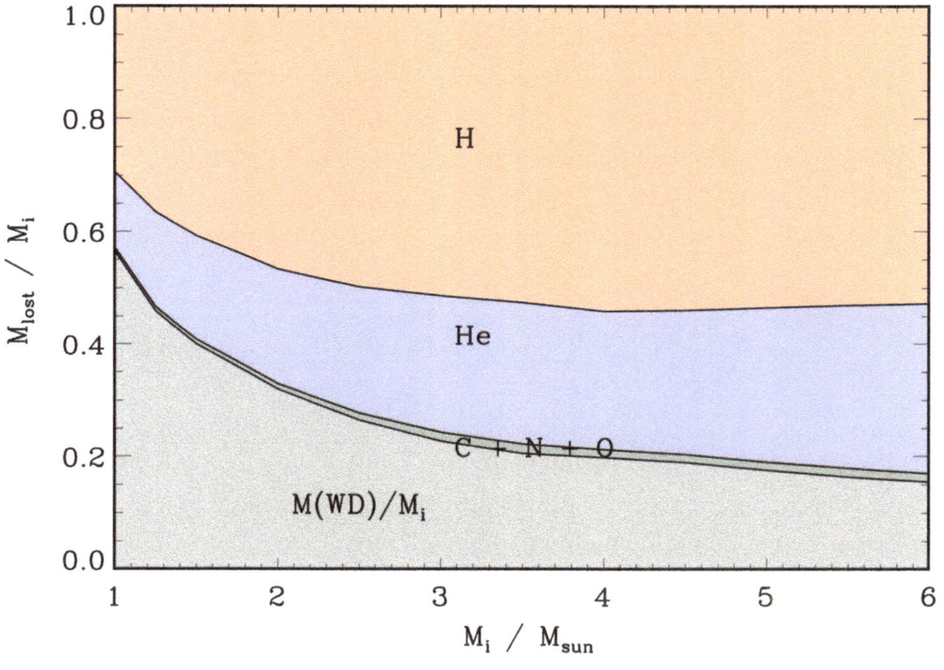

Figure 30.1. Amount of ejected mass and the remnant mass of low-mass stars, expressed as a fraction of the initial mass M_i. Red = hydrogen; blue = helium; green = CNO; and gray is the remnant mass. (Figure is based on data by Karakas 2010.)

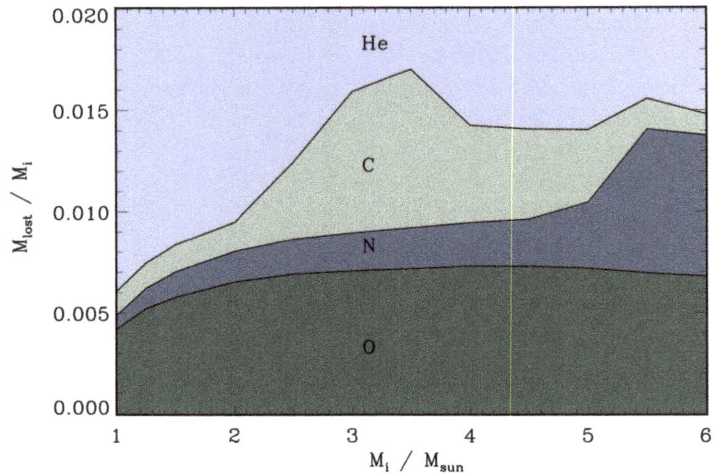

Figure 30.2. Amount of C, N, and O ejected during the evolution of low-mass stars. Notice the peak in the C production due to the loss of the C-rich envelope during the AGB phase and the peak in N production due to the enrichment of N in the CNO cycle. (Figure is based on data by Karakas 2010.)

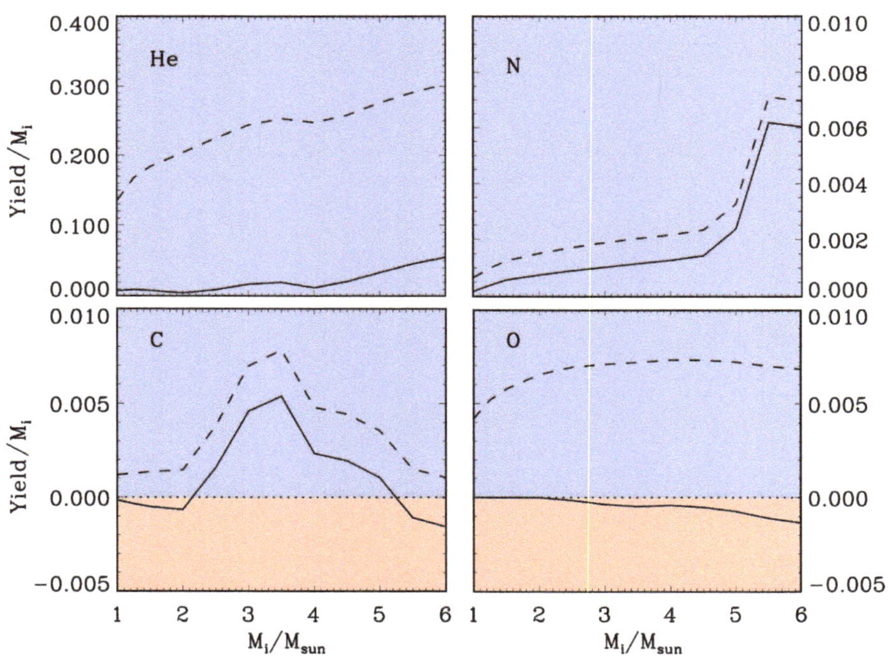

Figure 30.3. Ejected mass in M/M_i (dashed lines) and the yields in M/M_i (full lines) of He, C, N, and O. Notice that the yield of He is low, although a large amount of He was ejected. The yields of O and C are even negative (red) for stars with $M_i > 5M_{\odot}$. (Figure is based on data by Karakas 2010.)

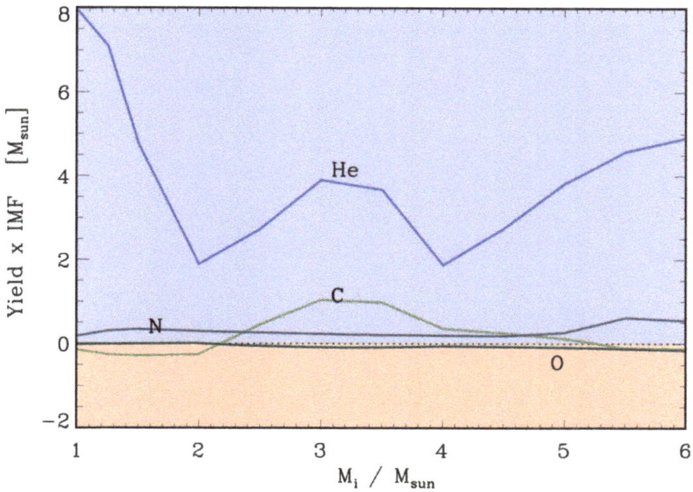

Figure 30.4. Total stellar yields, $Y_E \times N(M_i)$, of He, C, N, and O for an assumed initial mass function of $N(M_i) = 1000 \times M_i^{-2.35}$. The yields are positive in the blue area and negative in the red area. (Figure is based on data from Karakas 2010.)

situation occurs for O. In this case, the yield of O is even negative. This is because O has been converted into C and N in the CNO cycle in low-mass stars.

Figure 30.4 shows the **total yields** for an assumed Salpeter stellar mass function of $N(M_i) = 1000 \times M_i^{-2.35}$. The large increase in the total yield of He at low mass is due to the rapid increase in the number of these stars. Notice the peaks in the yields of C around $3M_\odot$ and N around $6M_\odot$.

30.2.2 Chemical Yields of Massive Stars

The top panels of Figure 30.5 show the yields of different chemical elements for massive stars, $9 \lesssim M_\text{i} \lesssim 120M_\odot$, with solar metallicity ($X = 0.705$, $Y = 0.275$, and $Z = 0.02$), as a function of their initial mass, for nonrotating and rapidly rotating models with $v_\text{init} = 300$ km s^{-1}. The yield of He is roughly constant at about 10% of the initial stellar mass, except for stars with $M_i > 60M_\odot$, which eject a larger fraction of He. Stars in the initial mass range of $M_i > 15M_\odot$ contribute strongly to the enrichment of the ISM with C, O, and α-elements like Ne, Mg, and Si.

The lower panels of Figure 30.5 show the *total yield* from the same models. It is the yield of each stellar mass, multiplied by the number of stars, using a Salpeter IMF of $N(M) = 1000 \times M^{-2.35}$. The figure shows that the contribution to the enrichment of the ISM by He increases strongly toward low-mass stars. The enrichment by C is about constant for all massive stars and the enrichment by α-elements is mainly due to stars in the range of $20 < M_i < 40M_\odot$. The much larger yield of O compared to C by massive stars explains why the cosmic abundance of C is about a factor of 3 lower than that of O.

Figure 30.5 shows that rapidly rotating massive stars contribute more to the enrichment of the ISM than nonrotating massive stars, and that their contribution to

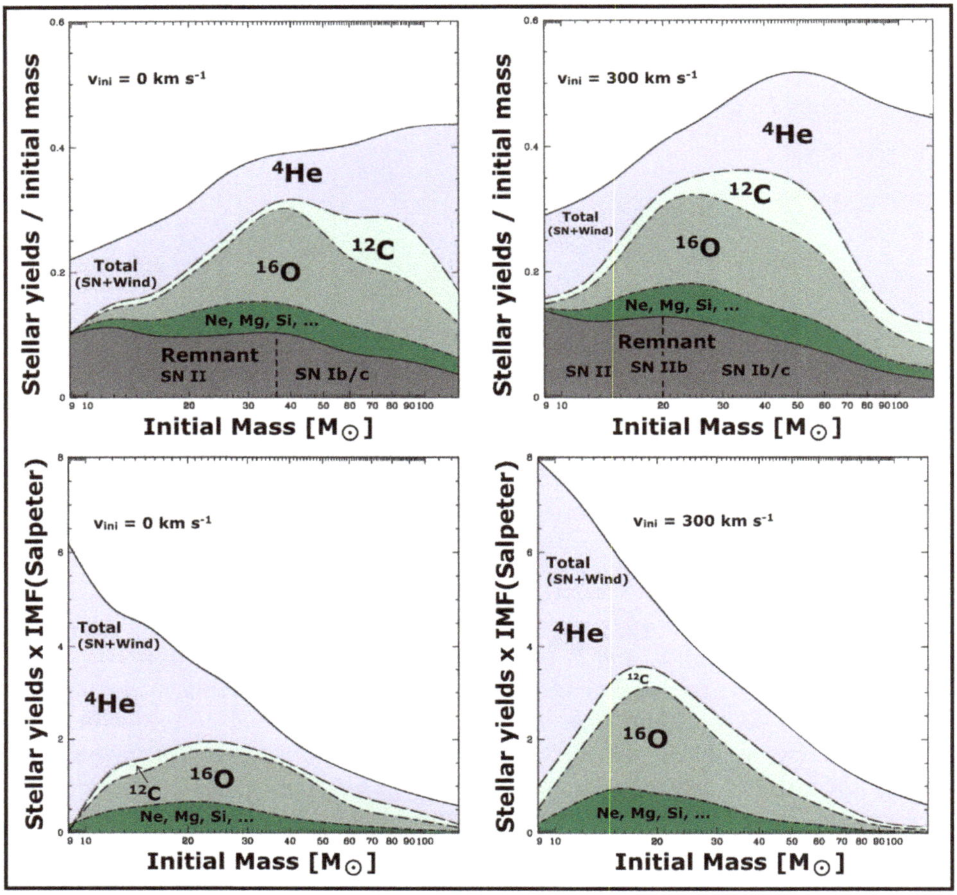

Figure 30.5. Chemical yields for nonrotating (left) and rapidly rotating models (right) of solar metallicity, Z = 0.02. Top panels: the yields are expressed as a fraction of the initial stellar mass. Colors indicate different chemical elements: H (top, white), He (blue), C (light green), O (middle green), and Ne, Mg, and Si (dark green). The mass fraction that remains in the remnant is shown in gray. The resulting supernova type is also indicated. Lower panels: the total stellar yields, $Y_E(M) \times N(M)$ in units of $M_\odot$, of massive stars of solar metallicity $Z = 0.02$, for a Salpeter IMF of $N(M) = 1000 \times M^{-2.35}$. The vertical scale is the same as for low-mass stars in figure 30.4. (Reproduced from Hirschi et al. 2005, with permission. © ESO.)

the He enrichment of the ISM is larger than for nonrotating models. At the same time, the rapidly rotating massive stars $M_i > 60\ M_\odot$ contribute less to the enrichment of C and O than nonrotating stars. This difference is mainly due to the different mass-loss histories of rotating and nonrotating stars. The rapidly rotating stars lose most of their mass early after the MS phase when nuclear fusion products have not yet appeared at the surface, whereas the mass loss of nonrotating stars occurs mainly later during the red supergiant phase or the WN and WC phases when enriched products have reached the surface.

30.3 The Main Producers of Various Elements

He:

- Most of the He was formed in the Big Bang.
- The most massive stars of 20–120$M_{\odot}$ and the least massive stars of <10$M_{\odot}$ eject He by their winds.
- Stars in the mass range of $10 < M < 20M_{\odot}$ eject most of their He in SNe.
- Although the lower-mass stars of $M < 10M_{\odot}$ eject a smaller fraction of their mass in the form of He than the massive stars, the large number of low-mass stars implies that even the stars with $M < 20M_{\odot}$ contribute significantly to the enrichment of He
- Most of the enrichment of He comes from stars with $1 < M < 30M_{\odot}$.

C:

- The most massive stars of $M > 60M_{\odot}$ eject C in their winds as WC stars.
- The stars in the range of 10–120$M_{\odot}$ also eject C by SNe explosions.
- The low-mass stars of ~2–5$M_{\odot}$ eject C in the form of C-rich AGB winds.
- More than half of the C enrichment is by low-mass stars.

O:

- Most of the O enrichment is by SNe of massive stars of $M > 10M_{\odot}$, with a peak contribution between ~20 and ~60$M_{\odot}$.
- The most massive stars of $M > 40M_{\odot}$ also lose O in the form of winds from WC and WO stars.
- Low-mass stars contribute very little to the enrichment of O.

α-elements, Ne-Si:

- The enrichment of these elements is due to SNe from massive stars of $M > 8\ M_{\odot}$.

Fe-peak elements and beyond:

- These elements are created and ejected in SN explosions. These could be due to core-collapse SNe of massive stars or white dwarfs exceeding the Chandrasekhar limit.

Neutron-rich s-process elements:

- Low-mass s-process elements such as Sr, Y, and Zr, are formed in massive stars and are ejected in SN explosions.
- The more massive s-process elements, such as Ba and Ce, are formed in low-mass AGB stars from neutrons that are formed during the He-fusion by the process $^{12}C + {}^{4}He \rightarrow {}^{16}O$ (Equation 8.18*b*).

Neutron-rich r-process elements:

- These are formed during the gravitational collapse of massive stars and ejected in SN explosions.

– Merging neutron stars could also provide an important source of n-rich r-process elements.

Two final comments:

1. The evolution of the stars described above is far from certain. For instance, meridional circulation, produced by rapid rotation, leads to more and deeper mixing than in slowly rotating stars. Rapid rotation also results in higher mass-loss rates by stellar winds. Moreover, the mass-loss rates in the RSG phase and the eruptions of LBVs are other uncertain factors in the evolution of stars. One might argue that this hardly matters for predicting stellar yields: if the products of nuclear fusion are not returned to the ISM by stellar winds, they will be released in the SN explosion at the end of the life of the star; however, this argument is too simplistic. The comparison between the yields of nonrotating and rapidly rotating stars shows that the yields depend on the timing of the mass loss by winds. If a large fraction of the mass is lost before the stellar envelope is mixed, the yield of fusion products will be smaller than if the matter is ejected later, after envelope mixing has been effective.
2. We only discussed the yields of single stars; however, many stars are in binaries, especially massive stars. We have seen in Sections 28 and 29 that mass transfer in binaries is a common phenomenon that can change the evolution of stars considerably. Part of the mass that is attracted by a binary companion may be lost from the system by nonconservative mass transfer. In this way, stars could lose a significant fraction of their mass before they are chemically enriched by nuclear fusion.

30.4 Summary

1. The evolution of stars results in chemical enrichment of the interstellar medium. The amount of enrichment of a given chemical element is called the "chemical yield." It is the mass of an element that is lost by the stellar wind or SN during the lifetime of the star, minus the initial mass of that element in the ejecta. The yield can be positive (e.g., for He) or negative (e.g., for H).
2. The chemical yield of a star depends strongly on its internal mixing and on the evolution phase when the mass is ejected. The yield of rapidly rotating stars is different from that of slow rotators.
3. The "total chemical yield" is the yield of individual stars of different masses, integrated over the initial mass function. Because the IMF is strongly biased toward low-mass stars, the contribution of low-mass stars is significant.
4. Most of the enrichment of the ISM by He is due to stars with $1 < M < 30M_{\odot}$. The enrichment of C is due to low- and high-mass stars. The enrichment of more massive elements is only due to stars more massive than about $10M_{\odot}$, except for the high-mass s-process elements, which are mainly formed during thermal pulses in low-mass stars.
5. Interacting binaries also contribute strongly to the enrichment of the ISM.

Exercises.

30.1 Figure 30.1 shows the amount of He (in M/M_i) that is lost during the evolution of stars as a function of their initial mass. Derive from this figure the net yield of He for stars of 3, 4, 5, and $6M_\odot$ and compare your results with Figure 30.3.

30.2 Derive the net yields of C, N, and O of a star of $4M_\odot$ from Figure 30.2 and compare your results with Figure 30.3.

30.3 Do you expect the net yields of He, C, N, and O in the mass range of 3 to 6 $M_\odot$ of very metal poor stars ($Z = 0.001Z_\odot$) to be higher, lower, or about the same as those of solar metallicity stars? Give arguments.

References

Hirschi, R., Meynet, G., & Maeder, A. 2005, A&A, **433, 1013**

Karakas, A. I. 2010, MNRAS, **433, 1013**

AAS | IOP Astronomy

Understanding Stellar Evolution

Henny J.G.L.M. Lamers and Emily M. Levesque

Appendix A

Physical and Astronomical Constants

A.1 Physical Constants in cgs Units

Speed of light	c	2.9979×10^{10}	cm s^{-1}
Gravitational constant	G	6.6726×10^{-8}	dyne cm^2 g^{-2}
Planck's constant	h	6.6261×10^{-27}	erg s
Boltzmann constant	k	1.3808×10^{-16}	erg K^{-1}
Atomic mass unit = amu	m_u	1.6605×10^{-24}	g
Mass of electron	m_e	9.1094×10^{-28}	g = 5.4858×10^{-4} amu
Mass of proton	m_p	1.6726×10^{-24}	g = 1.0073 amu
Mass of neutron	m_n	1.6749×10^{-24}	g = 1.0087 amu
Mass of He core (α-particle)	m_α	6.6447×10^{-24}	g = 4.0016 amu
Mass ratio proton/electron	m_p/m_e	1.8229×10^{3}	
Electron charge	e	1.6022×10^{-19}	C
	e	1.6022×10^{-20}	emu = 4.8032×10^{-10} esu
Electron volt	eV	1.6022×10^{-12}	erg
Gas constant	$\mathfrak{R}$	8.3145×10^{7}	erg K^{-1} $mole^{-1}$
Radiation density constant	a	7.5659×10^{-15}	erg cm^{-3} K^{-4}
Stefan–Boltzmann constant	$\sigma = ac/4$	5.6705×10^{-5}	erg cm^{-2} K^{-4} s^{-1}
Temperature corresponding to 1 eV	$T_{eV} = eV/k$	1.1604×10^{4}	K
Thomson scattering cross-section	σ_T	6.6525×10^{-25}	cm^2 $electron^{-1}$

A.2 Conversions from cgs to mks Units

erg → Joule	1 erg = 10^{-7} J
dyne → Newton	1 dyne = 10^{-5} N

A.3 Astronomical Constants in cgs Units

Solar mass	$M_\odot$	1.9891×10^{33}	g
Solar radius	$R_\odot$	6.9551×10^{10}	cm
Solar luminosity	$L_\odot$	3.8458×10^{33}	erg s^{-1}
Astronomical Unit	AU	1.4960×10^{13}	cm
Year	yr	3.1558×10^{7}	s
Lightyear	lyr	9.4607×10^{17}	cm
Parsec	pc	3.0857×10^{18}	cm
	pc	2.0626×10^{5}	AU = 3.2616 l yr
Arcsecond	arcsec	4.4816×10^{-6}	radian

Henny J.G.L.M. Lamers and Emily M. Levesque

Appendix B

Stellar Parameters

B.1 Main-sequence Stars (Luminosity Class V)

Spectral Type	T_{eff} (K)	log $L/L_{\odot}$	$R/R_{\odot}$	$M/M_{\odot}$	M_{bol} mag	B.C. mag	M_V mag	$U-B$ mag	$B-V$ mag
O5	44,500	5.90	15	60	−10.1	−4.40	−5.7	−1.19	−0.33
O8	35,800	5.23	11	23	−8.4	−3.54	−4.9	−1.14	−0.32
B0	30,000	4.72	8.4	17.5	−7.1	−3.16	−4.0	−1.08	−0.30
B5	15,400	2.92	4.1	5.9	−2.7	−1.46	−1.2	−0.58	−0.17
A0	9520	1.73	2.7	2.9	+0.3	−0.30	+0.6	−0.02	−0.02
A5	8200	1.15	1.9	2.0	+1.7	−0.15	+1.9	+0.10	+0.15
F0	7200	0.81	1.6	1.6	+2.6	−0.09	+2.7	+0.03	+0.30
F5	6440	0.46	1.4	1.4	+3.5	−0.14	+3.6	−0.02	+0.44
G0	6030	0.18	1.1	1.05	+4.2	−0.18	+4.4	+0.06	+0.58
Sun	5770	0.00	1.00	1.00	+4.74	−0.09	+4.83	+0.16	+0.64
G5	5770	−0.10	0.89	0.92	+4.9	−0.21	+5.1	+0.20	+0.68
K0	5250	−0.38	0.79	0.79	+5.6	−0.31	+5.9	+0.45	+0.81
K5	4350	−0.82	0.68	0.67	+6.7	−0.72	+7.4	+0.98	+1.15
M0	3850	−1.11	0.63	0.51	+7.4	−1.38	+8.8	+1.22	+1.40
M5	3 240	−1.96	0.33	0.21	+9.6	−2.73	+12.3	+1.24	+1.64
M8	2640	−2.92	0.17	0.06	+11.9	−4.1	+16.0	+1.53	+1.93

doi:10.1088/978-0-7503-1278-3ch32

B.2 Giants (Luminosity Class III

Spectral Type	T_{eff} (K)	log $L/L_\odot$	$R/R_\odot$	$M/M_\odot$	M_{bol} mag	B.C. mag	M_V mag	$U-B$ mag	$B-V$ mag
O5	42,500	6.00	18	60:	−10.3	−4.05	−6.3	−1.18	−0.32
B0	29,000	5.04	13	20	−8.0	−2.88	−5.1	−1.08	−0.29
A0	10,100	2.02	3.4	4.0	−0.4	−0.42	+0.0	−0.07	−0.03
F0	7150	1.30	2.9	3.0:	+1.4	−0.11	+1.5	+0.08	+0.30
G0	5850	1.53	5.7	1.0	+0.8	−0.20	+1.0	+0.21	+0.65
K0	4750	1.78	11	1.1	+0.2	−0.50	+0.7	+0.84	+1.00
M0	3800	2.52	42	1.2	−1.6	−1.25	−0.4	+1.87	+1.56
M6	3240	3.03	100	1.4:	−2.9	−2.73	−0.2	+1.16	+1.52

B.3 Supergiants (Luminosity Class Iab)

Spectral Type	T_{eff} (K)	log $L/L_\odot$	$R/R_\odot$	$M/M_\odot$	M_{bol} mag	B.C. mag	M_V mag	$U-B$ mag	$B-V$ mag
O5	40,300	6.04	22	70	−10.5	−3.87	−6.6	−1.17	−0.31
B0	26,000	5.52	25	25	−8.9	−2.49	−6.4	−1.06	−0.23
A0	9730	4.54	66	16	−6.7	−0.41	−6.3	−0.38	−0.01
F0	7700	4.50	100	12	−6.6	−0.01	−6.6	+0.15	+0.17
G0	5550	4.48	190	10	−6.6	−0.15	−6.4	+0.52	+0.76
K0	4420	4.46	290	13	−6.5	−0.50	−6.0	+1.17	+1.24
M0	3650	4.61	510	13	−6.9	−1.29	−5.6	+1.90	+1.67
M6	2600	5.65	3300	15:	−9.5	−3.90	−5.6	+1.80	+1.60

Data are from Carroll, B. W. and Ostlie, D. A. 1996 *An Introduction to Modern Astrophysics* and from Cox, A. N. 2000 *Allen's Astrophysical Quantities*
Values with colons are uncertain.

IOP Astronomy

Understanding Stellar Evolution

Henny J.G.L.M. Lamers and Emily M. Levesque

Appendix C

Solar Model

$r/R_\odot$	$M_r/M_\odot$	$L_r/L_\odot$	log T K	log P dyne cm^{-2}	log ρ g cm^{-3}	μ	ε_{nucl} erg g^{-1} s^{-1}	dT/dr K cm^{-1}	$(dT/dr)_{ad}$ K cm^{-1}	equil
0.0011	0.0000	0.0000	7.199	17.377	2.194	0.862	17.910	0.332	0.397	rad
0.0181	0.0007	0.0058	7.195	17.367	2.183	0.852	17.386	0.333	0.397	rad
0.0459	0.0098	0.0791	7.179	17.317	2.126	0.808	14.907	0.331	0.397	rad
0.0709	0.0322	0.2290	7.154	17.243	2.049	0.758	11.829	0.327	0.397	rad
0.0904	0.0601	0.3790	7.131	17.172	1.980	0.723	9.367	0.323	0.397	rad
0.1096	0.0959	0.5291	7.106	17.093	1.908	0.693	7.124	0.317	0.397	rad
0.1311	0.1448	0.6792	7.075	16.996	1.825	0.668	4.955	0.307	0.397	rad
0.1590	0.2183	0.8239	7.034	16.858	1.714	0.646	2.878	0.291	0.397	rad
0.1874	0.3001	0.9140	6.992	16.708	1.597	0.633	1.519	0.272	0.397	rad
0.2141	0.3795	0.9598	6.952	16.557	1.481	0.626	0.792	0.254	0.397	rad
0.2398	0.4541	0.9823	6.915	16.407	1.366	0.623	0.416	0.239	0.397	rad
0.2647	0.5228	0.9933	6.880	16.256	1.249	0.622	0.219	0.227	0.397	rad
0.2893	0.5851	0.9979	6.847	16.106	1.132	0.621	0.077	0.217	0.397	rad
0.3137	0.6410	0.9992	6.815	15.955	1.013	0.620	0.030	0.209	0.397	rad
0.3382	0.6907	0.9997	6.784	15.805	0.893	0.620	0.015	0.203	0.397	rad
0.3628	0.7346	0.9999	6.754	15.655	0.773	0.619	0.008	0.199	0.397	rad
0.3878	0.7732	1.0000	6.724	15.505	0.652	0.619	0.004	0.195	0.397	rad
0.4131	0.8070	1.0001	6.695	15.354	0.531	0.619	0.002	0.193	0.397	rad
0.4390	0.8365	1.0001	6.666	15.204	0.409	0.618	0.001	0.192	0.397	rad
0.4653	0.8620	1.0001	6.637	15.054	0.287	0.618	0.001	0.191	0.397	rad
0.4922	0.8842	1.0001	6.609	14.904	0.165	0.618	0.000	0.191	0.397	rad
0.5197	0.9032	1.0000	6.580	14.753	0.044	0.617	0.000	0.193	0.397	rad
0.5478	0.9196	1.0000	6.551	14.603	−0.077	0.617	0.000	0.196	0.397	rad

(*Continued*)

doi:10.1088/978-0-7503-1278-3ch33

(*Continued*)

$r/R_\odot$	$M_r/M_\odot$	$L_r/L_\odot$	log T K	log P dyne cm^{-2}	log ρ g cm^{-3}	μ	ε_{nucl} erg g^{-1} s^{-1}	dT/dr K cm^{-1}	$(dT/dr)_{ad}$ K cm^{-1}	equil
0.5764	0.9337	1.0000	6.521	14.453	−0.198	0.617	0.000	0.202	0.397	rad
0.6055	0.9456	1.0000	6.490	14.303	−0.317	0.617	0.000	0.212	0.397	rad
0.6349	0.9558	1.0000	6.456	14.152	−0.435	0.616	0.000	0.231	0.397	rad
0.6645	0.9644	1.0000	6.419	14.002	−0.549	0.615	0.000	0.269	0.396	rad
0.6939	0.9715	1.0000	6.376	13.852	−0.658	0.611	0.000	0.320	0.396	rad
0.7221	0.9775	1.0000	6.321	13.703	−0.755	0.607	0.000	0.396	0.396	conv
0.7490	0.9824	1.0000	6.262	13.553	−0.845	0.607	0.000	0.396	0.396	conv
0.7739	0.9863	1.0000	6.202	13.403	−0.935	0.608	0.000	0.396	0.396	conv
0.7970	0.9895	1.0000	6.143	13.253	−1.026	0.609	0.000	0.396	0.396	conv
0.8181	0.9920	1.0000	6.083	13.103	−1.116	0.609	0.000	0.396	0.396	conv
0.8550	0.9955	1.0000	5.964	12.803	−1.296	0.611	0.000	0.396	0.396	conv
0.8851	0.9975	1.0000	5.846	12.502	−1.476	0.613	0.000	0.395	0.395	conv
0.9094	0.9986	1.0000	5.727	12.202	−1.656	0.615	0.000	0.394	0.394	conv
0.9284	0.9993	1.0000	5.611	11.908	−1.833	0.618	0.000	0.392	0.392	conv
0.9436	0.9996	1.0000	5.494	11.608	−2.013	0.622	0.000	0.389	0.389	conv
0.9605	0.9999	1.0000	5.321	11.158	−2.284	0.630	0.000	0.381	0.381	conv
0.9813	1.0000	1.0000	4.964	10.158	−2.904	0.664	0.000	0.342	0.342	conv
0.9904	1.0000	1.0000	4.637	9.158	−3.542	0.720	0.000	0.276	0.276	conv
0.9972	1.0000	1.0000	4.258	7.158	−5.054	0.924	0.000	0.136	0.133	conv
0.9998	1.0000	1.0000	3.915	5.158	−6.581	1.247	0.000	0.578	0.249	conv
1.0000	1.0000	1.0000	3.705	4.483	−7.041	1.259	0.000	0.044	0.387	rad

Data are from: http://www.ap.stmarys.ca/~guenther/evolution/ssmyz47_0200.txt.

Appendix D

Main Sequence from ZAMS to TAMS

M_i	z=0.014 (solar)						z=0.002 (subsolar)					
	ZAMS		TAMS				ZAMS		TAMS			
1	2	3	4	5	6	7	8	9	10	11	12	13
0.8	-0.59	3.69	10.29	0.8	-0.26	3.71	-0.29	3.77	10.01	0.8	0.07	3.79
0.9	-0.36	3.72	10.09	0.9	-0.05	3.74	-0.05	3.80	9.82	0.9	0.28	3.81
1.0	-0.14	3.75	9.89	1.0	0.13	3.76	0.16	3.82	9.65	1.0	0.45	3.84
2.0	1.21	3.97	9.00	2.0	1.37	3.86	1.40	4.08	8.82	2.0	1.65	4.00
4.0	2.37	4.19	8.18	4.0	2.61	4.08	2.49	4.27	8.09	4.0	2.78	4.19
7.0	3.22	4.33	7.62	7.0	3.52	4.24	3.29	4.39	7.60	7.0	3.61	4.31
12.0	3.98	4.46	7.18	11.9	4.31	4.35	4.01	4.50	7.17	12.0	4.36	4.41
15.0	4.27	4.50	7.04	14.8	4.60	4.39	4.28	4.54	7.04	14.9	4.64	4.45
20.0	4.61	4.55	6.89	19.7	4.95	4.42	4.62	4.59	6.89	19.9	4.98	4.48
25.0	4.87	4.59	6.80	24.2	5.19	4.43	4.87	4.62	6.81	24.8	5.22	4.50
32.0	5.13	4.62	6.72	30.1	5.42	4.42	5.12	4.66	6.72	31.5	5.46	4.51
40.0	5.34	4.65	6.65	36.5	5.61	4.40	5.34	4.68	6.65	39.1	5.65	4.52
60.0	5.70	4.69	6.54	43.6	5.88	4.10	5.70	4.72	6.55	57.8	5.97	4.50
85.0	5.98	4.71	6.46	58.1	6.11	4.31	5.98	4.75	6.48	80.9	6.20	4.45
120.0	6.23	4.73	6.36	87.7	6.31	4.43	6.23	4.77	6.43	111.9	6.42	4.28

column 1 : M_i in $M_\odot$
columns 2 and 8 : log ($L/L_\odot$) at ZAMS
columns 3 and 9 : log (T_{eff}) at ZAMS in K
columns 4 and 10 : log (t) at TAMS in yrs
columns 5 and 11 : $M(t)$ at TAMS in $M_\odot$
columns 6 and 12 : log ($L/L_\odot$) at TAMS
columns 7 and 13 : log (T_{eff}) at TAMS

Data from: https://obswww.unige.ch/Recherche/evol/tables_grids2011/Z002/
https://obswww.unige.ch/Recherche/evol/tables_grids2011/Z014/

Based on:

Ekstrom, S., Georgy, C., and Eggenberger, P., *et al* 2012, A&A, **537,** 146
Georgy, C.,, Ekstrom, S., and Eggenberger, P., *et al* 2013, A&A, **558,** 103

Henny J.G.L.M. Lamers and Emily M. Levesque

Appendix E

Acronyms

AAVSO	American Association of Variable Star Observers
AGB	Asymptotic Giant Branch
BC	Bolometric Correction
BH	Black Hole
BSG	Blue Super Giant
CD	Complete Degeneracy
CMD	Color–Magnitude Diagram
CSPN	Central Star of Planetary Nebula
EoS	Equation of State
GRB	Gamma-Ray Burster
HB	Horizontal Branch
HBB	Hot Bottom Burning
HD-limit	Humphreys–Davidson limit
HE	Hydrostatic Equilibrium
HRD	Hertzsprung–Russell Diagram
IMF	Initial Mass Function
ISM	Interstellar Medium
ISZ	Intershell Zone
KD	Kippenhahn Diagram
LBV	Luminous Blue Variable
LMC	Large Magellanic Cloud
MESA	Modules for Experiments in Stellar Astrophysics
MS	Main Sequence
NRP	Non-Radial Pulsations
NS	Neutron Star
PD	Partial Degeneracy
PMS	Pre-Main Sequence
PN	Planetary Nebulae
RD	Relativistic Degeneracy
RE	Radiative Equilibrium
RGB	Red Giant Branch
RL	Roche Lobe
RSG	Red Super Giant
SC-limit	Schönberg–Chandrasekhar limit
SMC	Small Magellanic Cloud

doi:10.1088/978-0-7503-1278-3ch35

SN	Super Nova
TAMS	Terminal Age Main Sequence
TE	Thermal Equilibrium
TP-AGB	Thermal Pulsing Asymptotic Giant Branch
YSG	Yellow Super Giant
VT	Virial Theorem
WC	C-rich Wolf–Rayet Star
WD	White Dwarf
WN	N-rich Wolf–Rayet Star
WO	O-rich Wolf–Rayet Star
WR	Wolf–Rayet Star
ZAHB	Zero-Age Horizontal Branch
ZAMS	Zero-Age Main Sequence

www.ingramcontent.com/pod-product-compliance
Ingram Content Group UK Ltd.
Pitfield, Milton Keynes, MK11 3LW, UK
UKHW061858100926
8556IPUK00008B/918

9 780750 312790